Genomics and Genetic Engineering

Genomics and Genetic Engineering

Contributors

Sumiti Vinayaka, Carrie F. Brooks et al.

AURIS
Reference

www.aurisreference.com

Genomics and Genetic Engineering

Contributors: Sumiti Vinayaka, Carrie F. Brooks et al.

Published by Auris Reference Limited

www.aurisreference.com

United Kingdom

Genomics and Genetic Engineering

ISBN: 978-1-78154-956-8

British Library Cataloguing in Publication Data
A CIP record for this book is available from the British Library

Printed in the United Kingdom

Exclusively distributed by CBS Publishers & Distributors Pvt. Ltd.

Sales & Distribution Rights only for India, Pakistan, Bangladesh, Sri Lanka, Nepal and Bhutan. This book is not to be sold outside these territories.

Contents

List of Abbreviations

ARS	Autonomously Replicating Sequence
CAGE	Conjugative Assembly Genome Engineering
CAT	Chloramphenicol Acetyltransferase
CGM	Clostridium Growth Medium
CHEF	Clamped Homogeneous Electric Field
CNV	Copy Number Variation
CPK	Cryptic Polyketide
CRISPR	Clustered Regularly Interspaced Short Palindromic Repeat
DHR	Downstream Homology Region
GFP	Green Fluorescent Protein
GOI	Gene of Interest
GPI	Glycosyl Phosphatidyl Inositol
GS	Genome Shuffling
GWAS	Genome Wide Association Studies
IGV	Integrated Genome Viewer
ITR	Inverted Terminal Repeats
LMM	Linear Mixed Model
MAGE	Multiplex Automated Genome Engineering
NEGC	Non-Essential Gene Clusters
OD	Optical Density
OR	Open Reading Frames
PFF	Pig Fetal Fibroblasts
PFGE	Pulse Field Gel Electrophoresis
PNI	Pronuclear Injection
PTEF	Promoter for the Translation Elongation Factor
QTL	Quantitative Trait Loci
RCM	Reinforced Clostridial Medium
REML	Restricted Maximum Likelihood
ROS	Reactive Oxygen Species
RRS	Recombinase Recognition Sites
RS	Repeat Sequence
SBN	Sternopleural Bristle Number
SCNT	Somatic Cell Nuclear Transfer
SDS	Symmetric Differences Squared
SM	Synthetic Medium
SNP	Single Nucleotide Polymorphisms
TREC	Tandem Repeat Coupled With Endonuclease Cleavage
UHR	Upstream Homology Region
UT	Untranslated Regions
WLS	Weighted Least Square
YCP	Yeast Centromeric Plasmid
YEME	Yeast Extract Malt Extract

List of Contributors

Sumiti Vinayak
Center for Tropical and Emerging Global Diseases, University of Georgia, Athens, Georgia, USA

Carrie F. Brooks
Center for Tropical and Emerging Global Diseases, University of Georgia, Athens, Georgia, USA

Anatoli Naumov
Departments of Molecular Medicine & Global Health, University of South Florida, Tampa, Florida, USA

Elena S. Suvorova
bDepartments of Molecular Medicine & Global Health, University of South Florida, Tampa, Florida, USA

Michael W. White
Departments of Molecular Medicine & Global Health, University of South Florida, Tampa, Florida, USA

Boris Striepen
Center for Tropical and Emerging Global Diseases, University of Georgia, Athens, Georgia, USA
Department of Cellular Biology, University of Georgia, Athens, Georgia, USA

Deyao Du
State Key Laboratory of Microbial Resources, Institute of Microbiology, Chinese Academy of Sciences, Beijing, China
University of Chinese Academy of Sciences, Beijing, China

Lu Wang
State Key Laboratory of Microbial Resources, Institute of Microbiology, Chinese Academy of Sciences, Beijing, China
Key Laboratory of Industrial Fermentation Microbiology, Ministry of Education, College of Biotechnology, Tianjin University of Science and Technology, Tianjin, China.

Yuqing Tian
State Key Laboratory of Microbial Resources, Institute of Microbiology, Chinese Academy of Sciences, Beijing, China

Hao Liu
Key Laboratory of Industrial Fermentation Microbiology, Ministry of Education, College of Biotechnology, Tianjin University of Science and Technology, Tianjin, China.

Huarong Tan
State Key Laboratory of Microbial Resources, Institute of Microbiology, Chinese Academy of Sciences, Beijing, China

Guoqing Niu
State Key Laboratory of Microbial Resources, Institute of Microbiology, Chinese Academy of Sciences, Beijing, China

Zhiqiu Hu
Department of Agricultural, Food and Nutritional Science, University of Alberta, Edmonton, Alberta, Canada

Rong-Cai Yang
Department of Agricultural, Food and Nutritional Science, University of Alberta, Edmonton, Alberta, Canada
Alberta Agriculture and Rural Development, Edmonton, Alberta, Canada

Matthew L. Jones
Moritz Treeck Laboratory, The Francis Crick Institute, Mill Hill Laboratory, The Ridgeway, London NW71AA, United Kingdom

Sujaan Das
Michael J. Blackman Laboratory, The Francis Crick Institute, Mill Hill Laboratory, The Ridgeway, London NW71AA, United Kingdom

Hugo Belda
Moritz Treeck Laboratory, The Francis Crick Institute, Mill Hill Laboratory, The Ridgeway, London NW71AA, United Kingdom

Christine R. Collins
Michael J. Blackman Laboratory, The Francis Crick Institute, Mill Hill Laboratory, The Ridgeway, London NW71AA, United Kingdom

Michael J. Blackman
Michael J. Blackman Laboratory, The Francis Crick Institute, Mill Hill Laboratory, The Ridgeway, London NW71AA, United Kingdom

MoritzTreeck
Moritz Treeck Laboratory, The Francis Crick Institute, Mill Hill Laboratory, The Ridgeway, London NW71AA, United Kingdom

Suchismita Chandran
The J. Craig Venter Institute, 9704 Medical Center Drive, Rockville 20850, MD, USA

Vladimir N Noskov
The J. Craig Venter Institute, 9704 Medical Center Drive, Rockville 20850, MD, USA

Thomas H Segall-Shapiro
The J. Craig Venter Institute, 9704 Medical Center Drive, Rockville 20850, MD, USA

Li Ma
The J. Craig Venter Institute, 9704 Medical Center Drive, Rockville 20850, MD, USA

Caitlin Whiteis
The J. Craig Venter Institute, 9704 Medical Center Drive, Rockville 20850, MD, USA

Carole Lartigue
INRA, UMR 1332 de Biologie du Fruit et Pathologie, F-33140, Villenave d'Ornon Bordeaux, France
University Bordeaux, UMR 1332 de Biologie du Fruit et Pathologie, F-33140, Villenave d'Ornon Bordeaux, France

Joerg Jores
International Livestock Research Institute (ILRI), Old Naivasha Road, 00100 Nairobi, Kenya.

Sanjay Vashee
The J. Craig Venter Institute, 9704 Medical Center Drive, Rockville 20850, MD, USA

Ray-Yuan Chuang
The J. Craig Venter Institute, 9704 Medical Center Drive, Rockville 20850, MD, USA

Karl J Clark
Department of Animal Science, University of Minnesota, St. Paul, MN, USA
The Arnold and Mabel Beckman Center for Transposon Research, University of Minnesota, Minneapolis, MN, USA
The University of Minnesota Animal Biotechnology Center, University of Minnesota, St. Paul, MN, USA

Daniel F Carlson
Department of Animal Science, University of Minnesota, St. Paul, MN, USA
The Arnold and Mabel Beckman Center for Transposon Research, University of Minnesota, Minneapolis, MN, USA
The University of Minnesota Animal Biotechnology Center, University of Minnesota, St. Paul, MN, USA

Linda K Foster
Department of Animal Science, University of Minnesota, St. Paul, MN, USA

Byung-Whi Kong
Department of Animal Science, University of Minnesota, St. Paul, MN, USA

Douglas N Foster
Department of Animal Science, University of Minnesota, St. Paul, MN, USA
The University of Minnesota Animal Biotechnology Center, University of Minnesota, St. Paul, MN, USA

Scott C Fahrenkrug
Department of Animal Science, University of Minnesota, St. Paul, MN, USA
The Arnold and Mabel Beckman Center for Transposon Research, University of Minnesota, Minneapolis, MN, USA
The University of Minnesota Animal Biotechnology Center, University of Minnesota, St. Paul, MN, USA

Dominic Pinel
Department of Biology, Centre for Structural and Functional Genomics, Concordia University, 7141 Sherbrooke Street West, Montréal, Québec H4B 1R6, Canada
Energy Biosciences Institute, University of California, Berkeley, Berkeley, CA 94704, USA

David Colatriano
Department of Biology, Centre for Structural and Functional Genomics, Concordia University, 7141 Sherbrooke Street West, Montréal, Québec H4B 1R6, Canada

Heng Jiang
Department of Biology, Centre for Structural and Functional Genomics, Concordia University, 7141 Sherbrooke Street West, Montréal, Québec H4B 1R6, Canada
Crabtree Nutrition Laboratories, McGill University Health Center, Montreal, Quebec H3A 1A1, Canada.

Hung Lee
School of Environmental Sciences, University of Guelph, Guelph, Ontario N1G 2 W1, Canada

Vincent JJ Martin
Department of Biology, Centre for Structural and Functional Genomics, Concordia University, 7141 Sherbrooke Street West, Montréal, Québec H4B 1R6, Canada

Christian Croux
LISBP, INSA, University of Toulouse, 135 Avenue de Rangueil, 31077 Toulouse Cedex, France

Ngoc Phuong Thao Nguyen
LISBP, INSA, University of Toulouse, 135 Avenue de Rangueil, 31077 Toulouse Cedex, France

Jieun Lee
College of Life Sciences and Biotechnology, Korea University, Seoul, South Korea

Céline Raynaud
Metabolic Explorer, Saint Beauzire, France.

Florence Saint Prix
LISBP, INSA, University of Toulouse, 135 Avenue de Rangueil, 31077 Toulouse Cedex, France

Maria Gonzalez Pajuelo
LISBP, INSA, University of Toulouse, 135 Avenue de Rangueil, 31077 Toulouse Cedex, France

Isabelle Meynial Salles
LISBP, INSA, University of Toulouse, 135 Avenue de Rangueil, 31077 Toulouse Cedex, France

Philippe Soucaille
LISBP, INSA, University of Toulouse, 135 Avenue de Rangueil, 31077 Toulouse Cedex, France
Metabolic Explorer, Saint Beauzire, France.

Narayana Annaluru
Department of Environmental Health Sciences, Bloomberg School of Public Health, Johns Hopkins University, 615 North Wolfe Street, Baltimore, MD 21205, USA

Sivaprakash Ramalingam
Department of Environmental Health Sciences, Bloomberg School of Public Health, Johns Hopkins University, 615 North Wolfe Street, Baltimore, MD 21205, USA

Srinivasan Chandrasegaran
Department of Environmental Health Sciences, Bloomberg School of Public Health, Johns Hopkins University, 615 North Wolfe Street, Baltimore, MD 21205, USA

Preface

Genomics is an area within genetics that concerns the sequencing and analysis of an organism's genome. The genome is the entire DNA content that is present within one cell of an organism. Experts in genomics strive to determine complete DNA sequences and perform genetic mapping to help understand disease. Genetic engineering, also called genetic modification, is the direct manipulation of an organism's genome using biotechnology. It is a set of technologies used to change the genetic makeup of cells, including the transfer of genes within and across species boundaries to produce improved or novel organisms. New DNA may be inserted in the host genome by first isolating and copying the genetic material of interest using molecular cloning methods to generate a DNA sequence, or by synthesizing the DNA, and then inserting this construct into the host organism. Genomics and Genetic Engineering emphasizes on the development of genome-scale technologies and their application to all areas of biological investigation. First chapter focuses on genetic manipulation of the toxoplasma gondii genome by fosmid recombineering. In second chapter, a novel strategy based on phage BT1 integrase-mediated site-specific recombination was developed, and used for simultaneous *Streptomyces* genome engineering and cloning of antibiotic gene clusters. In third chapter, we advocate the use of a statistical procedure known as symmetric differences squared (SDS) as it may serve as a viable alternative when the LMM methods have difficulty or fail to work with large datasets. In fourth chapter, we describe a strategy for facile and rapid functional analysis of genes using an approach based on the Cre/ lox system and tailored for organisms with short and few introns. In fifth chapter, we demonstrate applications of the TREC-IN method in gene complementation and genome minimization studies in *Mmc*. Sixth chapter provides the basis for developing transposon and recombinase based tools for genetic engineering of the swine genome. In seventh chapter, we demonstrate that strain evolution by meiotic recombination-based genome shuffling coupled with deep sequencing can be used to deconstruct complex phenotypes and explore the nature of multigenic traits, while providing concrete targets for strain development. Construction of a restriction-less, marker-less mutant useful for functional genomic and metabolic engineering of the biofuel producer clostridium acetobutylicum is highlighted in eighth chapter. Last chapter reviews the current status of synthetic genomics, starting with a historical perspective that highlights the key milestones in the field and then continuing with a particular emphasis on the total synthesis of the first functional designer eukaryotic (yeast) chromosome.

Chapter 1

GENETIC MANIPULATION OF THE TOXOPLASMA GONDII GENOME BY FOSMID RECOMBINEERING

Sumiti Vinayak[a], Carrie F. Brooks[a], Anatoli Naumov[b], Elena S. Suvorova[b], Michael W. White[b], Boris Striepen[a,c]

[a]Center for Tropical and Emerging Global Diseases, University of Georgia, Athens, Georgia, USA

[b]Departments of Molecular Medicine & Global Health, University of South Florida, Tampa, Florida, USA

[c]Department of Cellular Biology, University of Georgia, Athens, Georgia, USA

ABSTRACT

Apicomplexa are obligate intracellular parasites that cause important diseases in humans and animals. Manipulating the pathogen genome is the most direct way to understand the functions of specific genes in parasite development and pathogenesis. In *Toxoplasma gondii*, nonhomologous recombination is typically highly favored over homologous recombination, a process required for precise gene targeting. Several approaches, including the use of targeting vectors that feature large flanks to drive site-specific recombination, have been developed to overcome this problem. We have generated a new large-insert repository of *T. gondii* genomic DNA that is arrayed and sequenced and covers 95% of all of the parasite's genes. Clones from this fosmid library are maintained at single copy, which provides a high level of stability and enhances our ability to modify the organism dramatically. We establish a robust recombineering pipeline and show that our fosmid clones can be easily converted into gene knockout constructs in a 4-day protocol that does not require plate-based cloning but can be performed in multiwell plates. We validated this approach to understand gene function in *T. gondii* and produced a conditional null mutant for a nucleolar protein belonging to the NOL1/NOP2/SUN family, and we show that this gene is essential for parasite growth. We also demonstrate a powerful complementation strategy in the context of chemical mutagenesis and whole-genome sequencing. This repository is an important new resource that will accelerate both forward and reverse genetic analysis of this important pathogen.

IMPORTANCE *Toxoplasma gondii* is an important genetic model to understand intracellular parasitism. We show here that large-insert genomic clones are effective tools that enhance homologous recombination and allow us to engineer conditional mutants to understand gene function. We have generated, arrayed, and sequenced a fosmid library of *T. gondii* genomic DNA in a copy control vector that provides excellent coverage of the genome. The fosmids are maintained in a single-copy state that dramatically improves their stability and allows modification by means of a simple and highly scalable protocol. We show here that modified and unmodified fosmid clones are powerful tools for forward and reverse genetics.

INTRODUCTION

Toxoplasma gondii is an obligate intracellular parasite that belongs to the phylum Apicomplexa, which includes numerous important pathogens, such as*Plasmodium*, *Cryptosporidium*, *Eimeria*, *Neospora*, and *Theileria*, that cause diseases in humans and animals. Among apicomplexans, *T. gondii* has emerged as the experimentally most tractable organism and is now used by many investigators as a genetic model to understand parasite biology (**1**). The ability to introduce transgenic reporters and to ablate or modify parasite genes has driven experimental work on apicomplexans over the last 2 decades. A variety of approaches have been developed to generate and introduce the DNA molecules that bring about these changes. Initially, this was based largely on mini-gene plasmids that place the coding sequence of a gene, typically obtained from cDNA, into the context of a promoter and suitable 5′ and 3′ untranslated regions (**2, 3**). These tools are easily constructed and allow researchers to study the expression and localization of proteins by appending an epitope tag, a fluorescent protein, or an enzyme reporter (**4**). These vectors can also be used for conditional gene expression in combination with regulatable promoters, such as those recognized by the tetracycline-regulated transactivator system or protein destabilization domains, which can be modulated with small-molecule ligands (**5, 6**). A limitation of this approach is that it removes the gene from its natural expression context in the genome. This can result in protein expression at an inappropriate level or time, which may obscure the true location or function of the protein or produce dominant negative effects that make it more difficult to interpret the results. Targeting the modification directly to the genomic locus of the gene can mitigate some of these problems. Typically, this is achieved by single- or double-crossover homologous recombination using sequences derived from genomic DNA to target the recombination event to the desired locus. *T. gondii* uses homologous as well as a nonhomologous end-joining DNA repair systems, and typically, nonhomologous insertion is highly favored, which can make gene targeting challenging for some genes. The development of ΔKu80 mutant strains overcomes this by drastically reducing nonhomologous recombination and thus increasing the proportion of transgenics derived by homologous recombination in a population of transfected cells. This allows gene localization and gene replacement to occur under the control of endogenous regulatory elements (**7, 8**). Another advancement has been the development of tetracycline-regulated transactivator TATi/ΔKu80 strains for creating conditional gene knockouts in the parasite (**9**). These combine superior efficiency of homologous recombination (due to deletion of Ku80) with the tetracycline-regulatable promoter system. Most recently, clustered regularly interspaced short palindromic repeat (CRISPR)/Cas9-induced double-strand breaks have also been shown to yield higher crossover frequencies (**10, 11**).

A third strategy uses the massive flanking sequences afforded by large-insert genomic constructs to enhance homologous-recombination events; this is independent of mutations in repair mechanisms or the induction of genome injury and can be used in wild-type (wt) parasite strains. These large genomic inserts are not amenable to restriction cloning but can readily be modified in *Escherichia coli* via recombination-based genetic engineering (recombineering) to convert them into gene-tagging or gene knockout constructs (**12, 13**). The two large-insert cosmid libraries (TOX and PSB) available for *T. gondii* have been used to study the functions of genes in various biological processes, such as cell division, egress from host cells, isoprenoid biosynthesis, fatty acid synthesis, and apicoplast and mitochondrial function (**1, 9, 14–20**). Recombineering is a widely used platform to quickly and cost-effectively modify large DNA to produce large numbers of vectors for genome-wide functional analysis in mice (**21, 22**). However, cosmids are maintained at 50 copies per bacterial cell, and the presence of multiple copies can result in the modification of only a subset of cosmids. Also, activation of the phage recombination system in the context of multicopy constructs can produce illicit recombination and rearrangements, creating deletions or chimeric molecules.

Here we report the construction of a new fosmid resource of genomic DNA from the highly virulent RH strain of *T. gondii*. The library uses a copy control vector that overcomes many of the above-mentioned complications and offers a robust and fast route to genetic modification. The fosmid clones are maintained as a single copy per bacterial cell, which dramatically improves their stability during storage and through the recombineering process. Upon completion of the modification, a second high-copy-number origin of replication can be triggered using an inducer molecule to produce bulk DNA. We describe a powerful recombineering approach using these fosmids to modify the parasite's genome by homologous recombination. Given our interest in understanding the role of nuclear compartmentalization in the regulation of the cell cycle in *T. gondii* (**23**), we decided to test the utility of the fosmid approach to understand the role of a previously uncharacterized parasite nucleolar protein belonging to the NOL1/NOP2/SUN family. We report here the creation of a conditional knockout mutant for this essential nucleolar protein via promoter replacement by the fosmid approach. The *T. gondii* model not only permits reverse genetic modification of parasites but also offers exciting forward genetic possibilities. Such approaches have been used to map the genetic loci underlying the differences in strain virulence and have led to the discovery of how secreted rhoptry kinases allow the parasite to evade attachment by innate and acquired host immunity (**24–26**). Similarly, chemical mutagenesis in combination with screens for temperature sensitivity has produced important insights into host cell invasion and parasite replication (**27, 28**). Here we show the potential of the new fosmid resource for genetic complementation analysis using the example of a cell cycle mutant. We tested fosmids covering six nonsynonymous single nucleotide polymorphisms (SNPs) predicted by whole-genome sequence analysis and identified the key mutation in the gene encoding regulator of chromatin condensation 1 (RCC1), responsible for the phenotype.

RESULTS

The Fosmid Library Provides High Coverage of the *Toxoplasma gondii* Genome

Our goal was to establish a resource for genome engineering in*T. gondii* that is stable, is amenable to efficient and scalable manipulation, and provides access to the entire genome. We constructed a fosmid library by hydroshearing *T. gondii* RH genomic DNA into ~40-kb fragments (**Fig. 1A**). Hydroshearing to produce a random unbiased library based on physical breakage and size fragmentation was used rather than the traditional approach of partial Sau3AI digestion of DNA. To construct the library, we chose a copy control vector (pCC2FOS) that contains an *E. coli* F-factor single origin of replication as well as an inducible high-copy-number *oriV* gene. This system offers the advantage of stably maintaining fosmid clones in EPI300-T1^R phage-resistant *E. coli* at a single copy per cell, thus avoiding the undesired recombination that we experienced with previously constructed cosmid libraries for a subset of clones. The fosmid clones could be amplified to high copy numbers (10 to 200 copies/cell) when desired since the EPI300-

T1[R] *E. coli* cells provided the product of the *trfA* replication initiation gene under the tight control of an arabinose-inducible promoter for initiation of replication from *oriV*.

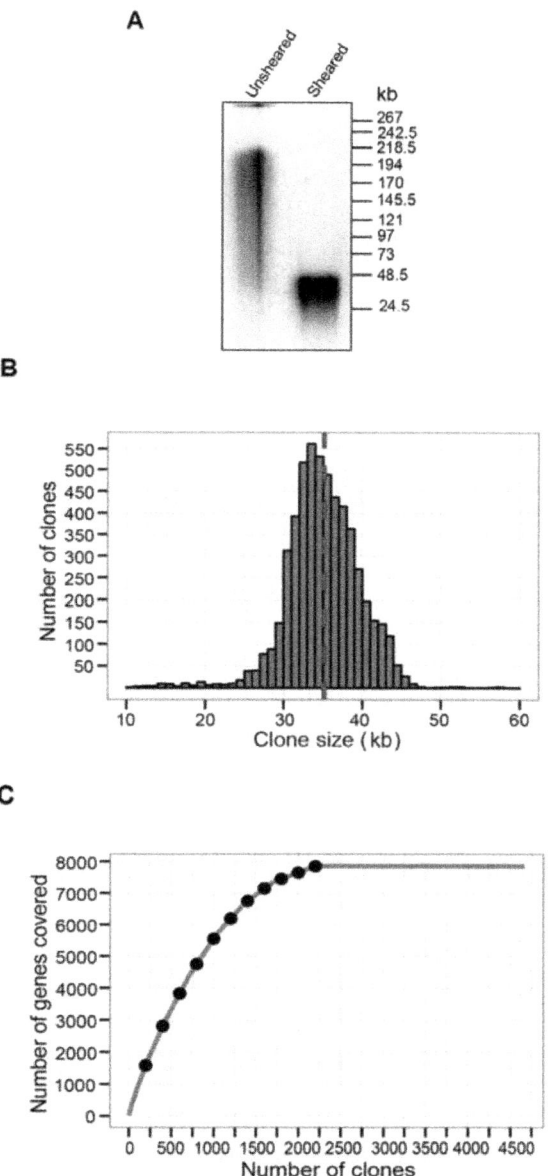

Figure 1: Characterization of the *Toxoplasma gondii* RH strain fosmid genomic DNA library. (A) CHEF gel showing the size comparison of native high-molecular-

weight *T. gondii* genomic DNA with hydrosheared DNA used for library preparation. The positions and sizes (kb) for the midrange II PFG marker are shown. (B) Size distribution (kb) of the 5,509 *T. gondii* fosmid library clones. The mean size of ~35 kb is indicated as a red dashed line. (C) Coverage of *T. gondii* genes as fosmid clone number increases. The line plateaus at 2,212 fosmids, covering 7,816 *T. gondii* genes.

The *T. gondii* fosmid library contained 200,000 independent clones that were maintained as a glycerol stock at −80°C. A random subset of 10,000 clones was picked and end sequenced using the vector-specific primers PC1F and PC1R (primer sequences are provided in **Table S1** in the supplemental material). Sequences from clones (n = 8,408) that produced high-quality reads from both ends were used as queries in a BLAST search against the *T. gondii* ME49 genome to obtain the start and end coordinates of each of the clones in the genome. Although all 8,408 clones mapped to the *T. gondii* genome, only 7,046 fosmids mapped to annotated chromosomes; the remaining (n = 1,362) mapped to unassembled regions, such as genomic scaffolds or assemblies. Of the 7,046 clones that mapped to *Toxoplasma* chromosomes, 1,537 clones showed multiple BLAST hits on different chromosomes and were excluded from our analysis to avoid ambiguity. The genome-wide set of the mapped fosmid clones (n = 5,509) is publicly available under the Genome Browser track of ToxoDB (**http://toxodb.org/cgi-bin/gbrowse/toxodb/**).

The insert size distribution of the sequenced and mapped fosmid library is shown in **Fig. 1B**, with 98.5% (5,426 out of 5,509) of the clone inserts in the 20- to 50-kb range; the average insert size is 35 kb. We next wanted to determine the number of genes that were covered by these fosmid clones (**Fig. 1C**). As expected, the number of genes covered increased with the number of clones, but the curve plateaus at 2,212 fosmids. These 2,212 fosmids covered ~95% (7,861 out of 8,317) of the genes on the *T. gondii* chromosomes (**Fig. 1C**), suggesting that we reached saturation with respect to the number of *T. gondii* genes covered.

We also wanted to compare the coverage of the fosmids to that of the existing clones from two cosmid libraries (**Fig. 2**). For this purpose, we used the start and end coordinates of the clones and generated a map of their coverage across all 14 *Toxoplasma* ME49 chromosomes, based on the current genome assembly and annotation. The 5,509 fosmid clones were found to provide deeper coverage of the 14 *T. gondii* chromosomes than did the 7,773 cosmid clones (**Fig. 2A**). We also performed an analysis to compare the numbers of genes covered per chromosome by cosmid (**Fig. 2B**) and fosmid clones (**Fig. 2C**). The fosmid clones, even though fewer, covered 79 more genes in total than the cosmids (also see **Table S2** in the supplemental material).

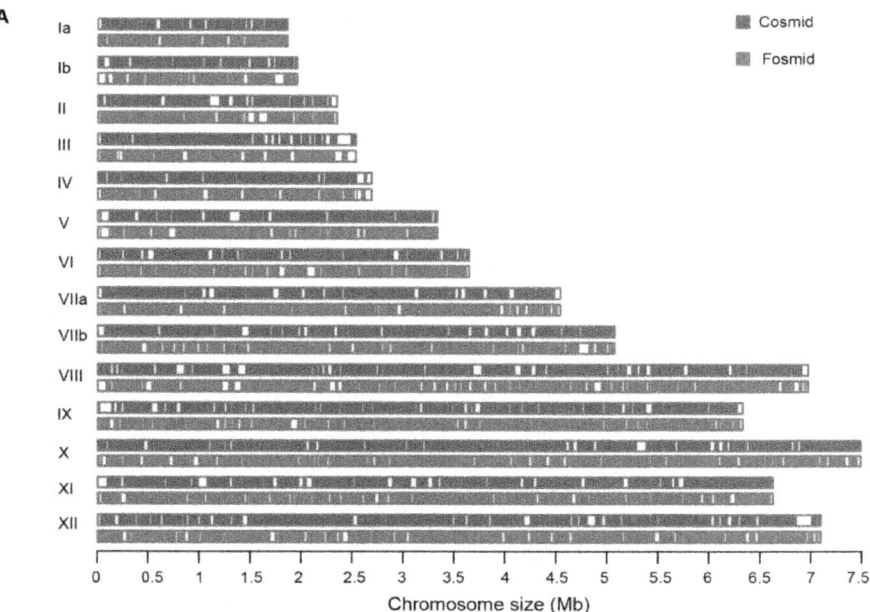

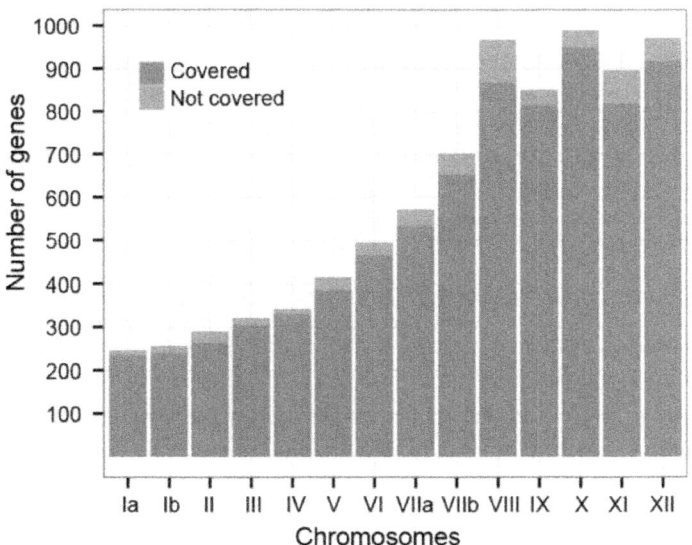

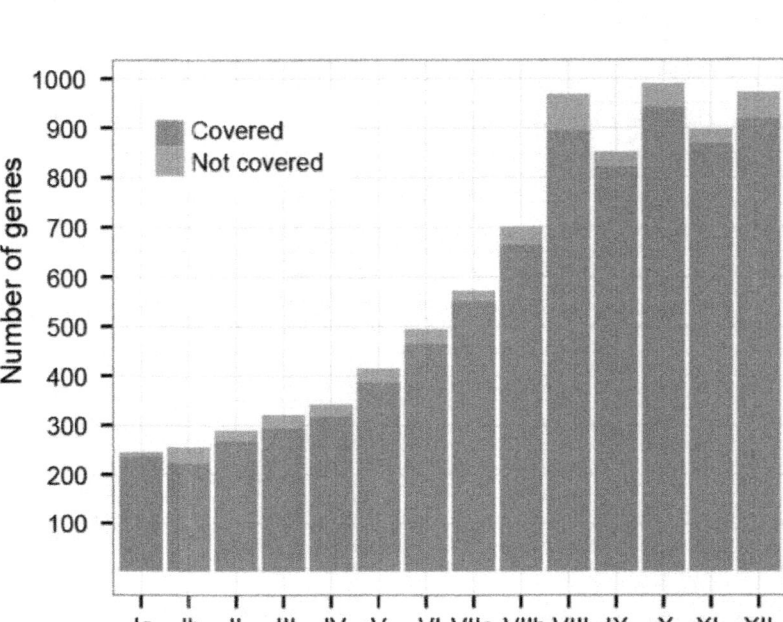

Figure 2 : Comparison of coverages of *T. gondii* chromosomes by cosmids and fos-mics. (A) Ideogram showing the regions of the *T. gondii* chromosomes covered by cosmids (red) and fosmids (blue). White spaces indicate no coverage. (B and C) Cosmids (B) and fosmids (C) mapping to genes on the 14 *T. gondii* chromosomes. The genes covered on each chromosome are shown in coral, whereas those that are not covered are shown in turquoise.

Modification of Fosmid DNA via Recombineering

We wanted to optimize a recombineering strategy to modify our set of arrayed and sequenced fosmids for transfection into the parasite to create gene knockouts. Given our interest in understanding the role of nucleolar proteins, we picked fosmid RHfos05J01 (TGME49_chrIX, positions 2582982 to

2625679; size, 42.69 kb) as an example. This clone covers the *T. gondii* gene (TGME49_288530) that encodes a protein belonging to the NOL1/NOP2/ SUN family. To create a conditional knockout for the SUN gene, we grew bacteria carrying fosmid RHfos05J01 recovered from our frozen clone collection and used the gentamicin-dihydrofolate reductase (DHFR)-T7S4 modification cassette (**9**) to replace the endogenous promoter in the fosmid with a tetracycline-regulatable promoter (**Fig. 3A**). This cassette was amplified with long primers (SUN_PR_F and SUN_PR_R) that amplify the cassette and also contain homology flanks that target 50 bp 5′ of the promoter and 3′ of the initiation codon of the SUN gene (primer sequences are provided in **Table S1** in the supplemental material). The RHfos05J01 fosmid was grown in chloramphenicol, and recombination was induced after electroporation with plasmid pSC101gbaArec. The amplified SUN promoter replacement cassette was then introduced, and recombination led to the replacement of the SUN gene promoter with the T7S4 promoter. The fosmids were selected on chloramphenicol and gentamicin plates, and the colonies that grew were screened by PCR using primers P3 and P4. No growth was observed on chloramphenicol and gentamicin plates for clones in which the recombination machinery was not switched on and for no-DNA controls. The copy number of the modified SUN fosmid clone was induced using L-arabinose to produce bulk DNA for parasite transfection.

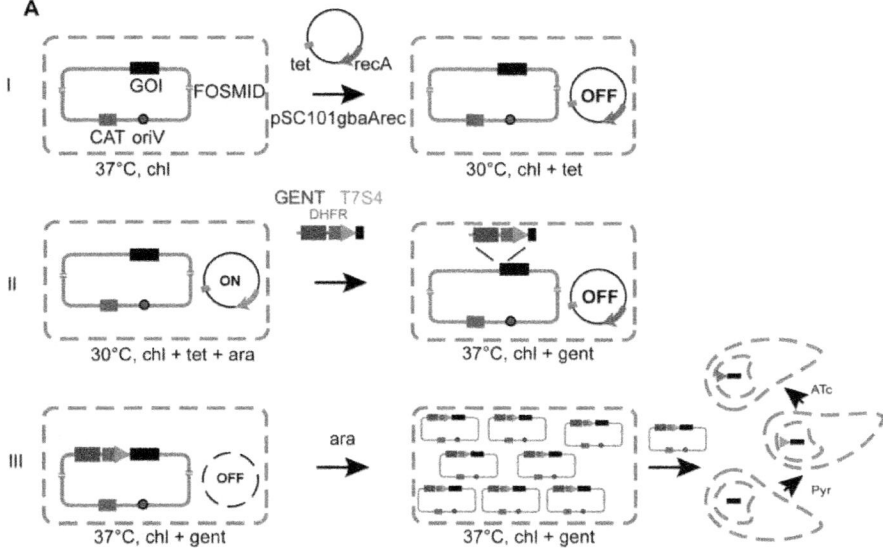

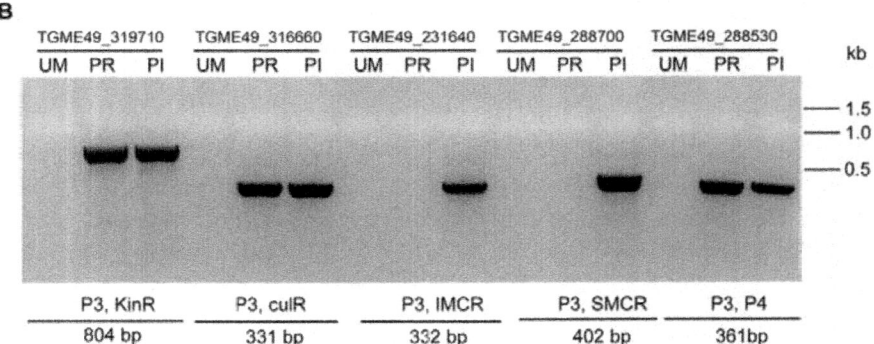

Figure 3: Fosmid modification by recombineering. (A) Steps of fosmid modification. In step I, the fosmid containing the gene of interest (GOI) is grown overnight in the presence of chloramphenicol (chl) at 37°C. Plasmid pSC101gbaArec, a plasmid carrying recombination Red γ, β, and α proteins is transformed into the cells. This plasmid has a tetracycline (tet) resistance marker and a temperature-sensitive origin of replication that allows growth only at 30°C, and the recombinase genes under the tight control of an arabinose-regulatable promoter. GENT, gentamicin. In step II, addition of L-arabinose (ara) switches on the recombination machinery. Cells are transformed with the gentamicin-DHFR-T7S4 promoter PCR cassette containing 50-bp regions of homology flanking the 5′ and 3′ ends of the cassette. In step III recombination occurs, and the cassette is integrated, leading to replacement of the endogenous promoter with the T7S4 promoter, at which time cells are grown with chloramphenicol and gentamicin. Overnight growth at 37°C leads to removal of the pSC101gbaArec recombination protein expression plasmid, preventing further rearrangement of the construct. The modified fosmid is induced to high copy numbers by addition of L-arabinose. Modified fosmid DNA can now be used directly for transfection into *T. gondii*, and the parasites are selected with pyrimethamine (Pyr). After homologous recombination in the parasite, the T7S4 promoter replaces the endogenous promoter and gene expression can be regulated upon addition of anhydrotetracycline (ATc). (B) PCR detection of modification of multiple fosmids performed using liquid recombineering. The results of PCR screening of five different fosmid clones to detect promoter replacement (PR) with the tetracycline-regulatable promoter or insertion (PI) are shown. No band is seen for the unmodified fosmids (UM). The PCR results shown here are an inverted image of an ethidium-bromide-stained agarose gel. The sequences of the primers used for PCR amplification are shown in **Table S1**, and further detail on the fosmid clones and the genes that they cover are provided in Table S3 in the supplemental material.

We also wanted to optimize a liquid recombineering protocol in deep 96-well plates. This protocol may be faster, as it avoids plates and colony picking, and the recombineering procedure can be performed in a high-throughput manner to allow modification of multiple fosmids in parallel. We repeated the modification of the SUN gene with two different targeting cassettes (resulting in promoter replacement or insertion) and picked four additional fosmids (see **Table S3** in the supplemental material). Ten independent fosmid modifications were conducted in parallel. Eight of the 10 targeted constructs were modified successfully using liquid recombineering in a 96-well plate in the first attempt (see **Table S3** in the supplemental material) as determined by PCR screening (**Fig. 3B**). No bacterial growth was observed for experiments in which the recombination machinery was not induced or in which a modification cassette was not supplied, indicating a negligible background. Our results suggest that the liquid recombineering procedure is very robust and fast. A targeting construct with large flanks on both sides could be obtained in 4 days; 7 days is required for modification when bacteria are plated and cloned at each step.

Modified Fosmid DNA Efficiently Replaces the Endogenous Promoter of the *T. gondii* SUN Gene

We tested the utility of modified fosmids to create conditional knockout mutants of *T. gondii* using the locus of the *T. gondii* SUN (TgSUN) protein. We first introduced an HA_3 epitope tag into this locus in a TATi ΔKu80 background using single homologous recombination (7) to be able to later follow the fate of the gene and protein. Analyzing the resulting drug-resistant clones by immunofluorescence assay, we found that TgSUN is a nucleolar protein that is robustly expressed in the tachyzoite stage and formed a ring-shape structure in the nucleolus of the parasite (**Fig. 4A**). To further describe the nucleolar compartment to which this protein localizes, we used an antibody to fibrillarin (to stain the dense fibrillar compartment) and found that TgSUN surrounds this protein (**Fig. 4B**). Thus, the SUN protein is localized to the outermost granular compartment of the nucleolus. Western blot analysis revealed that the tagged protein was of the predicted molecular mass of ~91 kDa (87.6-kDa SUN protein plus the 3.17-kDa epitope tag), and no reactive band was observed in the TATiΔKu80 parental line (**Fig. 4C**).

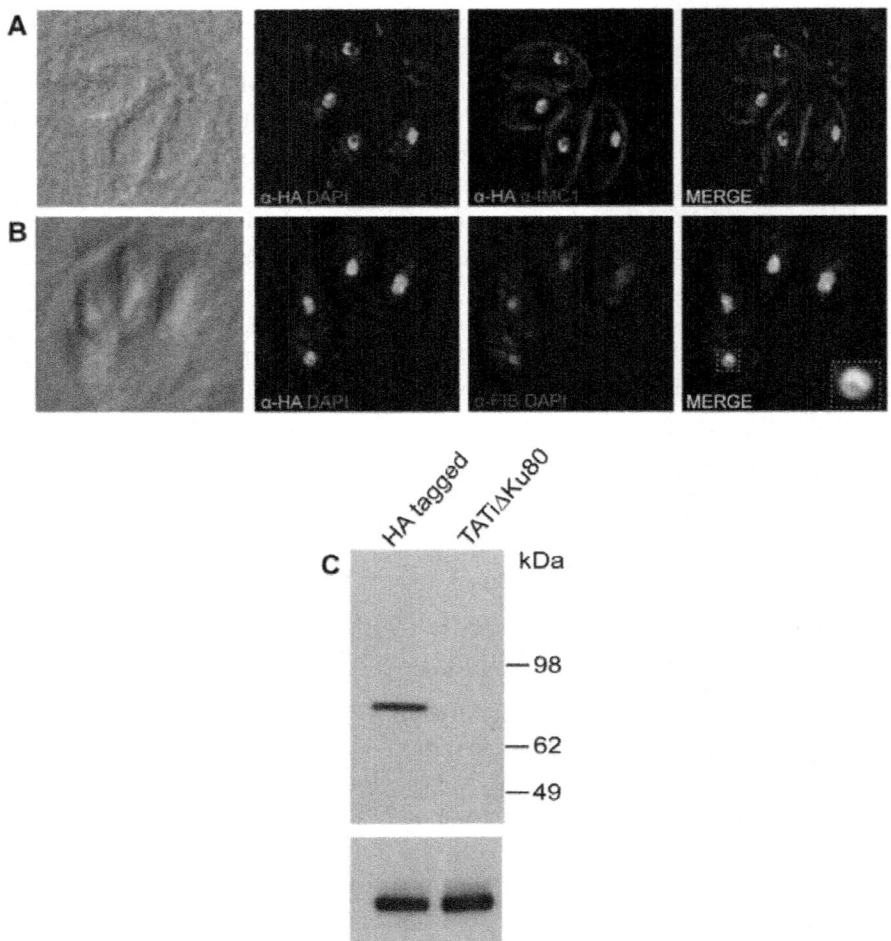

Figure 4: Localization and expression of the *T. gondii* SUN protein. (A) Fluorescence microscopy of the C-terminally HA-tagged SUN protein using anti-HA (green) and IMC1 (red [to outline parasite cells]) antibodies shows its nucleolar localization. Nuclei are stained with DAPI (4′,6-diamidino-2-phenylindole) (blue). (B) SUN protein is associated with the granular compartment (outer) of the nucleolus. Fibrillarin antibody staining of the dense fibrillar compartment is shown in red, and the HA-tagged SUN protein (green) forms a ring-shaped structure around the fibrillarin. (C) Western blot of parasite pellets showing expression of HA$_3$-tagged SUN protein using anti-HA antibody. The TATi ΔKu80 parental line showing no expression is included as a control. The lower panel shows the loading control using anti-α-tubulin antibody.

We then transfected this tagged parasite line with the modified fosmid DNA and selected for pyrimethamine resistance conferred by the DHFR marker that we introduced into the fosmid. Clones ($n = 48$) were screened by PCR for double homologous recombination and replacement of the endogenous promoter with the regulatable T7S4 promoter (**Fig. 5A**). Since the fosmid insert is large (42 kb), it was impractical to amplify a diagnostic PCR product that was anchored by a primer in the modification cassette on one side and by a primer in a region outside the fosmid on the other. Therefore, we performed a PCR screen to detect the loss of the endogenous promoter due to replacement by the modification cassette (**Fig. 5B**, amplification with primers P4 and P5) and found no amplification in the promoter replacement clone, while amplification was seen in the TATi ΔKu80 parental line. A success rate of 35% for homologous recombination was obtained, as 17 of the 48 clones screened showed replacement of the endogenous promoter with the T7S4 promoter. We also used primers flanking the bacterial chloramphenicol acetyltransferase (CAT) resistance marker found on the backbone of the fosmid. Double homologous recombination of the modification cassette into the chromosomal locus would eliminate this sequence; in contrast, episomal maintenance of the fosmid would preserve it. No amplification for the CAT gene using primers P6 and P7 was observed, and we therefore conclude that the promoter was successfully replaced in the parasite (**Fig. 5B**).

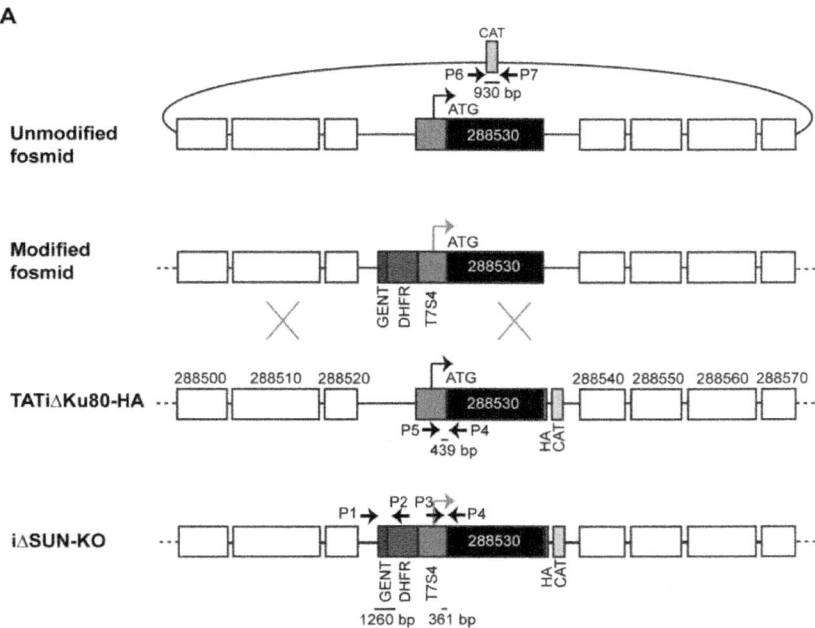

B

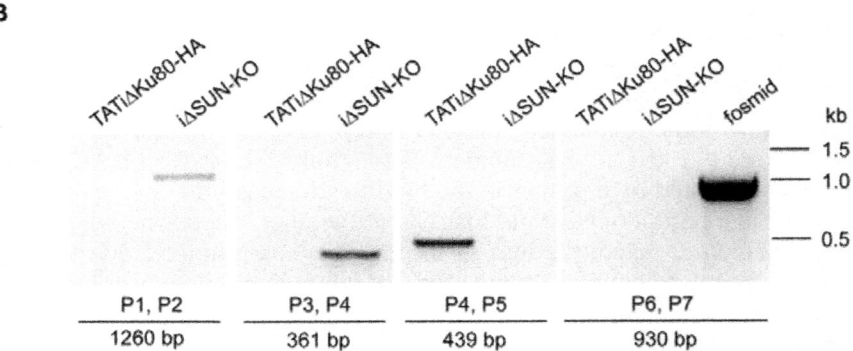

P1, P2	P3, P4	P4, P5	P6, P7
1260 bp	361 bp	439 bp	930 bp

Figure 5: Fosmid-based generation of the conditional SUN2 mutant by promoter replacement. (A) Scheme of replacement of the endogenous promoter with an inducible tetracycline-regulatable promoter (T7S4) via double homologous recombination in the TATi ΔKu80 HA-tagged line using fosmid DNA. The fosmid DNA is modified by recombineering in *E. coli* to introduce the gentamicin-DHFR-T7S4 cassette in place of the endogenous promoter for the SUN gene as described in the legend of **Fig. 2**. The modified fosmid DNA is then transfected into *T. gondii* for homologous recombination and promoter replacement to occur. The genes upstream (TGME49_288500, _288510, and _288520) and downstream (TGME49_288540, _288550, _288560, and _288570) of SUN are shown as unfilled boxes. The HA$_3$ tag and the chloramphenicol resistance marker in the parasite are displayed in pink and light blue, respectively. (B) PCR mapping of the SUN promoter replacement clone using primers indicated in panel A. The integration at the 5′ (P1 and P2 primers) and 3′ (P3 and P4 primers) ends of the promoter replacement is documented. PCR amplification with a primer set (P4, P5) that amplified within the endogenous promoter region in TATi ΔKu80, but not in the iΔSUN-KO line, is also shown. Primers (P6, P7) flanking the CAT resistance marker on the fosmid backbone resulted in a 930-bp band in the modified fosmid (shown as a control) but no amplification in the iΔSUN-KO and TATi ΔKu80 parasite lines. The primer sequences are provided in Table S1 in the supplemental material. The PCR results shown here are an inverted color image of an ethidium-bromide-stained agarose gel.

Modified Fosmid Provides Tight Regulation of Protein Expression

The tetracycline-inducible SUN knockout clone (iΔSUN-KO) was grown in the presence of anhydrotetracycline (ATc) to assess the regulation provided by the introduced modified fosmid cassette. Western blotting using anti-HA antibody revealed that the expression of the SUN protein was markedly downregulated over the course of 48 h of ATc treatment (**Fig. 6A**). This reduction was also apparent in immunofluorescence assays performed after 24 h of ATc treatment; these parasites also showed a more condensed labeling that had lost its typical ring shape. The remaining protein was now localized to the dense fibrillar component and colocalized with fibrillarin (**Fig. 6B**, +ATc, 24-h panels). After 48 h of ATc treatment, nucleolar TgSUN was no longer detectable (**Fig. 6B**, +ATc, 48-h panels). No change in nucleolar morphology was observed in the iΔSUN-KO and parental line in the absence of ATc (**Fig. 6B**, −ATc panels).

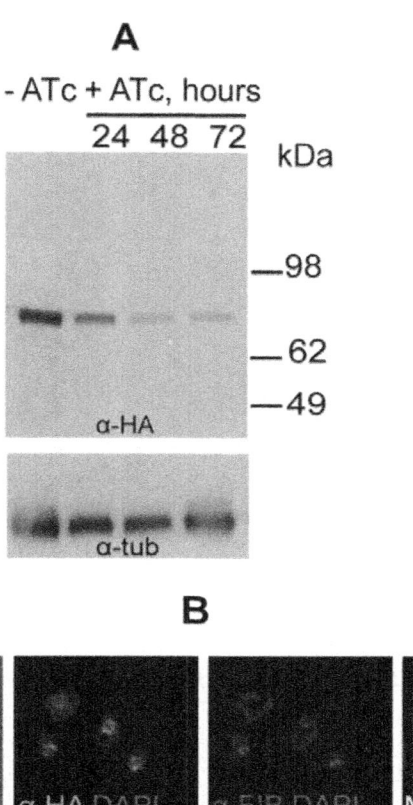

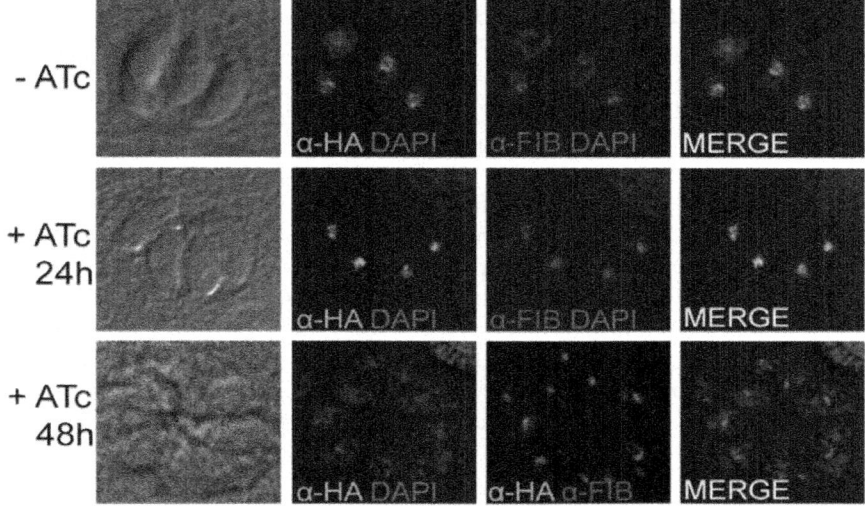

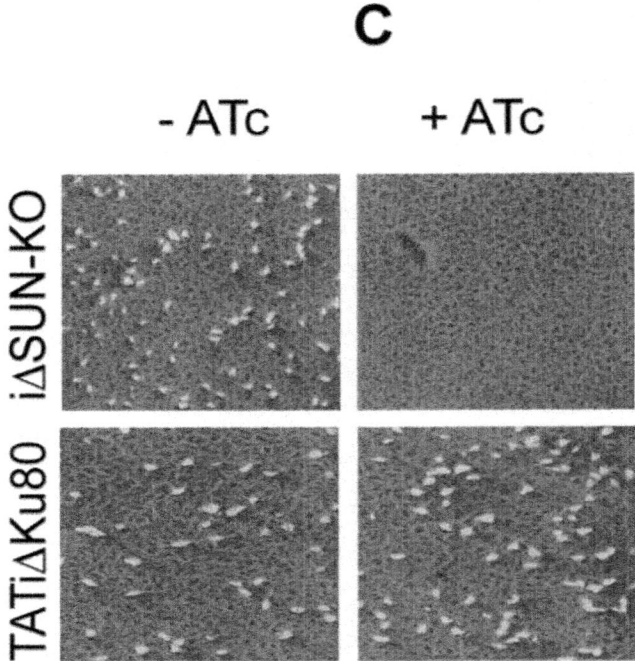

Figure 6: Regulation of SUN protein expression. (A) Western blot using anti-HA antibody protein lysates of the iΔSUN-KO line obtained from cells grown in the absence (−ATc) or presence (+ATc) of ATc for different lengths of time. Note the rapid decrease in expression of the SUN protein after the addition of ATc. The lower panel shows labeling with an anti-α-tubulin antibody as a loading control. (B) Fluorescence microscopy of *T. gondii* iΔSUN-KO parasites grown in the absence and presence of ATc for 24 and 48 h. The nucleus is stained with DAPI (blue); anti-HA (green) and anti-fibrillarin (red) antibodies show the nucleolar localization of the SUN protein. In the absence of ATc, the ring-like nucleolar localization pattern is indistinguishable from the native pattern shown in **Fig. 3B**, where the SUN protein surrounds the fibrillarin-labeled zone. In the presence of ATc, initially, the remaining SUN protein seems condensed, losing its ring-like pattern, and instead colocalizes with fibrillarin (24 h), whereas at 48 h of ATc treatment, staining for the SUN protein is no longer observed. (C) Plaque assays in the absence (−) or presence (+) of ATc. No plaques are seen in the iΔSUN-KO parasite line upon addition of ATc and after growth for 7 days, indicating that this gene is essential for parasite growth. The TATi ΔKu80 HA-tagged parental line grown in the presence or absence of ATc is shown as a control. Quantification of plaque size is shown in **Fig. S1** in the supplemental material.

Next we measured the growth of iΔSUN-KO and its TATi ΔKu80 HA-tagged parental line in the presence and absence of ATc by plaque assays. We did not observe plaques when the promoter replacement parasites were grown in the presence of ATc for 7 days, indicating that the protein is essential for parasite growth (**Fig. 6C**). The parental line continued to grow efficiently in the presence of ATc and is shown as a control (**Fig. 6C**). These results suggest that fosmid recombineering is suitable for modifying essential genes in the *T. gondii* parasite. In the absence of ATc, the plaques observed for the iΔSUN-KO promoter replacement line were smaller in size than those for the TATi ΔKu80-HA parental line (see **Fig. S1** in the supplemental material; the difference was moderate yet statistically significant). This may be due to differences between the level of expression of the SUN protein driven by the T7S4 promoter and the level of expression of the protein driven by its endogenous promoter.

Fosmid Complementation of a *ts* Mutant Identifies a New Variant of the *Toxoplasma* RCC1 Ortholog as the Key Protein Responsible for Conditional Growth Arrest

Large-insert clones of genomic DNA are ideal for complementation analysis. We explored the potential of this new library for genetic complementation of temperature-sensitive (*ts*) mutants. Specifically we tested the *ts* mutant 13-136A8 strain isolated in a previously described large chemical mutagenesis screen (**28**). This mutant was found to grow normally at 34°C but showed a severe growth defect when cultured at 40°C. Phenotypic characterization revealed that the majority of the cells were unable to complete a second round of the division (**Fig. 7A**, 40°C panel) and were arrested in the premitotic stage, as evidenced by costaining with the nuclear centrocone marker MORN1 and apicoplast protein Atrx1. Neither expansion of the centrocone that normally occurs in the S/M phase (**Fig. 7A**, 34°C, anti-MORN1 staining) nor duplication of the apicoplast was detected in the 13-136A8 mutant parasites grown at 40°C. These phenotypic features together with the absence of budding pointed toward a likely S-phase arrest of the *ts* mutant 13-136A8. We observed irregular DNA staining, with evident relaxation of the chromatin in the proximal part of the nucleus (**Fig. 7A**, graphs, blue DAPI line) and overcondensation in the apical region, although the over condensation region was close to but clearly segregated from plastid DNA (**Fig. 7A**). This mutant represents the first S-phase mutant of *Toxoplasma* characterized.

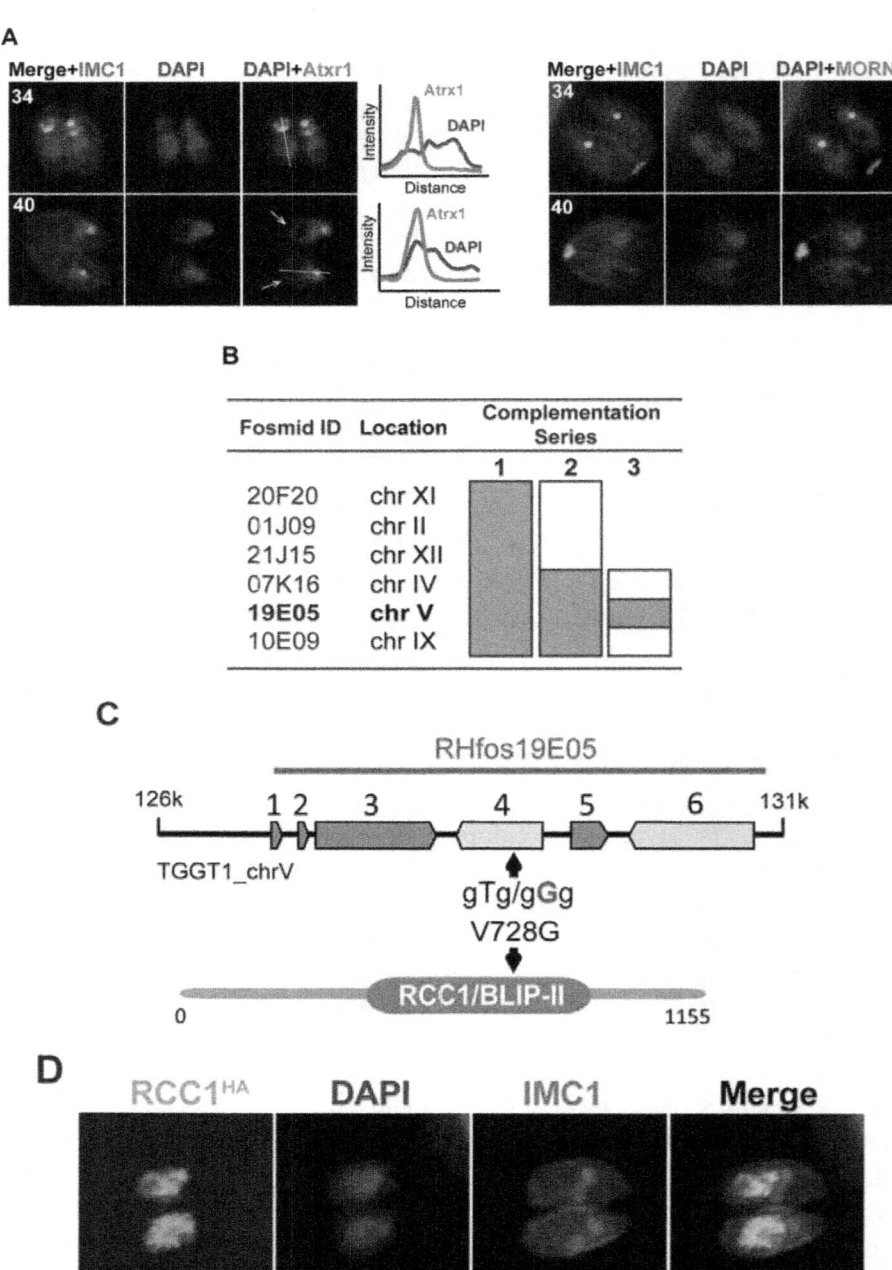

Figure 7: Growth defect and complementation of the *ts*mutant 13-136A8 using fosmids. (A) Mutant parasites were grown for 24 h at 34°C or 40°C and stained with

anti-Atrx1 (left panel) and anti-MORN1 (right panel) to detect apicoplasts and nuclear centrocone, respectively. Cells were costained with DAPI (blue) and anti-IMC1 (red). The white arrows indicate diffuse distribution of the nuclear chromatin in the parasites arrested at 40°C, whereas DNA condensation was not affected in the cells grown at 34°C. Quantification of the DAPI fluorescent signal across the nucleus (white line) is presented on the graphs. Incubation at 40°C resulted in the overcondensation of the chromatin at the apical end (overlapping green and blue peaks [40°C]) and DNA relaxation at the proximal end (blue line, right shoulder). Note that at 40°C, the mutant was unable to replicate the plastid or the nuclear centrocone, which is consistent with a premitotic arrest in S phase. (B) Genetic complementation of the *ts* mutant 13-136A8 was performed using fosmids in three sequential steps. A pool of fosmids covering predicted SNPs were used in the series 1 complementation. In series 2, the fosmid pool was divided into two sets, and in the third and final round, three fosmids were tested individually. Successful growth of the *ts* mutant at 40°C is shown in green. (C) The genomic locus spanned by the insert of fosmid clone RHfos19E05 resulted in rescue of the high temperature sensitivity of the *ts* mutant 13-136A8. Fosmid RH-fos19E05 contains six genes: TGGT1_213880, TGGT1_213885, TGGT1_213890, TGGT1_213900, TGGT1_213910, and TGGT1_213920. The V728G mutation identified by whole-genome sequencing in the RCC1/BLIP-II domain of TGGT1_213900 is indicated with an arrow. (D) HA-tagged TgRCC1 localizes to the parasite nucleus, consistent with the S-phase phenotype of the 13-136A8 *ts* mutant. Costaining of the TgRCC1 protein with anti-HA antibody (green), DAPI (blue) (nuclear staining), and anti-IMC1 antibody (red) is shown.

Previously, we developed a genetic protocol to identify the defective gene in *ts* mutants that employs primary complementation with a cosmid genomic library followed by marker rescue, secondary complementation with single cosmids to resolve the locus, and, finally, sequencing to identify the point mutation (**28**). While successful, this protocol is low throughput, labor-intensive, and time-consuming. To identify the mutation responsible for this growth defect, we developed an updated strategy taking advantage of the reduced cost of whole-genome sequencing and the fosmid library. We first performed whole-genome comparison of this mutant and its parent RH strain and identified SNPs resulting in nonsynonymous mutations in eight predicted proteins (see **Table S4** in the supplemental material). We prioritized candidates further by requiring genes to show expression in tachyzoites (>15th-percentile expression in tachyzoites) and a mutation leading to a nonconservative amino acid change. This narrowed the field to six genes that were selected for genetic complementation of the 13-136A8 mutant. Six corresponding fosmid clones spanning each gene were identified in the library and used for DNA preparation. We used a stepwise protocol to avoid the pitfall of reversion (**Fig. 7B**). The initial transformation was carried out using a pool of all six fosmids; this rescued the growth of the 13-136A8 mutant at 40°C.

Next, the mutant was transformed with two sets of three fosmids each (set 1, RHfos20F20, RHfos01J09, and RHfos21J15, and set 2, RHfos07K16, RHfos19E05, and RHfos10E09). Only set 2 was able to restore growth at the restrictive temperature. When fosmids from set 2 were individually tested, only fosmid RHfos19E05 resulted in phenotypic complementation (**Fig. 7B**). This 39.45-kb fosmid (chromosome V, positions 1401130 to 1440586) spans six genes, including TGGT1_213900, bearing the mutation V728G identified by whole-genome sequencing in a conserved RCC1/BLIP-II region (**Fig. 7C**). The complementing gene TgRCC1 encodes a predicted regulator of chromosome condensation, and the role of this factor in nuclear trafficking in *T. gondii* has been previously reported (**29**). Consistently with the S-phase phenotype of this mutant, endogenous HA_3 epitope tagging of RCC1 revealed a granular nuclear localization (**Fig. 7D**), which validates the results reported previously for this protein (**29**). Nuclear localization and conserved function in chromatin organization explain the DNA relaxation observed in the TgRCC1-deficient cells, which likely contributes to the temperature-sensitive arrest of parasites in the premitotic stage, which in other eukaryotes requires proper DNA condensation. Taken together, whole-genome sequencing followed by serial pool complementation with sets of fosmid clones rapidly reduces the effort and time required to pinpoint the mutations responsible for conditional growth in *ts*mutants.

DISCUSSION

We describe an improved system for genetic analysis and modification to study the biology of the parasite *T. gondii*. The main resource generated in this effort is an arrayed and end-sequenced fosmid library that was produced by random shearing, with an average insert size of 35 kb. This library yielded excellent coverage of the parasite's genome, exceeding and extending the available resources (**14, 30**). The sequenced clones cover ~95% of the genes that often provide researchers multiple fosmids, covering a particular gene of interest to choose from. We demonstrate several important applications of this resource.

The deep coverage of the library and the high stability of its clones make it an excellent resource for complementation assays. Forward genetic screens are very powerful tools of discovery, as they do not require preconceived notions of mechanism and function. While mutants are easily generated by chemical mutagenesis (**31**), linking mutation to phenotype has required significant effort (**28**). Improvements in sequencing technology now permit mutant identification by whole-genome sequencing (**27, 32**), and this technology will likely further improve and become more affordable. However, mutagenesis typically produces multiple changes per genome that have to be validated one by one. We demonstrate here that the arrayed genome resource can be used to quickly and rigorously reduce that complexity.

The arrayed genome also provides avenues for genome engineering and reverse genetic analysis. The optimized liquid recombineering pipeline in *E. coli* that we describe here for *T. gondii* fosmid modification is highly efficient and scalable and thus can be used to modify numerous clones in parallel (using the deep, 96-well plate format). This robust recombineering procedure also substantially reduces the time required to create constructs for gene knockouts. This platform can also be used for gene tagging, the introduction of point mutations at desired sites, and parasite complementation experiments. The power of liquid recombineering was first described to generate green fluorescent protein (GFP)-tagged transgenes from *Caenorhabditis briggsae* bacterial artificial chromosome (BAC) genomic clones (**33**). This technology has successfully been applied to engineer fosmid clones in *Drosophila* (**34**) and to the bacteriophage N15-based library for *Plasmodium berghei*, albeit with a smaller average insert size of 9 kb due to the AT-rich nature of the *Plasmodium* genome (**13**). Importantly, this resource is easily combined with other approaches. The use of CRISPR/Cas9 has recently been described for *T. gondii* for introducing double-strand breaks and allows efficient gene disruption (**10, 11**). The collection of the fosmid clones and the robust recombineering pipeline make the library an attractive system to provide homologous recombination donor constructs for modification experiments. Furthermore, while yielding exciting efficiency, CRISPR/Cas9 has some limitation stemming from significant off-target effects (**35–37**). Having available a technology that relies on simple homologous recombination without additional mutation may be a strength in certain settings.

In this study, we document the utility of the fosmid modification approach to create a conditional knockout for the essential nucleolar SUN gene in *T. gondii* via a promoter replacement strategy. The high efficiency of recombination that occurs due to the long homology arms of the fosmid made it easy to isolate stable transgenic clones in the parasite. This is the first SUN protein described for any apicomplexan parasite. The first SUN domain protein (Fmu) was described in *E. coli* and methylates C-967 in the 16S rRNA, while other members of the family containing this domain are involved in the methylation of tRNAs (**38**). The best studied among them is the Myc-induced SUN domain-containing protein (Misu or NSun2), a nucleolar tRNA methyltransferase important for c-Myc-induced proliferation in skin that is required for proper mitotic spindle assembly and cell cycle progression (**39–41**). Our results indicate the essential role of the *T. gondii* SUN protein in nucleolar stability and parasite growth.

In conclusion, the arrayed and sequenced *T. gondii* fosmid clone library is a valuable resource to the parasitology community, enabling and enhancing both forward and reverse genetic analysis.

MATERIALS AND METHODS

Toxoplasma gondii Fosmid Library Preparation

High-molecular-weight genomic DNA was extracted from *Toxoplasma gondii* RH strain parasites. Briefly, parasites from four T-175 parasite flasks were harvested and filtered through 3-μm polycarbonate membrane filters, followed by DNase I treatment to remove host cell DNA contamination. DNase I was heat inactivated at 75°C for 20 min, and the parasite pellet was suspended in 10 ml of 10 mM Tris, 1 mM EDTA (TE), pH 8.0. Parasites were lysed in 0.2% SDS and RNase A (0.04 μg/ml) for 4 h at 37°C. Proteinase K was added to the lysate and incubated at 37°C for 1 h, followed by overnight digestion at 56°C. DNA was extracted with phenol-chloroform-isoamyl alcohol (25:24:1) followed by chloroform extraction and precipitation using 0.2 volumes of 10 M ammonium acetate and 2 volumes of 100% ethanol. Precipitated DNA was spooled out of the tube and resuspended in TE overnight at 4°C.

The extracted DNA was aliquoted into tubes (20 μg per tube) and subjected to fragmentation using a HydroShear device employing a large shearing assembly (fragment size, 4 kb to >50 kb; Digilab Inc., MA). DNA shearing was monitored by pulse-field gel electrophoresis (PFGE) on a 1% agarose gel in 0.5× Tris-borate-EDTA (TBE) on a contour-clamped homogeneous electric field (CHEF) and crossed-field gel electrophoresis mapper system (Bio-Rad, CA) using the following parameters: 4 V/cm, an included angle of 120°, a switch time linearly ramped from 5 to 25 s, and 16 h at 16°C. The gel was stained with ethidium bromide and visualized on a UV transilluminator and compared with the Midrange II PFG marker (New England Biolabs, MA) to visualize the zone of DNA fragments in the ~40-kb range. Shearing was optimized in a series of pilot experiments and carried out at a speed code of 16 for 25 cycles, with a retraction speed of 20 for library construction.

DNA of the desired fragment length (~40 kb) was end repaired to generate blunt phosphorylated 5′ ends, ligated with the linearized pCC2Fos copy control vector, packaged using MaxPlax lambda packaging extracts, and transformed into phage EPI300-T1^R *E. coli* cells according to the manufacturer's instructions (Epicentre Biotechnologies, WI). Different dilutions of the package phage were plated to count colonies and determine the size of the library. The fosmid library contained 200,000 independent clones, and glycerol stocks of the library were prepared and frozen at −80°C.

Fosmid Library Clone Picking, Sequencing, and Mapping to the *T. gondii* genome

To characterize the library, 20 random fosmid clones were picked in 12.5 µg/ml chloramphenicol and induced to high copy numbers using 0.2% arabinose for miniprep DNA isolation. The clones were digested with NotI, recognition sites for which flank the insert on the fosmid, and insert size was determined by gel electrophoresis. The clones were also sequenced using vector-specific primers, PC1F and PC1R (primer sequences can be found in **Table S1** in the supplemental material). A dilution of library glycerol stock was plated onto large 22- by 22-cm rectangular QTrays containing 12.5 µg/ml chloramphenicol and incubated at 37°C overnight. A QBot robotic device (Genetix Inc., USA) was used to pick clones and array them into 384-well plates; the device was also used to replicate the plates. The clones were deposited into freezing medium [36 mM K_2HPO_4 (anhydrous), 13.2 mM KH_2PO_4, 1.7 mM sodium citrate, 0.4 mM $MgSO_4 \cdot 7H_2O$, 6.8 mM $(NH_4)_2SO_4$, 4.4% (vol/vol) glycerol in LB containing 12.5 µg/ml chloramphenicol], and plates were stored at −80°C. A set of plates was sent for end sequencing to Lucigen Corporation (WI) using primers PC1F and PC1R. The sequences obtained for each clone were subjected to BLAST search in the *Toxoplasma* database (http://www.toxodb. org), version 8.2, and mapped to the annotated ME49 genome.

High-Throughput Modification of Fosmids in *Escherichia coli* via Recombineering

The fosmid clones containing the gene of interest were picked from the 384-well plates and grown overnight at 37°C in 12.5 µg/ml chloramphenicol either in tubes (single fosmids) or in 2-ml-deep 96-well plates (for high-throughput modification of multiple fosmids). The overnight cultures were diluted, and a secondary inoculation was allowed to grow at 37°C with shaking at 250 rpm in the incubator (for tubes) or 900 rpm in a BioShake iQ ThermoMixer (Quantifoil Instruments GmBH, Germany) until the optical density at 600 nm (OD_{600}) reached 0.6 to 0.8. The tube or plate was chilled on ice for 15 min, and electrocompetent cells were prepared by pelleting and washing them three times in chilled sterile water. After the final washing step, 10 ng of recombination plasmid pSC101gbaArec (a kind gift from Oliver Billker, Wellcome Trust Sanger Institute, United Kingdom) was added. This is a low-copy-number plasmid and has a temperature-sensitive origin of replication and a tetracycline resistance marker. The resuspended pellet was transferred to a chilled 1-mm cuvette. Cells were electroporated with a BTX ECM 630 system (1,800 V, 25 µF, 200 Ω) and allowed to recover at 30°C for 70 min in antibiotic-free media. Chloramphenicol and tetracycline were added to the media to final concentrations of 12.5 μg/ml and 5 μg/ml, respectively, and cells were grown overnight at 30°C.

The overnight cultures were used for secondary inoculation and grown at 30°C to an OD of 0.3 to 0.4. The expression of recombination genes was induced by adding L-arabinose to a final concentration of 0.2%, and temperature was raised to 37°C for 40 min. The tubes/plates were chilled for 15 min, and electrocompetent cells were prepared. A recombineering PCR cassette was amplified with proofreading PrimeSTAR HS DNA polymerase (TaKaRa Bio Inc., Japan) using long primers (50-bp gene-specific overhangs and 25 bp matching the cassette). The PCR product was treated with DpnI to destroy the template plasmid prior to gel purification. Five hundred nanograms of this PCR recombineering cassette was delivered by electroporation as described above. After recovery, cultures were grown overnight at 37°C in media or plated on LB plates with 12.5 μg/ml chloramphenicol (for the fosmid) and 10 μg/ml gentamicin (for the cassette). The modification of fosmid clones was confirmed by PCR. Also, the confirmed modified fosmids were grown in chloramphenicol and gentamicin and induced with 0.2% arabinose (to increase fosmid copy number), and isolated miniprep DNA was subjected to Sanger sequencing.

Construction of a Tagged Reporter Parasite Line

T. gondii NOL1/NOP2/SUN (ToxoDB gene identifier, TGGT1_288530) and wild-type (wt) *T. gondii* RCC1 (TgRCC1) (ToxoDB gene identifier, TGGT1_213900) were tagged with a C-terminal 3-hemagglutinin (HA$_3$) epitope through modification of their respective genomic locus. For the SUN gene, a 2,726-bp region of the genomic sequence preceding the stop codon was amplified from *T. gondii* genomic DNA using primers SUN-LICF and SUN-LICR (see **Table S1** in the supplemental material). The amplified product was introduced by ligation-independent cloning (**42**) into the vector pLIC-HA$_3$-CAT (**7, 8**). *T. gondii* TATi ΔKu80 parasites (**9**) grown in human foreskin fibroblasts (HFF) were transfected with the linearized pLIC-SUN-HA$_3$-CAT plasmid and selected on chloramphenicol, and clonal parasite lines were isolated, as previously described (**4**). For the TgRCC1 gene, a 1,329-bp PCR DNA fragment was amplified using the RCC_LICF and RCC_LICR primers (see **Table S1** in the supplemental material), encompassing the 3' end of TgRCC1, to construct the plasmid pLIC-TgRCC1-HA3x/HxGPRC. This plasmid was electroporated into ΔKu80 parasites and subjected to positive selection on mycophenolic acid and xanthine.

Transfection of Modified Fosmids into Tagged Reporter Lines and Isolation of Conditional Mutants

The SUN fosmid was modified to replace the endogenous promoter with a tetracycline-regulatable promoter, tetO7sag4 (T7S4), using the recombineering procedure described above. The gentamicin-DHFR-T7S4 cassette described by Sheiner et al. (**9**) was used as a template to amplify the cassette with 50-bp homology flanks using primers SUN_PR_F and SUN_PR_R. The modified fosmid was transfected into TATi ΔKu80 parasites and selected on pyrimethamine. Clones were screened for promoter replacement, 5′ and 3′ integration, and the presence of the circulating fosmid. Growth of the conditional SUN mutant was measured by plaque assay in the presence and absence of 0.5 μM anhydrotetracycline (ATc). The area of plaques was quantified using ImageJ v1.84 software.

Whole-Genome Sequencing and Genetic Complementation

The *ts* mutant 13-136A8 and parental RH Δ*hxgprt* strain were grown at 34°C, and genomic DNA was extracted using the DNeasy blood and tissue kit (Qiagen GmbH, Hilden, Germany). Paired-end sequencing of the purified 13-136A8 and RH Δ*hxg prt* DNAs was performed at the Oklahoma Medical Research Foundation sequencing facility on the Illumina HiSeq 2500 platform using the chemistry and protocol recommended by the manufacturer (Illumina Inc., San Diego, CA). Coverages of 139 times (97 million reads) and 113 times (85 million reads), respectively, were achieved for the 13-136A8 and RH strains. The sequences were first filtered in PRINSEQ version 0.20.3 (**43**) to remove low-quality reads and were mapped against the *T. gondii*GT1 reference genome (ToxoDB version 9.0) using BWA short-read aligner version 0.7.2 (**44**). The single nucleotide polymorphisms (SNPs) were identified using the Genome Analysis tool kit (GATK) UnifiedGenotyper (**45, 46**). The SNPs detected in the 13-136A8 mutant were compared with SNPs in wild-type RH to identify the SNPs specific to the mutant. These SNPs were also manually verified by visualizing the alignment files in the Integrated Genome Viewer (IGV) tool (**47**).

Genetic complementation of the *ts* mutant 13-136A8 was performed using the pooled fosmid sets covering the SNPs identified by the whole-genome sequence comparison. In preliminary experiments, we measured the spontaneous frequency of temperature resistance in the 13-136A8 mutant to be well below 10^{-7}. In our complementation tests, we thus scored sustained growth at the restricted temperature of 40°C as a positive rescue. Differences between the abilities of fosmids to complement were readily observable as growth after transfection or a lack thereof. First, a complete set of six fosmids followed by two sets of three fosmids each were tested for complementation of the *ts* mutant 13-136A8. Finally, individual fosmids from the complemented set were used to confirm complementation and growth of the mutant at 40°C.

Fluorescence Microscopy

For immunofluorescence assays, HFF were grown on coverslips and infected with parasites. After 24 h of infection, coverslips were fixed with 4% paraformaldehyde, permeabilized with 0.25% Triton X-100 in phosphate-buffered saline (PBS), and blocked overnight in 4% bovine serum albumin (BSA) in PBS. Primary antibodies used were rat anti-HA (clone 3F10; Roche Applied Science, IN; 1:200 dilution), mouse monoclonal antifibrillarin (17C12, a gift from Michael Terns, University of Georgia; 1:1,000 dilution), mouse monoclonal anti-Atrx1 (a gift from Peter Bradley, University of California, Los Angeles, CA; 1:2,000 dilution), rabbit anti-MORN1 (kindly provided by Marc-Jan Gubbels, Boston, MA; 1:2,000 dilution), and mouse and rabbit anti-IMC1 (kindly provided by Gary Ward, University of Vermont, VT; 1:2,000 dilution). Secondary antibodies used were Alexa Fluor 488-, 546-, and 594-conjugated antibodies (Molecular Probes, Life Technologies, NY) at a dilution of 1:200. Images were collected on an Applied Precision Delta Vision inverted epifluorescence microscope using an Olympus UPlans APO 100×/1.40 oil lens and on a Carl Zeiss Axio Observer.Z1 inverted microscope with a Plan-Apochromat 100×/1.40 oil differential interference contrast (DIC) lens, deconvolved, and adjusted for contrast using SoftWoRx and AxioVs40 v4.8.1.0, respectively.

Western Blotting

Parasites were harvested, filtered, counted, washed with PBS, and pelleted by centrifugation. Parasites were suspended in lysis and loading buffer, heated to 100°C for 3 min, and run on a 4 to 12% Tris-glycine-SDS Mini-Protean precast gel (Bio-Rad, CA). Western blotting was performed as previously described (**48**). Monoclonal rat anti-HA tag (clone 3F10; Roche Applied Science, IN) and mouse anti-α-tubulin (12G10, a gift of Jacek Gaertig, University of Georgia) antibodies were used at 1:500 and 1:2,000 dilutions, respectively, to probe the blots. Horseradish peroxidase (HRP)-conjugated anti-rat or anti-mouse secondary antibody (Pierce, Thermo Scientific Inc., IL) was used at a 1:10,000 dilution and detected by chemiluminescence using the ECL Western blotting substrate (Pierce, Thermo Scientific Inc., IL).

SUPPLEMENTAL MATERIAL

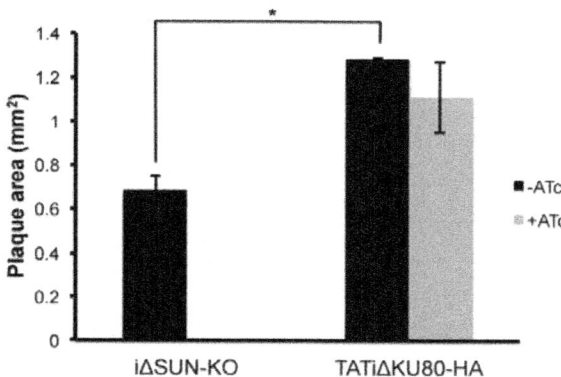

Figure S1: Quantification of the area of plaques formed by the iΔSUN-KO and TATi ΔKu80-HA parasite lines after 7 days of growth on HFF monolayers. The average area of 25 plaques was assessed for each line. The mean areas ± standard deviations (SD) are plotted for 3 independent experiments. In the absence of ATc, the plaques formed by the iΔSUN-KO line were significantly smaller in size than those of the parental TATi ΔKu80-HA line (*, $P < 0.0001$, Student's t test). ATc, anhydrotetracycline.

Table S1: Sequences of the primers used in the study

Primer	Sequence (5' to 3')
PC1F	GGATGTGCTGCAAGGCGATTAAGTTGG
PC1R	CTCGTATGTTGTGTGGAATTGTGAGC
SUN_LICF	TACTTCCAATCCAATTTAATGCACCACTGTACTGTTGATGCGT CTGTG
SUN_LICR	TCCTCCACTTCCAATTTTAGCCGCACGCTTCTTAGAAATACTT GC
SUN_PR_F	TGCGGCTGCCGCGAAATGCAGTGAAGGTATTTTCAGCAATTG GCACATGGAATGGTAACCGACAAACGCGTTC
RCC_LICF	TACTTCCAATCCAATTTAATGCACTAAGACAGACCT TGCCAAGGCCGC
RCC_LICR	TCCTCCACTTCCAATTTTAGCTCGACTGTTTGGGCGCC
SUN_PR_R	ATCCGTCCTCCCTTCTGGTCCAGTTGCCTCGACGCGGTTTTCT GTTCATAGATCTGGTTGAAGACAGACGAAAGC
P1	ATGCTAACAGCAGATGGCACGACTCTC
P2	GCACGGCAGTCAGATAACAGGTGTA
P3	CGCCTTGGCGAATGTTCATGAC
P4	GTAGCCGTCGTCCTGATCTTGGAC
P5	AACGCATGCACCTACAAATCGACAC
P6	ATCACTTATTCAGGCGTAGCAACC
P7	CGCGAATAAATACCTGTGACGGAAG
KinR	ATCTTTCCAGCTACACATCTGCGTC
culR	AAGACGTCGACTTCCGATGCAAAC
IMCR	CTGAGAAGCTTCGGCAGTCACGTG
SMCR	GTTTCTATGGCGGAAGTGACCATGT

Table S2: Number of genes covered by the mapped cosmids (n=7773) and fosmids (n=5509) per *Toxoplasma gondii* chromosome

Chromosome	Total number of genes	Number of genes covered by	
		Cosmids	Fosmids
Ia	246	234	239
Ib	256	241	224
II	290	263	270
III	321	304	293
IV	342	331	319
V	415	385	387
VI	495	466	466
VIIa	572	534	551
VIIb	702	653	666
VIIII	968	869	895
IX	851	816	823
X	990	949	942
XI	897	819	868
XII	972	918	918

Table S3: Fosmids modified using liquid recombineering

Gene ID	Chromosome	Product	Fosmid clone	Promoter Replacement	Promoter Insertion
TGME49_319710	IV	Kinesin motor domain-containing protein	RHfos24O11	Yes	Yes
TGME49_316660	XI	Cullin family protein	RHfos09P17	Yes	Yes
TGME49_231640	VIII	Alveolin domain containing intermediate filament IMC1 (ALV1)	RHfos22M15	No	Yes
TGME49_288700	IX	RecF/RecN/SMC N terminal domain-containing protein	RHfos05B19	No	Yes

Table S4: Single nucleotide polymorphisms (SNPs) in the Toxoplasma gondii ts-mutant 13-136A8 identified by whole genome sequencing

Gene ID	Product	SNP	Mutation	%mRNA[a]	mRNA[b]	Fosmid clone
TGGT1_311230	Hypothetical	Tgc/Agc	C3545S	70-95	M/C	RHfos20F20
TGGT1_204520	Hypothetical	aAT/aGt	N148S	5-15	M/C	RHfos10H15
TGGT1_255690	2-C-methyl-D-erythritol 2,4-cyclodiphosphate synthase	Aaa/Gaa	K75E	25-65	G1	RHfos01J09
TGGT1_248270	Zn-finger CCCH-type motif	Tcc/Gcc	S717A	55-70	n/p	RHfos2J15
TGGT1_301170	SAG-related sequence SRS19D	aAc/aGc	N282S	20-80	S	RHfos07K16
TGGT1_213900	Regulator of chromosome condensation RCC1	gTg/gGg	V728G	40-45	n/p	RHfos19E05
TGGT1_291050	Histone kinase SNF1	gTt/gCt	V1920A	15-35	S/M	RHfos10E09
TGGT1_206580	Formin FRM2	cAg/cTg	Q2795L	50-70	C/G1	-

[a]mRNA abundance in the tachyzoite stage (as percentile) from toxodb.org

[b]peak expression of the encoded mRNA from toxodb.org;

n/p: non-periodic (constitutive) expression

ACKNOWLEDGMENTS

This work was supported in part by U.S. National Institutes of Health RO1 grantsAI064671, AI084415 (to B.S.), AI077662, and A109843 (to M.W.W.). We thank Markus Meissner (Wellcome Trust Center for Molecular Parasitology, University of Glasgow, United Kingdom) for providing additional funding to sequence the fosmid library.

We thank Cornelia Lemke (Plant Genome Mapping Laboratory, University of Georgia) for her help with automated colony picking, replication, and arraying of the fosmid library and the EuPathDB team for adding the fosmid data set to the ToxoDB.

REFERENCES

1.   Weiss LM, Kim K. 2013. Toxoplasma gondii. The model apicomplexan: perspectives and methods, 2nd Éditions Elsevier Academic Press, Burlington, VT. Google Scholar

2.   Donald RG, Roos DS. 1993. Stable molecular transformation of Toxoplasma gondii: a selectable dihydrofolate reductase-thymidylate synthase marker based on drug-resistance mutations in malaria. Proc. Natl. Acad. Sci. U. S. A. 90:11703–11707. 10.1073/pnas.90.24.11703. Abstract/FREE Full Text

3.   Soldati D, Boothroyd JC. 1993. Transient transfection and expression in the obligate intracellular parasite Toxoplasma gondii. Science 260:349–352. 10.1126/science.8469986. Abstract/FREE Full Text

4.   Striepen B, Soldati D. 2007. Genetic manipulation of Toxoplasma gondii, p 391–418. In Weiss LM, Kim K (ed), Toxoplasma gondii. The model apicomplexan: perspectives and methods. Academic Press, Elsevier, London, United Kingdom. Google Scholar

5.   Herm-Götz A, Agop-Nersesian C, Münter S, Grimley JS, Wandless TJ, Frischknecht F, Meissner M. 2007. Rapid control of protein level in the apicomplexan Toxoplasma gondii. Nat. Methods 4:1003–1005. 10.1038/nmeth1134. CrossRefMedlineGoogle Scholar

6.   Meissner M, Schlüter D, Soldati D. 2002. Role of Toxoplasma gondii myosin A in powering parasite gliding and host cell invasion. Science 298:837–840. 10.1126/science.1074553. Abstract/FREE Full Text

7.   Huynh MH, Carruthers VB. 2009. Tagging of endogenous genes in a Toxoplasma gondii strain lacking Ku80. Eukaryot. Cell 8:530–539. 10.1128/EC.00358-08. Abstract/FREE Full Text

8.   Fox BA, Ristuccia JG, Gigley JP, Bzik DJ. 2009. Efficient gene replacements in Toxoplasma gondii strains deficient for nonhomologous end joining. Eukaryot. Cell 8:520–529. 10.1128/EC.00357-08. Abstract/FREE Full Text

9.   Sheiner L, Demerly JL, Poulsen N, Beatty WL, Lucas O, Behnke MS, White MW, Striepen B. 2011. A systematic screen to discover and analyze apicoplast proteins identifies a conserved and essential protein import factor. PLoS Pathog. 7:e1002392. 10.1371/journal.ppat.1002392. CrossRefMedlineGoogle Scholar

10. 10.↵ Shen B, Brown KM, Lee TD, Sibley LD. 2014. Efficient gene disruption in diverse strains of Toxoplasma gondii using CRISPR/CAS9. mBio 5(3):e01114-14. 10.1128/mBio.01114-14. Abstract/FREE Full Text

11. Sidik SM, Hackett CG, Tran F, Westwood NJ, Lourido S. 2014. Efficient genome engineering of Toxoplasma gondii using CRISPR/Cas9. PLoS One 9:e100450. 10.1371/journal.pone.0100450. CrossRefGoogle Scholar

12. Court DL, Sawitzke JA, Thomason LC. 2002. Genetic engineering using homologous recombination. Annu. Rev. Genet. 36:361–388. 10.1146/annurev.genet.36.061102.093104. CrossRefMedlineGoogle Scholar

13. Pfander C, Anar B, Schwach F, Otto TD, Brochet M, Volkmann K, Quail MA, Pain A, Rosen B, Skarnes W, Rayner JC, Billker O. 2011. A scalable pipeline for highly effective genetic modification of a malaria parasite. Nat.

Methods 8:1078–1082. 10.1038/nmeth.1742. CrossRefMedlineGoogle Scholar

14. Brooks CF, Johnsen H, van Dooren GG, Muthalagi M, Lin SS, Bohne W, Fischer K, Striepen B. 2010. The Toxoplasma apicoplast phosphate translocator links cytosolic and apicoplast metabolism and is essential for parasite survival. Cell Host Microbe 7:62–73. 10.1016/j. chom.2009.12.002. CrossRefMedlineGoogle Scholar

15. Ramakrishnan S, Docampo MD, Macrae JI, Pujol FM, Brooks CF, van Dooren GG, Hiltunen JK, Kastaniotis AJ, McConville MJ, Striepen B. 2012. Apicoplast and endoplasmic reticulum cooperate in fatty acid biosynthesis in apicomplexan parasite Toxoplasma gondii. J. Biol. Chem. 287:4957–4971. 10.1074/jbc.M111.310144. Abstract/FREE Full Text

16. Nair SC, Brooks CF, Goodman CD, Sturm A, Strurm A, McFadden GI, Sundriyal S, Anglin JL, Song Y, Moreno SN, Striepen B. 2011. Apicoplast isoprenoid precursor synthesis and the molecular basis of fosmidomycin resistance in Toxoplasma gondii. J. Exp. Med. 208:1547–1559. 10.1084/ jem.20110039. Abstract/FREE Full Text

17. Francia ME, Jordan CN, Patel JD, Sheiner L, Demerly JL, Fellows JD, de Leon JC, Morrissette NS, Dubremetz JF, Striepen B. 2012. Cell division in apicomplexan parasites is organized by a homolog of the striated rootlet fiber of algal flagella. PLoS Biol. 10:e1001444. 10.1371/journal. pbio.1001444. CrossRefMedlineGoogle Scholar

18. Brooks CF, Francia ME, Gissot M, Croken MM, Kim K, Striepen B. 2011. Toxoplasma gondii sequesters centromeres to a specific nuclear region throughout the cell cycle. Proc. Natl. Acad. Sci. U. S. A. 108:3767–3772. 10.1073/pnas.1006741108. Abstract/FREE Full Text

19. McCoy JM, Whitehead L, van Dooren GG, Tonkin CJ. 2012. TgCDPK3 regulates calcium-dependent egress of Toxoplasma gondii from host cells. PLoS Pathog. 8:e1003066. 10.1371/journal.ppat.1003066. CrossRefMedlineGoogle Scholar

20. Lin SS, Gross U, Bohne W. 2011. Two internal type II NADH dehydrogenases of Toxoplasma gondii are both required for optimal tachyzoite growth. Mol. Microbiol. 82:209–221. 10.1111/j.1365-2958.2011.07807.x. CrossRefMedlineGoogle Scholar

21. Copeland NG, Jenkins NA, Court DL. 2001. Recombineering: a powerful new tool for mouse functional genomics. Nat. Rev. Genet. 2:769–779. 10.1038/35093556. CrossRefMedlineGoogle Scholar

22. Adams DJ, Quail MA, Cox T, van der Weyden L, Gorick BD, Su Q, Chan WI, Davies R, Bonfield JK, Law F, Humphray S, Plumb B, Liu P,

Rogers J, Bradley A. 2005. A genome-wide, end-sequenced 129Sv BAC library resource for targeting vector construction. Genomics 86:753–758. 10.1016/j.ygeno.2005.08.003. CrossRefMedlineGoogle Scholar

23. Suvorova ES, Radke JB, Ting LM, Vinayak S, Alvarez CA, Kratzer S, Kim K, Striepen B, White MW. 2013. A nucleolar AAA-NTPase is required for parasite division. Mol. Microbiol. 90:338–355. 10.1111/mmi.12367. CrossRefGoogle Scholar

24. Taylor S, Barragan A, Su C, Fux B, Fentress SJ, Tang K, Beatty WL, Hajj HE, Jerome M, Behnke MS, White M, Wootton JC, Sibley LD. 2006. A secreted serine-threonine kinase determines virulence in the eukaryotic pathogen Toxoplasma gondii. Science 314:1776–1780. 10.1126/science.1133643. Abstract/FREE Full Text

25. Saeij JP, Boyle JP, Coller S, Taylor S, Sibley LD, Brooke-Powell ET, Ajioka JW, Boothroyd JC. 2006. Polymorphic secreted kinases are key virulence factors in toxoplasmosis. Science 314:1780–1783. 10.1126/science.1133690. Abstract/FREE Full Text

26. Hunter CA, Sibley LD. 2012. Modulation of innate immunity by Toxoplasma gondii virulence effectors. Nat. Rev. Microbiol. 10:766–778. 10.1038/nrmicro2858. CrossRefMedlineGoogle Scholar

27. Farrell A, Thirugnanam S, Lorestani A, Dvorin JD, Eidell KP, Ferguson DJ, Anderson-White BR, Duraisingh MT, Marth GT, Gubbels MJ. 2012. A DOC2 protein identified by mutational profiling is essential for apicomplexan parasite exocytosis. Science 335:218–221. 10.1126/science.1210829. Abstract/FREE Full Text

28. Gubbels MJ, Lehmann M, Muthalagi M, Jerome ME, Brooks CF, Szatanek T, Flynn J, Parrot B, Radke J, Striepen B, White MW. 2008. Forward genetic analysis of the apicomplexan cell division cycle in Toxoplasma gondii. PLoS Pathog. 4:e36. 10.1371/journal.ppat.0040036. CrossRefMedlineGoogle Scholar

29. Frankel MB, Mordue DG, Knoll LJ. 2007. Discovery of parasite virulence genes reveals a unique regulator of chromosome condensation 1 ortholog critical for efficient nuclear trafficking. Proc. Natl. Acad. Sci. U. S. A. 104:10181–10186. 10.1073/pnas.0701893104. Abstract/FREE Full Text

30. Behnke MS, Khan A, Wootton JC, Dubey JP, Tang K, Sibley LD. 2011. Virulence differences in Toxoplasma mediated by amplification of a family of polymorphic pseudokinases. Proc. Natl. Acad. Sci. U. S. A. 108:9631–9636. 10.1073/pnas.1015338108. Abstract/FREE Full Text

31. Pfefferkorn ER, Pfefferkorn LC. 1976. Toxoplasma gondii: isolation and preliminary characterization of temperature-sensitive mutants.

Exp. Parasitol. 39:365–376. 10.1016/0014-4894(76)90040-0. CrossRefMedlineGoogle Scholar

32. Brown KM, Suvorova E, Farrell A, McLain A, Dittmar A, Wiley GB, Marth G, Gaffney PM, Gubbels MJ, White M, Blader IJ. 2014. Forward genetic screening identifies a small molecule that blocks Toxoplasma gondii growth by inhibiting both host- and parasite-encoded kinases. PLoS Pathog. 10:e1004180. 10.1371/journal.ppat.1004180. CrossRefGoogle Scholar

33. Sarov M, Schneider S, Pozniakovski A, Roguev A, Ernst S, Zhang Y, Hyman AA, Stewart AF. 2006. A recombineering pipeline for functional genomics applied to Caenorhabditis elegans. Nat. Methods 3:839–844. 10.1038/nmeth933. CrossRefMedlineGoogle Scholar

34. Ejsmont RK, Bogdanzaliewa M, Lipinski KA, Tomancak P. 2011. Production of fosmid genomic libraries optimized for liquid culture recombineering and cross-species transgenesis. Methods Mol. Biol. 772:423–443. 10.1007/978-1-61779-228-1_25. CrossRefMedlineGoogle Scholar

35. Fu Y, Foden JA, Khayter C, Maeder ML, Reyon D, Joung JK, Sander JD. 2013. High-frequency off-target mutagenesis induced by CRISPR-Cas nucleases in human cells. Nat. Biotechnol. 31:822–826. 10.1038/nbt.2623. CrossRefMedlineGoogle Scholar

36. Hsu PD, Scott DA, Weinstein JA, Ran FA, Konermann S, Agarwala V, Li Y, Fine EJ, Wu X, Shalem O, Cradick TJ, Marraffini LA, Bao G, Zhang F. 2013. DNA targeting specificity of RNA-guided Cas9 nucleases. Nat. Biotechnol. 31:827–832. 10.1038/nbt.2647. CrossRefGoogle Scholar

37. Pattanayak V, Lin S, Guilinger JP, Ma E, Doudna JA, Liu DR. 2013. High-throughput profiling of off-target DNA cleavage reveals RNA-programmed Cas9 nuclease specificity. Nat. Biotechnol. 31:839–843. 10.1038/nbt.2673. CrossRefMedlineGoogle Scholar

38. Tscherne JS, Nurse K, Popienick P, Michel H, Sochacki M, Ofengand J. 1999. Purification, cloning, and characterization of the 16S RNA m5C967 methyltransferase from Escherichia coli. Biochemistry (Mosc.) 38:1884–1892. 10.1021/bi981880l. CrossRefGoogle Scholar

39. Hussain S, Benavente SB, Nascimento E, Dragoni I, Kurowski A, Gillich A, Humphreys P, Frye M. 2009. The nucleolar RNA methyltransferase Misu (NSun2) is required for mitotic spindle stability. J. Cell Biol. 186:27–40. 10.1083/jcb.200810180. Abstract/FREE Full Text

40. Frye M, Watt FM. 2006. The RNA methyltransferase Misu (NSun2) mediates Myc-induced proliferation and is upregulated in tumors. Curr.

Biol. 16:971–981. 10.1016/j.cub.2006.11.001. CrossRefMedlineGoogle Scholar

41. Sakita-Suto S, Kanda A, Suzuki F, Sato S, Takata T, Tatsuka M. 2007. Aurora-B regulates RNA methyltransferase NSUN2. Mol. Biol. Cell 18:1107–1117. 10.1091/mbc.E06-11-1021. Abstract/FREE Full Text

42. Aslanidis C, de Jong PJ. 1990. Ligation-independent cloning of PCR products (LIC-PCR). Nucleic Acids Res. 18:6069–6074. 10.1093/nar/18.20.6069. Abstract/FREE Full Text

43. Schmieder R, Edwards R. 2011. Quality control and preprocessing of metagenomic datasets. Bioinformatics 27:863–864. 10.1093/bioinformatics/btr026. Abstract/FREE Full Text

44. Li H, Durbin R. 2010. Fast and accurate long-read alignment with Burrows-Wheeler transform. Bioinformatics 26:589–595. 10.1093/bioinformatics/btp698. Abstract/FREE Full Text

45. DePristo MA, Banks E, Poplin R, Garimella KV, Maguire JR, Hartl C, Philippakis AA, del Angel G, Rivas MA, Hanna M, McKenna A, Fennell TJ, Kernytsky AM, Sivachenko AY, Cibulskis K, Gabriel SB, Altshuler D, Daly MJ. 2011. A framework for variation discovery and genotyping using next-generation DNA sequencing data. Nat. Genet. 43:491–498. 10.1038/ng.806. CrossRefMedlineGoogle Scholar

46. McKenna A, Hanna M, Banks E, Sivachenko A, Cibulskis K, Kernytsky A, Garimella K, Altshuler D, Gabriel S, Daly M, DePristo MA. 2010. The genome Analysis toolkit: a MapReduce framework for analyzing next-generation DNA sequencing data. Genome Res. 20:1297–1303. 10.1101/gr.107524.110. Abstract/FREE Full Text

47. Thorvaldsdóttir H, Robinson JT, Mesirov JP. 2013. Integrative Genomics viewer (IGV): high-performance genomics data visualization and exploration. Brief. Bioinform. 14:178–192. 10.1093/bib/bbs017. Abstract/FREE Full Text

48. Van Dooren GG, Tomova C, Agrawal S, Humbel BM, Striepen B. 2008. Toxoplasma gondii Tic20 is essential for apicoplast protein import. Proc. Natl. Acad. Sci. U. S. A. 105:13574–13579. 10.1073/pnas.0803862105.

Chapter 2

GENOME ENGINEERING AND DIRECT CLONING OF ANTIBIOTIC GENE CLUSTERS VIA PHAGE ΦBT1 INTEGRASE-MEDIATED SITE-SPECIFIC RECOMBINATION IN STREPTOMYCES

Deyao Du[1,2], Lu Wang[1,3], Yuqing Tian[1], Hao Liu[3], Huarong Tan[1] & Guoqing Niu[1]

[1] State Key Laboratory of Microbial Resources, Institute of Microbiology, Chinese Academy of Sciences, Beijing, China

[2] University of Chinese Academy of Sciences, Beijing, China

[3] Key Laboratory of Industrial Fermentation Microbiology, Ministry of Education, College of Biotechnology, Tianjin University of Science and Technology, Tianjin, China.

ABSTRACT

Several strategies have been used to clone large DNA fragments directly from bacterial genome. Most of these approaches are based on different site-specific recombination systems consisting of a specialized recombinase and its target sites. In this study, a novel strategy based on phage ΦBT1 integrase-mediated site-specific recombination was developed, and used for simultaneous *Streptomyces* genome engineering and cloning of antibiotic gene clusters. This method has been proved successful for the cloning of actinorhodin gene cluster from *Streptomyces coelicolor* M145, napsamycin gene cluster and daptomycin gene cluster from *Streptomyces roseosporus* NRRL 15998 at a frequency higher than 80%. Furthermore, the system could be used to increase the titer of antibiotics as we demonstrated with actinorhodin and daptomycin, and it will be broadly applicable in many *Streptomyces*.

INTRODUCTION

Streptomyces are high-GC Gram-positive bacteria well known for their ability to produce a wide variety of medically and agriculturally useful antibiotics and related compounds[1]. Genes responsible for the biosynthesis of a specific secondary metabolite are usually arranged in clusters that vary in size from a few to over 100 kb[2]. To gain insight into the biosynthesis and regulation of antibiotics in *Streptomyces*, it is of great importance to clone their gene clusters. Recently, various approaches have been developed to clone gene clusters directly from bacterial genomic DNA. These methods include RecET-mediated linear-plus-linear homologous recombination (LLHR)[3], *oriT*-directed capture system[4] and transformation-associated recombination (TAR)[5]. The RecET-mediated LLHR was successful in cloning gene clusters (10 to 52 kb in length) from the genome of *Photorhabdus luminescens* into expression vectors in *Escherichia coli*3. The *oriT*-directed capture system has been used to clone regions up to 140 kb from the genome of*Burkholderia pseudomallei*[6] and 200 kb from megaplasmid of *Sinorhizobium meliloti*[4]. However, the use of this system was limited to Gram-negative bacteria that can be established as conjugation donors[6]. Taking advantage of the natural *in vivo* homologous recombination of*Saccharomyces cerevisiae*, TAR cloning strategy was used to capture a 21.3 kb enterocin gene cluster from *Salinispora pacifica* CNT-150[7] and a 67 kb taromycin A biosynthetic gene cluster from *Saccharomonospora* sp. CNQ-490[8].

The ability to delete large genomic fragments within *Streptomyces* genome is of great interest for genetic manipulations of *Streptomyces*. Several strategies have been developed for a number of bacteria. Some methods are based on the meganuclease I-SceI system which involves the meganuclease I-SceI of *Saccharomyces cerevisiae* and its 18 bp recognition sequence[9,10]. Many of them are based on site-specific recombination systems consisting of a specialized recombinase and its target sites. Nearly all site-specific recombinases fall into two families, the tyrosine recombinases and the serine recombinases[11]. The recombination systems of the tyrosine recombinase family include Cre/loxP from the P1 phage[12], Dre/rox from the P1-like transducing phage D6[13] and the Flp/FRT from yeast[14]. The Cre, Dre and Flp proteins are the tyrosine recombinases which catalyze reciprocal site-specific recombination of DNA at loxP, rox and FRT sites, respectively. Integrases (Int) from *Streptomyces* temperate phage φC31 and φBT1 belong to serine recombinase family. They catalyze site-specific recombination of the phage attachment site (*attP*) with the bacterial attachment site (*attB*), resulting in the formation of two hybrid sites (*attL* and *attR*)[15,16]. Both φC31 and φBT1 *attP-int* loci have been used to construct versatile vectors which can integrate into different *attB* sites in *Streptomyces*[15,16]. To increase the diversity of *attP-attB* pair of φBT1, 15 mutated *attP-attB* pairs (*attP$_{01}$-attB$_{01}$* → *attP$_{15}$-attB$_{15}$*) were generated by PCR mutagenesis of the central dinucleotide sequence of *attB* and*attP*[17]. The Cre/loxP system was successfully used for the deletion of large fragments in *Magnetospirillum gryphiswaldense* and several *Streptomyces* species[18,19,20]. However, the use of φC31 and φBT1 integrase in this aspect has not been exploited.

We devised a novel strategy for *Streptomyces* genome engineering and cloning of antibiotic gene clusters. This method is based on phage φBT1 *attP-attB-int* system and requires two single crossovers for targeted integration of mutated *attB* and *attP* into the recipient chromosome. Using the system, we easily cloned 25 kb fragment containing actinorhodin (*act*) gene cluster from *S. coelicolor* M145, 45 kb fragment containing napsamycin (*nap*) gene cluster and 157 kb fragment containing daptomycin (*dap*) gene cluster from *S. roseosporus* NRRL 15998. In addition, this method could be used to improve the titer of antibiotics by increasing copy numbers of antibiotic gene clusters.

RESULTS

Construction of pUC119- and pKC1139-Based Plasmids

Our strategy used in this study requires both homologous and site-specific recombinations. The homologous recombinations were used for targeted integration of the mutated *attB* and *attP* into *Streptomyces* chromosome, while the φBT1 integrase-mediated site-specific recombination was employed to excise targeted region of interest from the chromosome (Fig. 1). The mutated *attB* and *attP* sites were chosen to avoid site specific recombination with the endogenous *attB* site in *Streptomyces* genome and consequently undesirable DNA rearrangements. Sites of $attB_6$ and $attP_6$ were randomly chosen from the 15 mutated *attP-attB* pairs. For the integration of $attB_6$ into *Streptomyces* chromosome, pUC119-based suicide plasmids (pSV::$attB_6$-*act*, pSV::$attB_6$-*nap* and pSV::$attB_6$-*dap*) were constructed. These plasmids are derivatives of pUC119 containing the kanamycin-resistance gene (*neo*), the origin of transfer (*oriT*) from plasmid RK2 (for the intergeneric conjugation between *E. coli* and *Streptomyces*), $attB_6$ and a 2.0 kb homologous region flanking 5′ end of the targeted regions (Fig. S1a). We also constructed pKC1139-based plasmids (pKC1139::$attP_6$-*act*, pKC1139::$attP_6$-*nap* and pKC1139::$attP_6$-*dap*) for the integration of $attP_6$ into *Streptomyces* chromosome. These plasmids are derivatives of pKC1139 containing $attP_6$ and a 2.0 kb homologous region flanking 3′ end of the targeted regions (Fig. S1b).

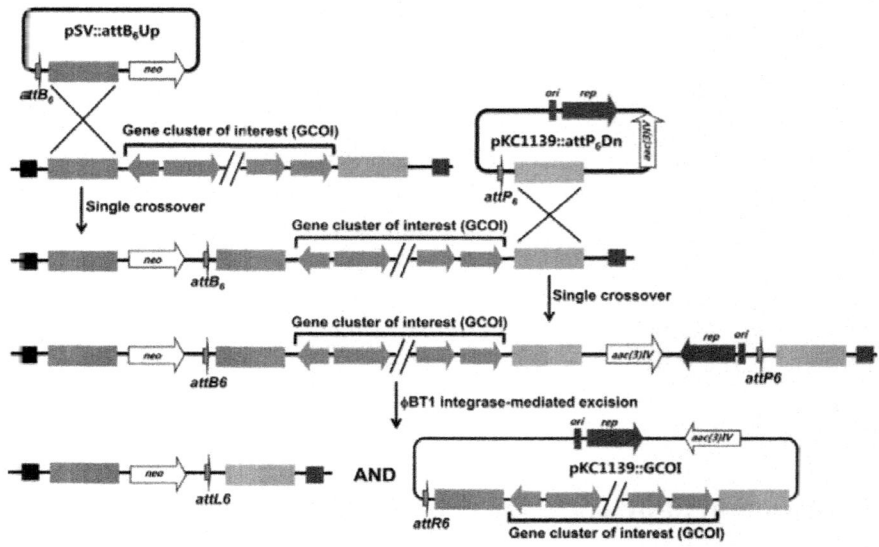

Figure 1: Schematic diagram of antibiotic gene cluster cloning from *Streptomyces* **chromosome.**

Initially, a pUC119-based suicide plasmid (pSV::attB$_6$Up) carrying *attB$_6$* and a region homologous to 5' end of the cluster is introduced into the chromosome by a single crossover. A second plasmid pKC1139::attP$_6$Dn is based on pKC1139 carrying *attP$_6$* and a region homologous to 3' end of the cluster. When the incubation temperature is higher than 34°C, pKC1139::attP$_6$Dn turns into a non-replicating plasmid and then is integrated into the chromosome by a single crossover. Expression of φBT1 integrase (encoded in the plasmid pIJ10500) leads to excision of the pKC1139 backbone with gene cluster of interest, leaving behind the suicide vector pUC119::*neo* and 42 bp *attL6* site. *aac(3)IV*: apramycin resistance gene; *neo*: kanamycin resistance gene; *ori*: temperature-sensitive origin of replication from pSG5; *rep*: *rep* encoding a replication initiator protein from pSG5.

Cloning of *act* Gene Cluster from *S. coelicolor* M145

To test this strategy, we first chose to clone the well-studied *act* gene cluster (*SCO5070-SCO5092*) from *S. coelicolor* M145. For this purpose, pSV::*attB$_6$*-*act* and pKC1139::*attP$_6$*-*act* were introduced into the recipient chromosomes via single-crossover homologous recombination to obtain double-cointegrate strain Sco-actB$_6$P$_6$ (Fig. 1). Further introduction of pIJ10500 (an integrative plasmid containing the φBT1 integrase gene) into Sco-actB$_6$P$_6$ allowed subsequent excision of 23 kb *act* gene cluster from *S. coelicolor* M145, leaving behind the suicide vector pUC119::*neo*, a scar of 42 bp *attL* site and pIJ10500 integrated within SCO4848. Excision of the gene cluster was confirmed by PCR analysis using both genomic and plasmid DNA as templates (Fig. 2). For 9 out of 10 exconjugates tested, the φBT1 integrase-mediated excision of *act* gene cluster occurred at a frequency of 90%. Furthermore, the presence of *attL* and *attR* in the amplified fragments was confirmed by DNA sequencing (Fig. 3). To recover the plasmid containing the entire *act* gene cluster (pKC1139::*act*) from *Streptomyces*, the DNA extract containing pKC1139::*act* from M145-MCact was used to transform *E.coli* Top10. Plasmid DNA from four apramycin resistant *E. coli* colonies was confirmed by BamHI and NotI digestion, respectively. The restriction fragments showed correct band patterns (Fig. S2a).

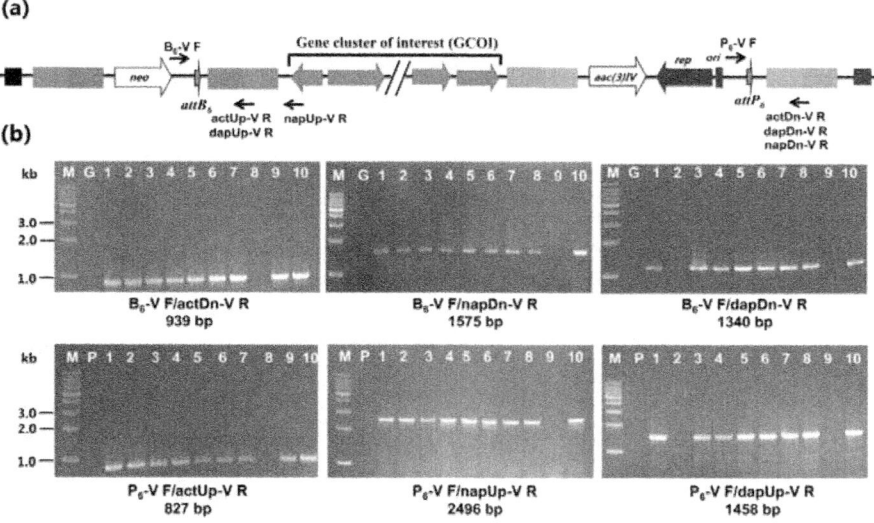

Figure 2: Confirmation of the excision events by PCR amplifications. (A) The schematic diagram showing the position of primers in the chromosome of double-cointegrate strains. (B) Agarose gel electrophoresis showing PCR amplified fragments. PCR templates in the upper panels are genomic DNAs from *S. coelicolor* M145 or *S. roseosporus* NRRL 15998 (G) and ten randomly selected double-cointegrate strains with pIJ10500 (M145-MCact, Sro-MCnap or Sro-MCdap), while PCR templates in the lower panels are plasmid DNAs including pKC1139 (P) and ten different clones of pKC1139::*act*, pKC1139::*nap* and pKC1139::*dap*. The primers used and the expected size of amplification fragments were indicated.

(a)
attB₆: CCAGGTTTTTGACGAAA**CT**GATCCAGATGATCCAGC
attP₆: TGCTGGGTTGTTGTCTCTGGACA**CT**GATCCATGGGAAACTACTCAGCA
attL₆: CCAGGTTTTTGACGAAA**CT**GATCCATGGGAAACTACTCAGCA
attR₆: TGCTGGGTTGTTGTCTCTGGACA**CT**GATCCAGATGATCCAGC

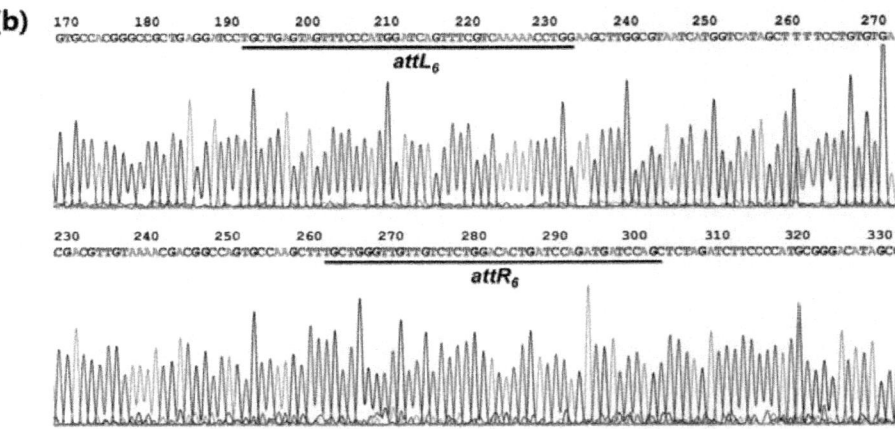

Figure 3: Representative excision of the *act* **gene cluster from** *S. coelicolor* **M145.** (A) Nucleotide sequence of *attB₆*, *attP₆*, *attL₆* and *attR₆*. The mutated core dinucleotide (CT) at which the crossover occurs is in bold. (B) Verification of *attL₆* and *attR₆* by DNA sequencing. Sequences of *attL₆* and *attR₆* from DNA sequencing are underlined.

To verify the cluster is complete, the recombinant plasmid pKC1139::*act* was introduced into *S. coelicolor* M1146 (M1146) to obtain M1146-MCact. Unlike M1146 and M1146-pKC1139 (*S. coelicolor* M1146 containing empty vector pKC1139), M1146-MCact regained the ability to produce the blue pigment actinorhodin (Fig. 4a). These results showed that the cloned *act* gene cluster was complete and functional.

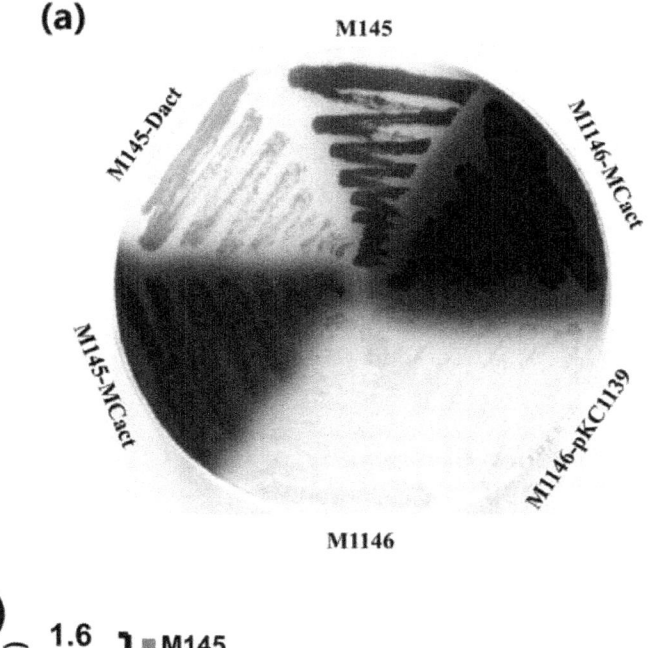

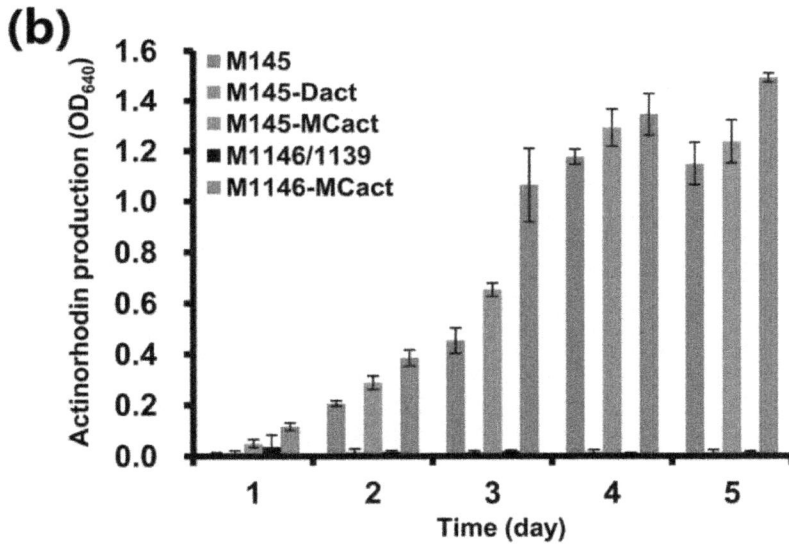

Figure 4: Comparison of actinorhodin production in *S. coelicolor*M145 and its derivatives. (A) Comparison of actinorhodin production (blue pigment) of *S. coelicolor* M145 and its derivatives. Photograph was taken from the bottom of the plate after grown on R5MS agar medium for 4 days at 28°C. Representative image of three independent experiments with similar results was shown. (B) Actinorhodin titers of *S. coelicolor* M145 and its derivatives grown in 50 ml of R5MS at 28°C. Error bars show standard deviations.

Deletion of *act* Gene Cluster from *S. coelicolor* M145

To delete the *act* gene cluster from *S. coelicolor* M145, a single colony of M145-MCact was randomly chosen for the removal of pKC1139::*act*. After three rounds of nonselective growth at 28°C and subsequent cultivation at 40°C, approximately 5% of M145-MCact colonies lost pKC1139::*act*. Strains lacking the *act* gene cluster (M145-Dact) were first confirmed by PCR (data not shown), and then patched on R5MS solid agar plate for visual comparison of actinorhodin production. Unlike *S. coelicolor* M145 that could produce both blue pigment actinorhodin and red pigment undecylprodigiosins, M145-Dact could only produce the red pigment undecylprodigiosins (Fig. 4a). This was further validated by no actinorhodin production of M145-Dact in R5MS liquid culture (Fig. 4b).

Cloning and Deletion of *nap* and *dap* Gene Cluster from *S. roseosporus* NRRL 15998

To clone gene cluster of medium and large sizes, we used the same strategy to clone *nap* and *dap* gene cluster from *S. roseosporus* NRRL 15998. Excision of *nap* gene cluster from *S. roseosporus* NRRL 15998 occurred in 9 out of 10 exconjugates, and excision of *dap* gene cluster from *S. roseosporus* NRRL 15998 occurred in 8 out of 10 exconjugates (Fig. 2). Like pKC1139::*act*, plasmid containing *nap* gene cluster (pKC1139::*nap*) was passed through *E. coli* Top10 and isolated plasmid DNA was confirmed by BglII and EcoRI digestion, respectively (Fig. S2b). The cloned fragment covers a contiguous DNA region of 45 kb from*SSGG02973* to *SSGG03009*. For pKC1139::*dap*, the plasmid was isolated directly from *Streptomyces* and confirmed with restriction digestion (Fig. S2c). The 157 kb fragment covering *SSGG00215-SSGG00287*contains the complete *dap* gene cluster. Similar to that of *act* gene cluster, the removal of pKC1139::*nap* and pKC1139::*dap* from Sro-MCnap and Sro-MCdap generated strains lacking *nap* and *dap* gene clusters (Sro-Dnap and Sro-Ddap).

Improvement of Antibiotic Titers

The pKC1139 contains a temperature-sensitive origin of replication from pSG5, which is a medium copy plasmid with an approximate 20–50 copy numbers per chromosome[21]. When cultured at 28°C, pKC1139 exists as autonomous plasmid in *Streptomyces*. In *S. coelicolor* M145, there is only one copy of *act* gene cluster in the chromosome. After the φBT1 integrase-mediated excision, the *act* gene cluster was transferred into pKC1139. An increase in copy number of *act* gene cluster will improve actinorhodin production. This was confirmed both on R5MS agar plate (Fig. 4a) and in R5MS liquid culture (Fig. 4b). It should be noted that the titer of actinorhodin in M1146-MCact was even higher than that of M145-MCact. Similarly, daptomycin titer could also increase after the excision of *dap* gene cluster from its chromosome location in *S. roseosporus* NRRL 15998. Cultures of Sro-MCdap and *S. roseosporus* NRRL 15998 were subjected to bioassay against *S. aureus*, the results showed that Sro-MCdap exhibited bigger inhibition zones against *S. aureus* than *S. roseosporus* NRRL 15998 at time intervals from 2–5 days (Fig. 5a). This was further verified by comparison of daptomycin from fermentation broth of *S. roseosporus* NRRL 15998 and Sro-MCdap by high-performance liquid chromatography (HPLC) analysis (Fig. 5b). In addition, we noticed that existence of extra copy numbers of antibiotic gene clusters caused a slowdown in growth of*Streptomyces*. When cultured on AS-1 agar medium, growth of Sro-MCnap and Sro-MCdap are severely impaired, especially at earlier stages of cultivation (Fig. S3). This phenotype was most likely attributed to the metabolic burden of extra copy numbers of antibiotic gene clusters. This assumption is based on the observation that growth of Sro-Dnap (devoid of *nap* gene cluster) and Sro-Ddap (devoid of *dap*gene cluster) are converted back to that of *S. roseosporus* NRRL 15998 (Fig. S3).

(a)

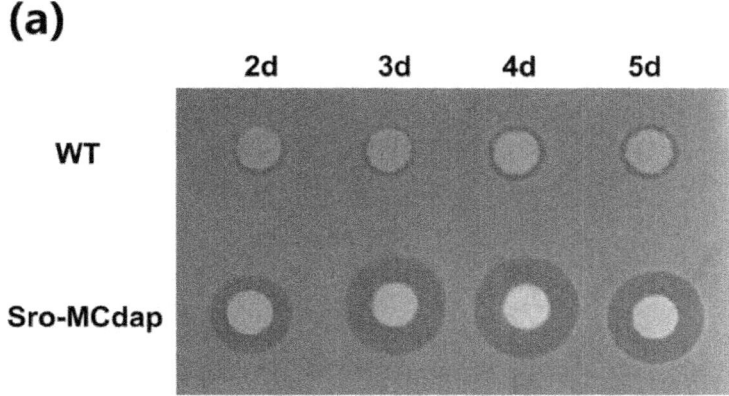

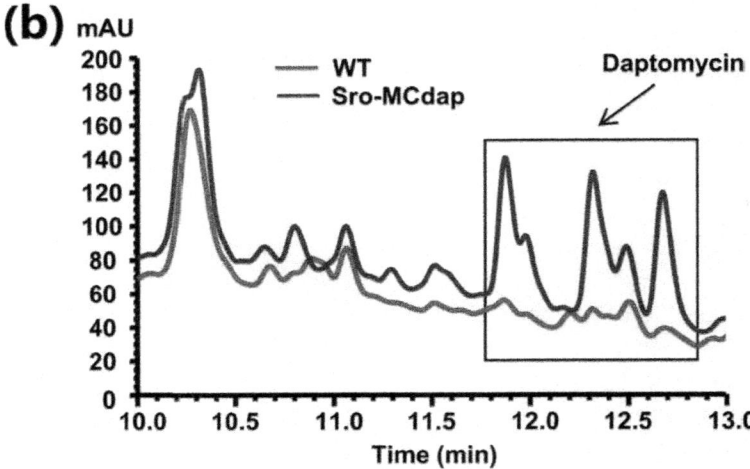

Figure 5: Analysis of daptomycin production in *S. roseosporus*NRRL 15998 (WT) and Sro-MCdap. (A) Bioassay of daptomycin against *S. aureus*. After grown on AS-1 agar for 2–5 days at 28°C, the patches of WT and Sro-MCdap were overlaid with cultures of *S. aureus* and the zone of inhibition was assessed after overnight incubation at 37°C. Representative images of three independent experiments with similar results are shown. (B) HPLC analysis of fermentation filtrates from WT and Sro-MCdap after incubation for 4 days. Components of daptomycin were indicated by comparison with standards.

To examine the stability of multiple copy plasmids in *Streptomyces*, two randomly chosen strains of Sro-MCdap were passed consecutively for five or ten times on AS-1 plates supplemented with or without apramycin. Biological activities of these stains (G_5 and G_{10}) were compared with that of the original Sro-MCdap (G_0). All Sro-MCdap strains exhibited similar inhibitory activity against *S. aureus* (Fig. S4), suggesting that pKC1139-derived large plasmids are stable in the engineered *Streptomyces* in the presence or absence of selective pressure.

DISCUSSION

We have established an efficient method for genome engineering and direct cloning of gene clusters in *Streptomyces*. The strategy is based on phage φBT1 *attP-attB-int* system and provides several advantages over similar methods. First, it can be used for the deletion of large fragment (up to 157 kb) from *Streptomyces* genome. In the meantime, the large fragment containing gene cluster of interest was cloned into pKC1139 simultaneously. Another advantage of our strategy is that it could clone gene cluster in size up to 157 kb.

This is the largest size ever reported in Gram positive bacteria and should be good enough for most antibiotic gene clusters. Last, our strategy can be used to improve the titer of industrial important antibiotics by creating strains with extra copy numbers of antibiotic biosynthetic gene clusters.

The ϕBT1 *attP-attB-int* system is helpful for genetic modifications of *Streptomyces* genome at multiple sites. In addition to the intact *attP-attB* pair, there are 15 mutated *attP-attB* pairs which can be recognized by ϕBT1 integrase[17]. Multiple rounds of large fragment deletion can be achieved with the following modifications. (1) Relocation of the $attB_6$ sequence (or any other mutated *attB*) to the downstream of the 2.0 kb homologous fragment in pSV::attB$_6$Up. This change will allow the excision of the pUC119::*neo* backbone from *Streptomyces* genome together with pKC1139. (2) Construction of an autonomous helper plasmid containing a temperature-sensitive origin of replication from pSG5, the origin of transfer (*oriT*) from plasmid RK2 and ϕBT1 integrase gene. This plasmid can ensure the high efficient excision of large fragment from *Streptomyces* genome and subsequent removal of ϕBT1 integrase. With these modifications, there is only 42 bp *attL* site left in the chromosome of *Streptomyces*.

Genome analysis suggested that *S. roseosporus* NRRL 15998 has potential capacity to produce napsamycins[22,23]. However, the production of napsamycins in *S. roseosporus* NRRL 15998 has not been reported. With this strategy, we cloned *nap* gene cluster in pKC1139 to generate pKC1139::*nap*. It can be manipulated extensively in *E. coli*. These manipulations include replacement of vector backbone with integrative plasmid and deletion or constitutive expression of regulatory gene by PCR targeting[24]. The modified gene cluster can be transferred into heterologous hosts for expression. It can also be transferred back into the mutant devoid of *nap* gene cluster after removal of pKC1139::*nap*. Detection of napsamycin in these strains will shed light on the activation of cryptic gene clusters in *Streptomyces*.

In some industrial overproducing strains generated by traditional mutagenesis, amplification of biosynthetic gene cluster has been observed[25,26,27]. Based on these observations, controlled amplification of gene cluster was used to increase the productivity of commercially important antibiotics. Integration of an additional copy of gene cluster for nikkomycin and gougerotin biosynthesis led to an increased production of nikkomycin and gougerotin by *Streptomyces ansochromogene*[28] and *Streptomyces graminearus*[29], respectively. The *zouA*-mediated gene amplification of *act* gene cluster in *S. coelicolor* M145 led to a 20-fold increase in actinorhodin production[30]. The *zouA* encodes a site-specific relaxase similar to TraA protein which catalyzes RecA-independent site-specific recombination. The recombination sites of ZouA are *oriT*-like

RsA and RsB[30]. In this study, we reported the amplification of gene clusters mediated by phage φBT1 integrase and improved antibiotic titers in the engineered *Streptomyces* strains. We believe that the system described here could be used readily to increase antibiotic titers in many *Streptomyces* and possible other actinomycetes.

METHODS

Bacterial Strains, Plasmids, Primers and Growth Conditions

Bacterial strains and plasmids used in this study are listed in Table 1, and primers are listed in Table S1. *S. coelicolor* M145 and *S. roseosporus* NRRL 15998 were used for cloning of *act*, *nap* and *dap* gene clusters. *S. coelicolor* M1146 is an engineered derivative of *S. coelicolor* M145 that lacks gene clusters for actinorhodin (ACT), undecylprodigiosins (RED), cryptic polyketide (CPK) and calcium-dependent antibiotic (CDA) biosynthesis[31]. *Staphylococcus aureus* was used as an indicator strain for daptomycin bioassay. *E. coli* Top10 was used as a general host for propagating plasmids. *E. coli* ET12567 (pUZ8002) was used as a host for transferring DNA from *E. coli* to *Streptomyces* by intergeneric conjugation[32].

Table 1: Strains and plasmids used in this study

Strains/plasmids	Genotype/description	Reference/source
E. coli		
Top10	F⁻ *mcrA* Δ(*mrr-hsdRMS-mcrBC*) Φ80*lacZ*ΔM15 Δ*lacX74 recA1 araD139* Δ(*ara leu*)7697 g *alU galK rpsL* (StrR) *endA1 nupG*	Invitrogen
ET12567	F⁻ *dam-13*::Tn*9 dcm-6 hsdM hsdR zjj-202*::Tn*10 recF143 galK2 galT22 ara-14 lacY1 xyl-5 leuB6 thi-1 tonA31 rpsL136 hisG4 tsx-78 mtl-1 glnV44*	40
Staphylococcus		
S. aureus	A indicator strain	41
Streptomyces		
S. coelicolor M145	Prototrophic; SCP1⁻ SCP2⁻ Pgl⁺	32

S. coelicolor M1146	Δact Δred Δcpk Δcda	31
Sco-actB$_6$P$_6$	A derivative of *S. coelicolor* M145 with *attB$_6$* and *attP$_6$* flanking *act* gene cluster	This study
S. roseosporus NRRL 15998	A daptomycin-producing strain	Broad Institute
Sro-pKC1139	A derivative of *S. roseosporus* NRRL 15998 containing pKC1139	This study
Sro-napB$_6$P$_6$	A derivative of *S. roseosporus* NRRL 15998 with *attB$_6$* and *attP$_6$* flanking *nap*-gene cluster	This study
Sro-dapB$_6$P$_6$	A derivative of *S. roseosporus* NRRL 15998 with *attB$_6$* and *attP$_6$* flanking *dap*-gene cluster	This study
M145-Dact	Δact	This study
M145-MCact	A derivative of *S. coelicolor* M145 containing multicopy of *act* gene cluster	This study
M1146-pKC1139	*S. coelicolor* M1146 containing pKC1139	This study
M1146-MCact	*S. coelicolor* M1146 containing pKC1139::*act*	This study
Sro-MCnap	A derivative of *S. roseosporus* NRRL 15998 containing multicopy of *nap*gene cluster	This study
Sro-MCdap	A derivative of *S. roseosporus* NRRL 15998 containing multicopy of *nap*gene cluster	This study
Plasmids		
pUZ8002	*tra neo* RP4	42
pUC119::*neo*	pUC119 containing kanamycin resistance gene (*neo*)	43
pKC1139	*E.coli-Streptomyces* shuttle plasmid contains a *Streptomyces* temperature-sensitive origin of replication	15
pIJ10500	A derivative of pMS82 containing φBT1 integrase gene	36
pUC119::*neo-attB$_6$-act*	A derivation of pUC119::*neo* containing *attB$_6$* and 2.0 kb homologous region flanking the 5′ end of *act* gene cluster	This study

pSV::$attB_6$-act	A derivation of pUC119::*neo* containing the origin of transfer (*oriT*) from plasmid RK2, $attB_6$ and 2.0 kb homologous region flanking the 5′ end of *act* gene cluster	This study
pSV::$attB_6$-nap	A derivation of pUC119::*neo* containing the origin of transfer (*oriT*) from plasmid RK2, $attB_6$ and 2.0 kb homologous region flanking the 5′ end of *nap* gene cluster	This study
pSV::$attB_6$-dap	A derivation of pUC119::*neo* containing the origin of transfer (*oriT*) from plasmid RK2, $attB_6$ and 2.0 kb homologous region flanking the 5′ end of *dap* gene cluster	This study
pKC1139::$attP_6$-act	A derivation of pKC1139 containing $attP_6$ and 2.0 kb homologous region flanking the 3′ end of *act* gene cluster	This study
pKC1139::$attP6$-nap	A derivation of pKC1139 containing $attP_6$ and 2.0 kb homologous region flanking the 3′ end of *nap* gene cluster	This study
pKC1139::$attP6$-dap	A derivation of pKC1139 containing $attP_6$ and 2.0 kb homologous region flanking the 3′ end of *dap* gene cluster	This study
pKC1139::*act*	A derivation of pKC1139 containing 25 kb fragment including actinorhodin gene cluster and its flanking sequences	This study
pKC1139::*nap*	A derivation of pKC1139 containing 45 kb fragment including napsamycin gene cluster and its flanking sequences	This study
pKC1139::*dap*	A derivation of pKC1139 containing 157 kb fragment including daptomycin gene cluster and its flanking sequences	This study

For general purpose, *S. coelicolor* M145 and its derivatives were grown on mannitol soya flour medium (MS) agar or in yeast extract-malt extract (YEME) liquid medium[32]. For actinorhodin production, *S.coelicolor* M145 and its derivatives were grown on R5MS agar or in R5MS liquid medium[33]. *S. roseosporus* NRRL 15998 was cultured on AS-1 agar medium or in tryptic soy broth (TSB) liquid medium[34]. All *Streptomyces* stains were maintained at 28°C unless specified otherwise. General approaches for *E. coli* or *Streptomyces* manipulations were performed according to standard protocols[32,35]. When necessary, the final antibiotic concentrations used for selection of *E. coli* transformants were as follows: ampicillin, 100 µg ml⁻¹; apramycin, 100 µg ml⁻¹; kanamycin, 100 µg ml⁻¹; chloramphenicol, 12.5 µg ml⁻¹. For selection of *Streptomyces* transformants, the final antibiotic concentrations were, kanamycin, 50 μg ml^{-1} in MS for *S. coelicolor* and 20 μg ml^{-1} in AS-1 for *S. roseosporus*; apramycin, 50 μg ml^{-1} in MS for *S. coelicolor* and 10 μg ml^{-1}in AS-1 for *S. roseosporus*; hygromycin, 50 μg ml^{-1} in MS or AS-1 for*Streptomyces*; nalidixic acid, 25 μg ml^{-1} in MS or AS-1 for *Streptomyces*.

Construction of Plasmids

Of the 15 mutated *attP-attB* pairs[17], *attP$_6$-attB$_6$* was randomly chosen for this experiment. The sequences of *attB$_6$* and *attP$_6$* were obtained by overlapping PCR. For construction of pSV::*attB$_6$-act*, a 2.0 kb fragment flanking 5′ end of the *act* gene cluster was amplified from genomic DNA of *S. coelicolor* M145 with primer pair act-Up F/act-Up R. The amplicon was diluted 1:100 and used as templates for the second round of PCR with primer pair attB$_6$-in F/act-Up R. The product from the second amplification reaction was diluted again and underwent a third run with primer pair attB$_6$-out F/act-Up R. The final product was digested with HindIII/BamHI and then inserted into the corresponding sites of pUC119::*neo* to generate pUC119::*neo-attB$_6$-act*. The origin of transfer (*oriT*) from plasmid RK2 was amplified from pKC1139 with primer pair oriT F and oriT R, subsequently digested with EcoRI and inserted into the EcoRI site of pUC119::*neo-attB$_6$-act* to generate pSV::*attB$_6$-act* (Fig. S1a). For construction of pKC1139::*attP$_6$-act*, a 2.0 kb fragment flanking 3′ end of the *act* gene cluster was amplified from genomic DNA of *S. coelicolor* M145 with primer pair act-Dn F/act-Dn R. The amplicon with 1:100 dilution served as templates for the second round of PCR with primer pair attP$_6$ F/act-Dn R. The final product was digested with HindIII/EcoRI and then inserted into the corresponding sites of pKC1139 to generate pKC1139::*attP$_6$-act* (Fig. S1b).

For construction of pSV::*attB$_6$-nap* and pSV::*attB$_6$-dap*, a 2.0 kb fragment flanking 5′ end of the *nap* and *dap* gene cluster was amplified from genomic DNA of *S. roseosporus* NRRL 15998 with primer pairs nap-Up F/nap-Up R and dap-UpF/dap-Up R, respectively. The product was digested with XbaI/BamHI and used to replace the 2.0 kb fragment upstream of the *act* gene cluster in pSV::*attB$_6$-act*. For construction of pKC1139::*attP$_6$-nap* and pKC1139::*attP$_6$-dap*, a 2.0 kb fragment flanking 3′ end of the *nap* and *dap* gene cluster was amplified from genomic DNA of *S. roseosporus* NRRL 15998 using primer pairs nap-Dn F/nap-Dn R and dap-Dn F/dap-Dn R, respectively. The product was digested with BamHI/EcoRI and used to replace the 2.0 kb fragment downstream of the *act* gene cluster in pKC1139::*attP$_6$-act*. To ensure the authenticity of DNA sequences, all PCR products were verified by sequencing.

Construction of Double-Cointegrate Strains

To insert $attB_6$ and $attP_6$ at sites flanking the *act* gene cluster of *S. coelicolor*, pSV::$attB_6$-*act* and pKC1139::$attP_6$-*act* were conjugated into *S. coelicolor* M145. The pSV::$attB_6$-*act* is unable to replicate alone in *Streptomyces* and selection with kanamycin allows to select exconjugants in which pSV::attB6-act is inserted into the *S. coelicolor* genome. The pKC1139::$attP_6$-*act* is a derivative of the *E. coli–Streptomyces* shuttle vector pKC1139 that contains a *Streptomyces* temperature-sensitive origin of replication from pSG5[15]. When the incubation temperature is higher than 34°C, pKC1139::$attP_6$-*act* turns into non-replicating plasmid and $attP_6$ was then inserted into *S. coelicolor* genome with selection of apramycin to obtain double-cointegrate strain Sco-actB$_6$P$_6$. Similar strategy was used for the construction of double-cointegrate strains Sro-napB$_6$P$_6$ and Sro-dapB$_6$P$_6$.

Excision of Targeted Regions

The integrative plasmid pIJ10500[36] is a derivative of pMS82 which contains the phage φBT1 integrase gene and integrates intragenically into *SCO4848* encoding a putative integral membrane protein[16]. It was conjugated into a randomly selected strain Sco-actB$_6$P$_6$, Sro-napB$_6$P$_6$ and Sro-dapB$_6$P$_6$, respectively. The exconjugants were initially selected with hygromycin and ten randomly chosen exconjugants were passed twice on MS or AS-1 plates supplemented with kanamycin and apramycin, and subject to genomic and plasmid extraction. Excision of targeted region from *Streptomyce* genome was analyzed by PCR amplifications using genomic DNA templates and primer pairs B$_6$-VF/actDn-VR, B$_6$-VF/napDn-VR and B$_6$-VF/dapDn-VR. In the meantime, PCR amplifications were performed with plasmid DNA template by using primer pairs P$_6$-VF/actUp-VR, B$_6$-VF/napUp-VR and B$_6$-VF/dapUp-VR. Strains with excision of the targeted regions were designated as M145-MCact, Sro-MCnap and Sro-MCdap, respectively.

Deletion of the Antibiotic Gene Clusters from *Streptomyces*

A single colony of M145-MCact was randomly chosen and passed three times on nonselective MS plates at 28 °C. Spores were harvested, serially diluted, and then spread on MS agar. After growing for 4 days at 40°C, colonies were replicated on MS agar plates containing kanamycin or apramycin. Strains lacking the *act* gene cluster (M145-Dact) are apramycin sensitive (Apr^s) and kanamycin resistant (Kan^r). Apr^s and Kan^r strains were further verified by PCR. In a similar way, strains lacking the *nap* and *dap* gene clusters (Sro-Dnap and Sro-Ddap) were obtained by the removal of pKC1139::*nap* and pKC1139::*dap* from Sro-MCnap and Sro-MCdap.

Actinorhodin Quantification

To quantitate actinorhodin production, *S. coelicolor* M145 and its derivatives were grown in 50 ml of R5MS at 28°C. 1 ml culture was harvested in a time-course and treated with KOH (1 N final concentration), and titer was calculated by measuring the absorbance at 640 nm[37].

Production and Analysis of Daptomycin

Small-scale fermentation of daptomycin was carried out by following the procedures described previously[38,39] with minor modifications. In brief, starter culture was grown in TSB for 48 h, 1 ml of starter culture was transferred to A355 (1% [wt/vol] glucose, 1.5% [vol/vol] glycerol, 1.5% [wt/vol] soya peptone, 0.3% [wt/vol] NaCl, 0.5% [wt/vol] malt extract, 0.5% [wt/vol] yeast extract, 0.1% [vol/vol] Tween 80 and 2% [wt/vol] MOPS, pH 7.0) and grown for 36 h as seed culture, and 1 ml of seed culture was transferred into a shake flask containing 50 ml A346 (1% [wt/vol] glucose, 2% [wt/vol] soluble starch, 0.5% [wt/vol] yeast extract, 0.5% [wt/vol] casein and 4.6% [wt/vol] MOPS, pH 7.0). The cultures were incubated for different time points at 28°C before fermentation broths were collected by centrifugation.

For daptomycin analysis, culture broths were centrifuged at 13,000 × g for 10 min to remove the mycelia. The supernatants were filtered through a Millipore membrane (pore diameter, 0.22 μm) and 50 μl of sample was used for HPLC analysis. Separation of daptomycin was achieved with an Agilent 1100 HPLC system and a ZORBAX SB-Aq column (5 μm pore size, 4.6 by 250 mm). HPLC conditions were described as follows: gradient elution with buffer A (0.01% [vol/vol] trifluoroacetic acid in acetonitrile) and buffer B (0.01% [vol/vol] trifluoroacetic acid in ddH$_2$O), flow rate at 1.0 ml/min, ultraviolet detection at wavelength of 224 nm. The elution profile was a linear gradient of 10%–100% buffer A over 22 min, a hold at 100% buffer A over 3 min, a linear gradient of 100%–10% buffer A over 2 min and a final hold at 10% buffer A over 3 min.

Bioassay against *S. aureus* was performed as previously described with modifications[38]. In brief, *S. roseosporus* and its derivatives were patched on AS-1 agar. After incubation for 2–5 days at 28°C, agar plugs were prepared from the patches, placed on the surface of an empty Petri dish, and overlaid with culture of indicator strain in soft nutrient agar containing 5 mM CaCl$_2$. The zone of inhibition was assessed after overnight incubation at 37°C.

ACKNOWLEDGEMENTS

This work was supported by grants from the Ministry of Science and Technology of China (grant nos. 2012CB721103 and 2013CB734001) and the National Natural Science Foundation of China (grant nos. 31270110 and 31171202). We would like to thank Professor Mervyn Bibb and Dr. Chris D. Den Hengst (John Innes Centre, Norwich, UK) for providing *S. coelicolor* M1146 and pIJ10500, respectively. We also thank Dr. Guojian Liao (Southwest University, Chongqing, China) for the gift of daptomycin standard.

CONTRIBUTIONS

D.D. and W.L. performed the experiments. T.Y. assisted with design of the project. L.H. assisted with the primary data analysis. N.G. conceived and designed the project, and wrote the manuscript. T.H. supervised the project and revised the manuscript.

REFERENCES

1. Niu, G. & Tan, H. Biosynthesis and regulation of secondary metabolites in microorganisms. Sci. China Life Sci. 56, 581–583 (2013).

2. Bibb, M. J. Regulation of secondary metabolism in streptomycetes. Curr. Opin. Microbiol. 8, 208–215 (2005).

3. Fu, J. et al. Full-length RecE enhances linear-linear homologous recombination and facilitates direct cloning for bioprospecting. Nat. Biotechnol. 30, 440–446 (2012).

4. Chain, P. S., Hernandez-Lucas, I., Golding, B. & Finan, T. M. oriT-directed cloning of defined large regions from bacterial genomes: identification of the Sinorhizobium meliloti pExo megaplasmid replicator region. J. Bacteriol. 182, 5486–5494 (2000).

5. Shao, Z., Luo, Y. & Zhao, H. Rapid characterization and engineering of natural product biosynthetic pathways via DNA assembler. Mol. Biosyst. 7, 1056–1059 (2011).

6. Kvitko, B. H., McMillan, I. A. & Schweizer, H. P. An improved method for oriT-directed cloning and functionalization of large bacterial genomic regions. Appl. Environ. Microbiol. 79, 4869–4878 (2013).

7. Bonet, B., Teufel, R., Crusemann, M., Ziemert, N. & Moore, B. S. Direct capture and heterologous expression of Salinispora natural product genes for the biosynthesis of enterocin. J. Nat. Prod. 10.1021/np500664q (2014).

8.   Yamanaka, K. et al. Direct cloning and refactoring of a silent lipopeptide biosynthetic gene cluster yields the antibiotic taromycin A. Proc. Natl. Acad. Sci. U.S.A. 111, 1957–1962 (2014).

9.   Fernandez-Martinez, L. T. & Bibb, M. J. Use of the meganuclease I-SceI of Saccharomyces cerevisiae to select for gene deletions in actinomycetes. Sci. Rep. 4, 7100, 10.1038/srep07100 (2014).

10. Lu, Z., Xie, P. & Qin, Z. Promotion of markerless deletion of the actinorhodin biosynthetic gene cluster in Streptomyces coelicolor. Acta Biochim. Biophys. Sin. (Shanghai) 42, 717–721 (2010).

11. Grindley, N. D., Whiteson, K. L. & Rice, P. A. Mechanisms of site-specific recombination. Annu. Rev. Biochem. 75, 567–605 (2006).

12. Sternberg, N. & Hamilton, D. Bacteriophage P1 site-specific recombination. I. Recombination between loxP sites. J. Mol. Biol. 150, 467–486 (1981).

13. Sauer, B. & McDermott, J. DNA recombination with a heterospecific Cre homolog identified from comparison of the pac-c1 regions of P1-related phages. Nucleic Acids Res. 32, 6086–6095 (2004).

14. Schweizer, H. P. Applications of the Saccharomyces cerevisiae Flp-FRT system in bacterial genetics. J. Mol. Microbiol. Biotechnol. 5, 67–77 (2003).

15. Bierman, M. et al. Plasmid cloning vectors for the conjugal transfer of DNA from Escherichia coli to Streptomyces spp. Gene 116, 43–49 (1992).

16. Gregory, M. A., Till, R. & Smith, M. C. Integration site for Streptomyces phage φBT1 and development of site-specific integrating vectors. J. Bacteriol. 185, 5320–5323 (2003).

17. Zhang, L. et al. DNA cleavage is independent of synapsis during Streptomyces phage φBT1 integrase-mediated site-specific recombination. J. Mol. Cell. Biol. 2, 264–275 (2010).

18. Ullrich, S. & Schuler, D. Cre-lox-based method for generation of large deletions within the genomic magnetosome island of Magnetospirillum gryphiswaldense. Appl. Environ. Microbiol. 76, 2439–2444 (2010).

19. Komatsu, M., Uchiyama, T., Omura, S., Cane, D. E. & Ikeda, H. Genome-minimized Streptomyces host for the heterologous expression of secondary metabolism. Proc. Natl. Acad. Sci. U.S.A. 107, 2646–2651 (2010).

20. Herrmann, S. et al. Site-specific recombination strategies for engineering actinomycete genomes. Appl. Environ. Microbiol. 78, 1804–1812 (2012).

21. Muth, G., Wohlleben, W. & Puhler, A. The minimal replicon of the Streptomyces ghanaensis plasmid pSG5 identified by subcloning and Tn5 mutagenesis. Mol. Gen. Genet. 211, 424–429 (1988).

22. Kaysser, L. et al. Identification of a napsamycin biosynthesis gene cluster by genome mining. Chembiochem 12, 477–487 (2011).

23. Niu, G. & Tan, H. Nucleoside antibiotics: biosynthesis, regulation, and biotechnology. Trends Microbiol. 23, 110–119 (2015).

24. Gust, B., Challis, G. L., Fowler, K., Kieser, T. & Chater, K. F. PCR-targeted Streptomyces gene replacement identifies a protein domain needed for biosynthesis of the sesquiterpene soil odor geosmin. Proc. Natl. Acad. Sci. U.S.A. 100, 1541–1546 (2003).

25. Peschke, U., Schmidt, H., Zhang, H. Z. & Piepersberg, W. Molecular characterization of the lincomycin-production gene cluster of Streptomyces lincolnensis 78-11. Mol. Microbiol. 16, 1137–1156 (1995).

26. Fierro, F. et al. The penicillin gene cluster is amplified in tandem repeats linked by conserved hexanucleotide sequences. Proc. Natl. Acad. Sci. U.S.A. 92, 6200–6204 (1995).

27. Yanai, K., Murakami, T. & Bibb, M. Amplification of the entire kanamycin biosynthetic gene cluster during empirical strain improvement of Streptomyces kanamyceticus. Proc. Natl. Acad. Sci. U.S.A. 103, 9661–9666 (2006).

28. Liao, G. et al. Cloning, reassembling and integration of the entire nikkomycin biosynthetic gene cluster into Streptomyces ansochromogenes lead to an improved nikkomycin production. Microb. Cell Fact. 9, 6 (2010).

29. Jiang, L., Wei, J., Li, L., Niu, G. & Tan, H. Combined gene cluster engineering and precursor feeding to improve gougerotin production in Streptomyces graminearus. Appl. Microbiol. Biotechnol. 97, 10469–10477 (2013).

30. Murakami, T., Burian, J., Yanai, K., Bibb, M. J. & Thompson, C. J. A system for the targeted amplification of bacterial gene clusters multiplies antibiotic yield in Streptomyces coelicolor. Proc. Natl. Acad. Sci. U.S.A. 108, 16020–16025 (2011).

31. Gomez-Escribano, J. P. & Bibb, M. J. Engineering Streptomyces coelicolor for heterologous expression of secondary metabolite gene clusters. Microb. Biotechnol. 4, 207–215 (2011).

32. Kieser, T., Bibb, M. J., Buttner, M. J., Chater, K. F. & Hopwood, D. A. Practical Streptomyces genetics. (John Innes Foundation, Norwich, U.K., 2000).

33. Okamoto, S., Taguchi, T., Ochi, K. & Ichinose, K. Biosynthesis of actinorhodin and related antibiotics: discovery of alternative routes for quinone formation encoded in the act gene cluster. Chem. Biol. 16, 226–236 (2009).

34. Nguyen, K. T. et al. A glutamic acid 3-methyltransferase encoded by an accessory gene locus important for daptomycin biosynthesis in Streptomyces roseosporus. Mol. Microbiol. 61, 1294–1307 (2006).

35. Sambrook, J. & Russell, D. W. Molecular cloning: a laboratory manual, 3rd ed. (Cold Spring Harbor Laboratory Press, NY, 2001).

36. Pullan, S. T., Chandra, G., Bibb, M. J. & Merrick, M. Genome-wide analysis of the role of GlnR in Streptomyces venezuelae provides new insights into global nitrogen regulation in actinomycetes. BMC Genomics 12, 175 (2011).

37. Bystrykh, L. V. et al. Production of actinorhodin-related "blue pigments" by Streptomyces coelicolor A3(2). J. Bacteriol. 178, 2238–2244 (1996).

38. Miao, V. et al. Genetic engineering in Streptomyces roseosporus to produce hybrid lipopeptide antibiotics. Chem. Biol. 13, 269–276 (2006).

39. Miao, V. et al. Daptomycin biosynthesis in Streptomyces roseosporus: cloning and analysis of the gene cluster and revision of peptide stereochemistry. Microbiology 151, 1507–1523 (2005).

40. MacNeil, D. J. et al. Analysis of Streptomyces avermitilis genes required for avermectin biosynthesis utilizing a novel integration vector. Gene 111, 61–68 (1992).

41. Yang, H. et al. Autoregulation of hpdR and its effect on CDA biosynthesis in Streptomyces coelicolor. Microbiology 156, 2641–2648 (2010).

42. Paget, M. S., Chamberlin, L., Atrih, A., Foster, S. J. & Buttner, M. J. Evidence that the extracytoplasmic function sigma factor σE is required for normal cell wall structure in Streptomyces coelicolor A3(2). J. Bacteriol. 181, 204–211 (1999).

43. Li, R. et al. polR, a pathway-specific transcriptional regulatory gene, positively controls polyoxin biosynthesis in Streptomyces cacaoi subsp. asoensis. Microbiology 155, 1819–1831 (2009).

Chapter 3

MARKER-BASED ESTIMATION OF GENETIC PARAMETERS IN GENOMICS

Zhiqiu Hu[1], Rong-Cai Yang[1,2]

[1] Department of Agricultural, Food and Nutritional Science, University of Alberta, Edmonton, Alberta, Canada

[2] Alberta Agriculture and Rural Development, Edmonton, Alberta, Canada

ABSTRACT

Linear mixed model (LMM) analysis has been recently used extensively for estimating additive genetic variances and narrow-sense heritability in many genomic studies. While the LMM analysis is computationally less intensive than the Bayesian algorithms, it remains infeasible for large-scale genomic data sets. In this paper, we advocate the use of a statistical procedure known as symmetric differences squared (SDS) as it may serve as a viable alternative when the LMM methods have difficulty or fail to work with large datasets. The SDS procedure is a general and computationally simple method based only on the least squares regression analysis. We carry out computer simulations and empirical analyses to compare the SDS procedure with two commonly used LMM-based procedures. Our results show that the SDS method is not as good as the LMM methods for small data sets, but it becomes progressively better and can match well with the precision of estimation by the LMM methods for data sets with large sample sizes. Its major advantage is that with larger and larger samples, it continues to work with the increasing precision of estimation while the commonly used LMM methods are no longer able to work under our current typical computing capacity. Thus, these results suggest that the SDS method can serve as a viable alternative particularly when analyzing 'big' genomic data sets.

INTRODUCTION

Recent surge of genome-wide association studies (GWAS) based largely on the use of single nucleotide polymorphisms (SNPs) has enabled plant/animal breeders and human geneticists to identify hundreds of SNPs responsible for the genetic variation of quantitative traits or complex diseases. While such identification contributes to an in-depth understanding of the genetic architecture of complex traits, the numerous SNPs identified have collectively accounted for only a small amount of genetic variation in many studies. Several possible explanations have been put forward to explain this phenomenon known as the "missing heritability" [1]–[4]. Meanwhile, several studies [5]–[8] suggested the use of Henderson's [9] linear mixed models (LMM) to estimate the total genetic variation captured by all SNPs by replacing the pedigree-based relationship matrix with the marker-based relationship matrix in the mixed-model equations. As usual, restricted maximum likelihood (REML) method is used for the estimation of variance components. The use of LMM uncovers a substantial amount of hidden (rather than missing) heritability. However, the LMM-REML analysis is computationally very demanding particularly when there are a large number of individuals. For this reason, there are now a whole array of software packages including ASREML [10], EMMA [6], [11], FaST-LMM [12], GEMMA [8], GCTA [13], rrBLUP [14] and TASSEL [15] that implement the REML in the LMM analysis.

Since Meuwissen et al. [16], the use of DNA markers over the whole genome for prediction of unobserved phenotypes (genome-wide prediction or genomic selection, GS) has been extensively exploited in animal and plant breeding [17], [18]. While the Bayesian-based analyses are often used to simultaneously estimate variance parameters and marker effects in GS studies, it is recommended [17] that with moderate to large sample sizes, a two-step approach should be employed:

- to estimate the variance components using a non-Bayesian algorithm, usually REML-based algorithm; and

- to compute BLUP of genetic effects from standard mixed-model equations. This two-step approach is computationally less intensive than the Bayesian algorithms.

In order to achieve sufficient statistical power of detecting numerous variants with small-sized effects that collectively contribute to the genetics of most complex traits, large sample sizes are often used [19]–[23] though it is most convenient to scan the whole genome one marker at a time. For example, a recent large-scale (126,559 individuals) GWAS [23] is able to detect very small allelic effect with R^2 of $\sim 0.02\%$. However, the LMM analysis of such large data sets creates a much heavier computational burden because the computing time required for constructing the genetic relationship matrix and solving LMM equations increases with the cube of the number of individuals fit as a random effect. The computing time is further increased because iteration is needed to estimate population parameters, such as variance components that are due to the effects of individual tested markers. Despite recent proposed improvements [e.g., 8,12], the computational burden with the LMM analysis will remain to be a major issue in anticipation of larger GWAS or GS studies in the future.

It has been well known since the founding days of quantitative genetics [24] that genetic and environmental variance components can be estimated directly from phenotypic resemblance between related individuals. Koch [25] proposed the use of symmetric differences squared (SDS) as a general statistical procedure to compute the phenotypic resemblance between relatives. The SDS procedure or its variant based on the phenotypic similarity index have been subsequently used to estimate genetic and environmental variances for linkage analysis [24], animal breeding [25], estimation of heritability in natural populations [26]–[29], and recently in the genetic analysis of complex traits in humans [3], [5], [30] and QTL mapping in plants [31],[32]. Since the SDS procedure is a general and computationally simple method based only on the least squares regression analysis, it may serve as a viable alternative when the REML-based methods have difficulty or fail to work with large datasets. However, the SDS procedure has not always been employed appropriately and its correct usage needs to be clarified. The objectives of this study are (i) to provide a comprehensive evaluation of the SDS procedure in terms of its statistical properties and computational efficiencies and (ii) to compare it with two commonly used REML-based methods implemented in the software packages, rrBLUP and GCTA.

RESULTS

Simulation Results

The effects of sample size and marker density on the correlation between the actual and theoretical genetic relatedness are depicted in Figure 1. The correlations are nearly perfect ($r > 0.98$) with the high marker density ($m = 20000$) regardless of sample sizes. However, the situation is different for lower marker densities: while there is little change in the correlations for smaller samples, the correlations for larger samples can be substantially reduced. For example, at the marker density of $m = 200$, the correlation for the population of size $n = 200$ is $r = 0.95$, but the correlation for the sample of size $n = 10000$ is nearly halved at $r = 0.48$.

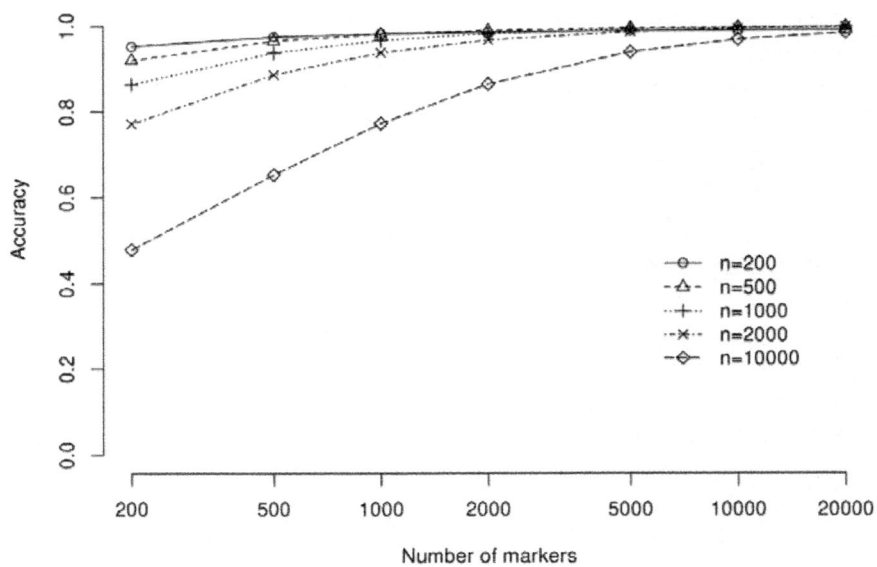

Figure 1. Effects of marker density on correlation between realized genetic related-ness and its expected value under the AR1 model.

Presented in Table 1 are the means and ranges of realized heritability values ($\hat{h}^2$) and SDS and REML heritability estimates ($\hat{h}^2$) from 100 replicated samples for each of the 27 simulation combinations consisting of theoretical heritability, sample size and marker density. A realized heritability is calculated as the ratio of the variance components directly from simulated genetic and residual effects and thus it gauges the variation in genetic sampling across different simulated samples. The effect of genetic sampling is evident as the $\hat{h}^2$ values are distributed around their respective true values (h^2). Such effect is more evident with smaller samples but the impact of marker density is less obvious. Since the estimates by rrBLUP and GCTA are identical, the mean values and ranges of estimated heritability under the header of REML are those of GCTA estimates. Both REML and SDS approaches give the mean values that are close to the corresponding true values of heritability for all the parameter combinations. The t-tests show that with a few exceptions (seven in the REML estimates and one in the SDS estimates), the means of the estimates are not significantly different from their theoretical values. However, while the ranges of the estimated heritability by the REML approach are relatively consistent over different sample sizes and marker densities, the ranges by the SDS approach are much larger for n=500 than for n=5000 under different h^2 and m values. For example, for the parameter combination of h^2=0.8 and m=200, the ranges of REML estimates are 0.075 (=0.828–0.753) for n=500 and 0.014

(0.806–0.792) for $n=5000$, but the ranges of SDS estimates are 0.558 (1.096–0.538) for $n=500$ and 0.268 (0.955–0.687) for $n=5000$. Unlike the sample size, the marker density has a minor effect on the estimates of heritability though there are larger standard deviations and ranges of heritability estimates when too many markers ($m=20000$) are used.

Table 1. Means and 90% ranges of 100 SDS and REML estimates of narrow-sense heritability (h^2) for 27 simulation trials[a]

| | | | Realized h^2 $(\hat{h}^2)$ | | SDS | | REML | |
h^2	n	m	$\hat{h}^2 \pm$ SD	$\hat{h}^2$ Range of	$\hat{h}^2 \pm$ SD[b]	$\hat{h}^2$ Range of	$\hat{h}^2 \pm$ SD	$\hat{h}^2$ Range of
0.2	500	200	0.187+0.030	0.146-0.243	0.193±0.078	0.081-0.313	0.187+0.047*	0.121-0.249
0.2	500	2000	0.190±0.032	0.140-0.264	0.197±0.077	0.082-0.342	0.199±0.065	0.095-0.303
0.2	500	20000	0.186±0.031	0.136-0.238	0.184±0.088	0.054-0.319	0.192±0.069	0.095-0.300
0.2	1000	200	0.194±0.029	0.153-0.243	0.198±0.069	0.106-0.337	0.197±0.039	0.134-0.261
0.2	1000	2000	0.189±0.022	0.157-0.223	0.192±0.045	0.118-0.262	0.193±0.040	0.132-0.263
0.2	1000	20000	0.191±0.024	0.155-0.228	0.201±0.086	0.091-0.342	0.194±0.048	0.113-0.273
0.2	5000	200	0.199±0.011	0.181-0.216	0.196±0.041	0.136-0.275	0.198±0.011	0.179-0.216
0.2	5000	2000	0.199±0.013	0.181-0.218	0.198±0.039	0.149-0.269	0.199±0.016	0.173-0.223
0.2	5000	20000	0.200±0.011	0.184-0.217	0.204±0.050	0.124-0.297	0.201±0.019	0.174-0.233
0.5	500	200	0.476±0.049	0.407-0.562	0.492±0.138	0.296-0.725	0.482±0.048*	0.415-0.561
0.5	500	2000	0.480±0.052	0.394-0.564	0.495±0.119	0.300-0.702	0.499±0.070	0.384-0.602
0.5	500	20000	0.474±0.051	0.386-0.556	0.480±0.150	0.271-0.694	0.480±0.085*	0.337-0.602
0.5	1000	200	0.487±0.046	0.419-0.562	0.494±0.114	0.344-0.683	0.495±0.035	0.435-0.548
0.5	1000	2000	0.481±0.036	0.428-0.535	0.482±0.072*	0.374-0.613	0.491±0.047	0.417-0.585
0.5	1000	20000	0.485±0.038	0.423-0.542	0.511±0.144	0.343-0.774	0.490±0.050	0.408-0.563
0.5	5000	200	0.498±0.017	0.470-0.525	0.497±0.063	0.408-0.615	0.497±0.010*	0.479-0.513
0.5	5000	2000	0.497±0.020	0.468-0.528	0.498±0.052	0.428-0.595	0.498±0.016	0.472-0.521
0.5	5000	20000	0.499±0.017	0.474-0.526	0.507±0.069	0.411-0.609	0.501±0.022	0.464-0.534
0.8	500	200	0.782±0.034	0.733-0.837	0.802±0.177	0.538-1.096	0.790±0.022*	0.753-0.828
0.8	500	2000	0.785±0.035	0.722-0.838	0.801±0.135	0.592-1.022	0.797±0.044	0.725-0.860
0.8	500	20000	0.781±0.035	0.715-0.833	0.797±0.202	0.536-1.114	0.783±0.064*	0.670-0.874
0.8	1000	200	0.790±0.031	0.742-0.837	0.797±0.137	0.612-1.008	0.797±0.014	0.775-0.819
0.8	1000	2000	0.766±0.024	0.749-0.821	0.785±0.084	0.652-0.953	0.795±0.027	0.754-0.842
0.8	1000	20000	0.789±0.026	0.746-0.825	0.829±0.191	0.573-1.189	0.796±0.032	0.736-0.841
0.8	5000	200	0.799±0.011	0.780-0.815	0.799±0.081	0.687-0.955	0.798+0.004*	0.792-0.806
0.8	5000	2000	0.798±0.013	0.779-0.817	0.800±0.057	0.715-0.899	0.799±0.007	0.787-0.809
0.8	5000	20000	0.799±0.011	0.783-0.816	0.809±0.088	0.665-0.937	0.801±0.013	0.778-0.819

*Indicates a significant deviation from h^2 according to a t-test: $t = (\hat{h}^2 - h^2)/(SD/\sqrt{100})$.
[a]The 27 simulation trials consist of three levels of true narrow-sense heritability (h^2), three sample sizes (n) and three marker densities (m). In each simulation sample, h^2 is estimated by the symmetric difference squared (SDS) method implemented in our R package, SDS/R and by residual maximum likelihood (REML) method implemented in the GCTA software.
[b]SD = standard deviation.
doi:10.1371/journal.pone.0102715.t001

The computational time and memory requirements for the two REML-based analyses (rrBLUP and GCTA) and the SDS analysis are recorded in Table 2 for seven sample sizes from $n=500$ to $n=40000$. These records of time and memory requirements are taken from the runs of the analyses on Dell Precision T1650 with Intel Xeon E3-1280 (3.8 GHz) and 32GB of RAM under the Redhat Linux 6.4 operating system. In anticipation of insufficient RAMs required for the REML-based analyses of large data sets, we decide that any analysis is terminated if its RAM usage exceeds 30GB. Thus, the recorded memory for a given analysis is the maximum memory usage over the entire process of the analysis if the memory usage at any given time of the analysis is within 30GB RAMs, or the maximum memory usage during the time period from the beginning to the termination of the analysis if the memory usage is beyond

30GB RAMs. In this study, the time period for terminating an analysis is set to be one hour after the 30GB RAM usage criterion is reached. For GCTA, only one CPU core is used in each analysis run to measure the computational time and memory usage for a fair comparison with the computational efficiencies for rrBLUP and SDS. For rrBLUP, the memory usage is measured in a clean R environment with only the Q matrix and the vector of phenotypic values being loaded and the time of loading the data is not included in the computational time recorded.

Table 2. Actual computational efficiency by SDS and REML procedures under samples of seven sizes $(n)^a$

| | REML | | | | SDS | |
| | GCTA | | rrBLUP | | | |
n	Time (s)	RAM(GB)	Time (s)	RAM(GB)	Time (s)	RAM(GB)
500	0.235	<0.01	1.438	0.27	0.054	0.26
1000	0.746	0.02	17.856	0.42	0.061	0.32
2000	3.881	0.33	112.780	0.94	0.084	0.42
5000	58.451	1.91	2738.159	4.77	0.231	0.59
10000	226.827	7.46	9054.193	19.33	0.756	1.14
20000	1610.518	28.90	NA^b	65.60	2.851	2.05
40000	NA	NA	NA	NA	11.286	5.13

^{a}The computational times in seconds (s) and memory requirements in gigabites (GB) that are required to run a simulated sample of size n with the true heritability of $h^2 = 0.5$ and marker density of $m = 2000$ by the symmetric difference squared (SDS) and two residual maximum likelihood (REML) methods, GCTA and rrBLUP. Each of these times and memory requirements is an average over five simulation samples.
^{b}NA indicates that no information is available due to termination of the analysis.
doi:10.1371/journal.pone.0102715.t002

It is evident from Table 2 that the SDS analysis succeeds with all sample sizes and it requires far less computational time and memories than either the GCTA or rrBLUP analysis. For the two REML-based methods, while the rrBLUP and GCTA analyses give identical estimates of variance components and heritability, the GCTA analysis requires much less computational time and memories than the rrBLUP analysis. When the sample size is increased to $n=20000$, the rrBLUP analysis is terminated because its memory requirement (65.6GB) is far beyond our criterion of 30GB RAMs. With this same sample size, the GCTA analysis is successfully completed in ~27 minutes but its memory requirement (28.9GB) is near the RAM capacity of our computer. Obviously, both rrBLUP and GCTA are terminated for $n=40000$ because the memory requirement certainly exceeds our criterion of 30GB RAMs.

Empirical Study

Wheat Data

We will analyze two empirical examples. The first empirical data set is taken from Crossa et al.[33] and it includes 599 wheat inbred lines developed by the CIMMYT Global Wheat Breeding program and tested in four target sets of environments (E1–E4). The pedigree data and 1279 polymorphic Diversity Array Technology (DArT) markers for these lines are provided as well. Grain yield (GY) was measured in all the environments. Crossa et al. [33] compared the estimates of variance components and heritability under different schemes ranging from pedigree-based information only to the combination of both pedigree-based and marker information. Since the genotype of each inbred line can only be one of the two homozygotes at a marker locus (essentially a haploid model), the ilth element of $\mathbf{W}$ matrix for the genotype of the ith individual at lth locus is coded as, $w_{il} = (z_{il} - \hat{p}_l)/\sqrt{\hat{p}_l(1-\hat{p}_l)}$, where the indicator variable z_{il} is 1 for the reference homozygote and 0 for the alternative homozygote, and $\hat{p}_l$ is the estimated frequency of the reference homozygote at the lth locus. Both rrBLUP and GCTA do not allow for inputting haploid data for constructing $\mathbf{W}$ matrix and thus the GRM, but GCTA does have an option to allow for inputting the GRM. Thus the comparison will be between the SDS and GCTA analyses for this empirical data.

The estimates of heritability for the GY in E1-E4 are presented in Table 3. The SDS estimates are comparable with the GCTA estimates, judging from the standard deviations and the 95% confidence intervals generated by bootstrapping. The bootstrapping procedure used here is somewhat different from the usual bootstrapping in that any duplicates in a bootstrap sample from sampling with replacement are excluded to avoid the problem of the GRM being singular in the REML-based GCTA estimation.

Table 3. SDS and REML estimates of heritability for wheat grain yield in four environments (E1–E4)[a]

Environment	SDS			REML		
	h^2	SD	CI95 [b]	h^2	SD	CI95
E1	0.564	0.086	0.362–0.706	0.498	0.049	0.399–0.589
E2	0.452	0.114	0.192–0.629	0.448	0.048	0.324–0.520
E3	0.379	0.070	0.223–0.497	0.423	0.061	0.305–0.544
E4	0.481	0.071	0.304–0.587	0.430	0.058	0.292–0.524

[a] The estimates of heritability for the wheat data set taken from Crossa et al. [33] by the symmetric difference squared (SDS) method and a residual maximum likelihood (REML) method, GCTA.

[b] The 95% confidence intervals (CI 95) are constructed based on 1000 bootstrap samples.

Drosophila Data

The second empirical data set is taken from Macdonald et al. [34]. These authors examined associations of 203 SNPs at the *Enhancer of split* Complex [(E(spl)-C] with sternopleural bristle number (SBN) and abdominal bristle number (ABN) for 2000 *Drosophila melanogaster* individuals (1000 males and 1000 females) sampled from a location in Napa Valley, California. The 203 genotyped polymorphisms actually consisted of 191 SNPs and 12 insertion/ deletion events, but for simplicity, they were all referred to as SNPs. There was little evidence of associations between individual SNPs and SBN or ABN in the female or male population. In fact, Macdonald et al. [34] concluded that individual SNPs in the E(spl)-C gene region contributed little to phenotypic variation in SBN and ABN. For this reason, we use the LMM-REML and SDS methods to determine if the joint effect of all SNPs in the E(spl)-C would make a detectable contribution to the variation in SBN and ABN. Prior to our analysis, we remove 50 individuals with missing phenotypic values or genotype scores in both female and male populations. In addition, a total of 41 SNPs are removed due to (i) minor allele frequency (MAF) of <0.05 and (ii) significant ($p<0.01$) departure from Hardy-Weinberg proportions of three genotypes at each locus. Thus, 950 individuals with 162 SNPs in the E(spl)-C are retained for the analysis. To be consistent with the analysis of the first data set, only the GCTA is used for comparison with the SDS.

As shown in Table 4, the joint effect of 162 SNPs in the E(spl)-C makes a small contribution to the phenotypic variation in SBN and ABN in both female and male populations. The percentages of the contribution range from 0.5% to 9% by the SDS method and 0% to 1.4% by the GCTA. These are reasonable values given that (i) this is the joint contribution of potential causal variants from one major gene complex for the traits; and (ii) the previous gene-wide scan by Macdonald et al. [34] revealed only marginally significant associations of three SNPs with SBN and of two SNPs with ABN regardless of sexes. The lower limits of the 95% confidence intervals are all bound to 0 because, in a REML-based analysis, a constraint of $\hat{\sigma}_a^2 \geq 0$ and thus $\hat{h}^2 \geq 0$ is imposed.

Table 4. SDS and REML estimates of heritability for bristle number in a large wild-caught cohort of fruit fly (*Drosophila melanogaster*)[a]

		SDS			REML		
Sex	Trait[b]	h^2	SD	95% CI[c]	h^2	SD	95% CI
Female	ABN	0.035	0.027	−0.028–0.074	0.000	0.004	0.000–0.015
Female	SBN	0.008	0.023	−0.034–0.053	0.000	0.006	0.000–0.020
Male	ABN	0.090	0.039	−0.015–0.136	0.014	0.013	0.000–0.043
Male	SBN	0.005	0.027	−0.033–0.072	0.000	0.009	0.000–0.031

[a]The estimates of heritability for the fruit fly data set taken from Macdonald et al. [34] by the symmetric difference squared (SDS) method and a residual maximum likelihood (REML) method. GCTA.
[b]ABN = abdominal bristle number and SBN = sternopleural bristle number.
[c]The 95% confidence intervals (CI 95) are constructed based on 1000 bootstrap samples.
doi:10.1371/journal.pone.0102715.t004

DISCUSSION

The computational burdens of many LMM applications including heritability estimation in large-scale genomic studies have recently stimulated a huge amount of research interest in the development of faster and memory-efficient algorithms for feasible and successful analyses of large-scale genomic data. Despite these efforts, the computational challenges remain. In this study, we investigate the statistical properties and computational efficiency of the least-squares-based SDS method in comparison to the two LMM methods (rrBLUP and GCTA) to see the feasibility of the SDS method as a viable alternative to LMM-based methods. Our results (Table 1) show that the SDS method is inferior to the REML methods for small sample sizes, but it becomes progressively better and can match well with the precision of estimation by the REML methods for large sample sizes. Thus, these results suggest that the SDS method can serve as a viable alternative particularly when analyzing 'big' genomic data sets. Its major advantage is that with larger and larger data sets, it continues to work with the increasing precision of estimation while the other current commonly used methods are no longer able to work with our current computing capacity. To illustrate this point, we go beyond the parameter combinations set in Table 1 to simulate a larger data set with sample size of $n=50000$ and marker density of $m=2000$ and heritability of $h^2=0.8$. We analyze this data set using the SDS only because the other methods have already stopped to work with a sample of smaller size $n=40000$ and the same marker density ($m=2000$) as shown in Table 2. The means of estimates from 100 simulation samples are very similar and are closely around the true value of 0.8 is 0.805 with SD being 0.026, and the 90% range is 0.072 (0.838–0.766). This range is much narrower than that for $n=5000$. On average, the per-sample time requirement for constructing the GRM and the SDS analysis is 6355.3 seconds or about 1 hour 46 minutes.

It is evident from Table 2 that the SDS requires far less computational time and memory than the LMM methods. Henderson [9] and recently others [5]–[8], [12] have shown that the LMM estimates the variance components through simultaneously estimating the variance parameters and marker effects in an iterative manner. Usually, the iterative process starts with an initial (guessed) set of variance values or their ratio to provide the first round of estimates of random additive genetic effects (cf. equation A5 in the Text S1). These estimates of random effects are in turn used to estimates additive and residual variances using equation (A8) in the Text S1. The iteration continues until the successive rounds of estimates of variance components are stabilized. On the other hand, the SDS estimation of variance components or heritability is based directly on the linear regression of two sets second-order statistics, phenotypic and genetic similarities or dissimilarities. The SDS approach is much simpler and thus computationally much less demanding than the LMM approaches because it does not require (i) iteration and (ii) computing the inverse of GRM. The Bayesian analysis was not used in our study because in our initial investigation of different estimation methods, it took the Bayesian LASSO [17] 304.161 seconds or >5 minutes to complete the analysis of a single simulated sample of size $n=1000$ with $m=2000$, comparing to only 17.856 seconds by rrBLUP, 0.746 seconds by GCTA and 0.061 seconds by the SDS as shown in Table 2 for the same n and m. It would have been hardly feasible to run the data sets from a number of replicated simulations. Furthermore, de los Campos et al. [17] suggested the Bayesian analysis would be computationally even more demanding than the LMM analysis.

Computational efficiency has become an emerging issue particularly with increasing availability of larger and larger data sets from genomic studies and it has been recently investigated [8],[35]. However, these investigations focus on the computational complexity that is platform-independent, but ignore the implementation issues that may affect the actual computational efficiency in reality. In other words, the actual computational efficiency needs to be evaluated by considering the memory capacity and management under a given operating and software environment. Thus, in this study, we instead emphasize the *actual* computational time and memory requirements under different sample sizes (Table 2). Should the computational efficiency be based solely on the computational complexity [8], [35], the two REML implementations, rrBLUP and GCTA, would have had the same computational complexity. However, because rrBLUP is a cross-platform R package that is implemented strictly in the R environment and GCTA is a program that was written in C and has routinely run in the Linux operating system, the two programs are obviously different in terms of their actual computational efficiency (Table 2). This discrepancy is due mainly to the differences in the platforms and programming languages inherent in the two software packages.

In our study, any analysis would be terminated if its RAM usage exceeds 30GB, the maximum allowable RAMs after accounting for the RAM requirements by the operating system and other essential utility programs. It may be argued that we can increase the memory for the current computer or tap into supercomputing resources to address the issue of insufficient memory. However, neither solution is very feasible in our situation and perhaps in many other situations. Our workstation has a motherboard that can only support the maximum RAMs of 32GB and thus there is no room to add more memory. Many supercomputing servers such as the Westgrid in Canada (https://www. westgrid.ca/) have provided excellent computing resources to researchers with larger computing needs. However, access to these servers is a complicated process. It involves (i) submission of separate proposals for those systems with considerably large computing capacity in the supercomputing network; and (ii) a long waiting time after submitting batch jobs. In addition, some servers only serve certain countries or regions. Another possibility is the use of GPU (graphics processing unit)-accelerated computer for more efficient and even faster solutions to the large-scale GRM and its inverse essential in GWAS and genomic prediction. There are recent studies on the use of GPU-based parallel computing in GWAS [36]–[38], but these studies focus on detecting individual gene or gene-gene epistatic effects through genome-wide scan using single-marker analysis. Thus, it remains to be developed the GPU-based parallel computing algorithms for calculating the GRM and its inverse needed for genomic and marker-based estimation of variance components or heritability.

Our choice of rrBLUP and GCTA for comparison with the SDS method is somewhat arbitrary and it is due to our own familiarity and experiences with these software packages. Other LMM-REML software packages as mentioned in the Introduction section would have been equally effective for the comparison. ASREML [10] is widely used in agriculture community and it is implemented based on the average information algorithm as GCTA is. TASSEL [15] is very popular in plant breeding because it is able to accommodate both intra- and inter-population genetic relatedness in the LMM analysis. EMMA [6], [11] and its more efficient version GEMMA[8] have been extensively used in detection of causal variants responsible complex traits and diseases in human. FaST-LMM [12] uses a low-rank approximation to the GRM and thus it greatly improves computational efficiency and reduces memory requirement. For example, the computing time for the FaST-LMM analysis of a simulated data set with n=20000 and m=2000 is only 50.1 seconds and the maximum RAM usage for this run is 2.8 GB. These computational performance and memory requirement by FaST-LMM are much better than the GCTA analysis (1610.5 seconds and 29.0 GB), but are still not as good as the SDS analysis (2.9 seconds and 2.0 GB). Despite FaST-LMM's superior computational performance and low memory requirement, it remains to be determined the optimal number of eigenvalues and corresponding eigenvectors that should be retained for a good low-rank approximation to the GRM for a given data set.

The analysis of the wheat yield data supplied by Crossa et al. [33] (Table 3) shows the similar estimates of heritability by the SDS and REML methods in all four environments, despite a relatively small sample size (n=599) and low marker density (m=1279). At the first glance, this result is somewhat surprising because the SDS estimates would have been more fluctuating than the REML at this level of sample size and marker density as can be inferred from the simulation results of Table 1. However, it should be remembered that this wheat population consists of 599 recombinant inbred lines (i.e., they are essentially 599 haplotypes) whereas our simulation data are all based on diploid individuals. The estimation of genetic relationship between haploids would be obviously more accurate than that between diploids at the same level of sample size and marker density. We also employ the GCTA analysis to confirm the estimates of genetic and residual variances as in Table 1 of Crossa et al. [33] based on their P model (i.e., the **A** matrix obtained from the pedigree information only) and the M-RKHS model (i.e., the **K** matrix obtained using the reproducing kernel Hilbert space nonlinear function of marker distance). The heritability estimates are subsequently calculated. Our GRM-based estimates of heritability, by either SDS or GCTA, are close to or slightly higher than those based on the P model, but they are all smaller than those based on the M-RKHS model. The elements of the K matrix derived from the M-RKHS model are a nonlinear (exponential) function of squared-Euclidean distance between markers in pairs of individuals, thereby probably capturing more genetic variation. It remains to be investigated how the use of nonlinear functions of marker distances generally affects the estimation of the genetic relationship between individuals, thereby influencing the estimation of variance components or heritability. The analysis of the drosophila data taken from Macdonald et al. [34] (Table 4) shows that the joint effect of all potential causal variants in in the E(spl)-C gene region remains small. This reinforces the earlier conclusion [34] that individual SNPs contribute little to the phenotypic variation in the bristle number.

A key part of both LMM and SDS analyses is to estimate the GRM based on a set of SNPs or other genetic markers. The use of GRM is more advantageous over the use of the traditional pedigree-based relationship matrix (**A** matrix) for two major reasons. First, the GRM would capture much of the

Mendelian sampling variation that is missing in the **A** matrix. Second, the use of marker SNP data (rather than pedigree information) allows for estimation of relatedness of distantly related individuals, thereby controlling confounding effects from the environmental correlation between relatives due to their shared (common) environment. However, the unbiased estimate of GRM is achieved only if all the markers used for the estimation are the causal variants [4], [39], [40]. This of course is not true almost in all the cases. In our simulation studies, we use 10% of the total markers as the 'causal' (relevant) variants and the remaining 90% of the markers as irrelevant variants. It is evident from Table 1 that the accuracy of the estimated heritability is the best with the medium marker density ($m=2000$) for most simulation trials. With the high marker density ($m=20000$), there are more irrelevant markers, thereby costing some accuracy in the estimation of heritability. Similar results were observed in other simulation studies [e.g., 39].

In applying the SDS method to the actual analysis, we propose a modification to the Ritland's estimation procedure. Instead of directly regressing the phenotypic similarity indexes on the corresponding genetic relatedness values between pairs of individuals as originally suggested in Ritland [26] and Lynch and Walsh [29, p. 800–803], we propose that the regression analysis should be based on the averages of bins to improve the feasibility and accuracy of the estimation in large genomic studies. The bins are constructed as a complete set of nonoverlapping intervals covering the entire range of genetic relatedness coefficients (i.e., all the elements of lower triangle of the GRM). The basis of our binning (grouping) procedure is simple: the neighboring values of genetic relatedness should contribute very similarly to the shape and pattern of the linear regression line in equation (9) and thus they can be bracketed into the same group or bin. A question would naturally arise: how many bins should we have to achieve the best estimates of variance components or heritability? In our own simulation and empirical data analysis, we use a set of 1000 equal-width bins for all the data. While this set is somewhat arbitrary, we feel it suffices for the data sets we have analyzed. Nevertheless, it is an interesting issue that needs to be further examined particularly when there are nonlinear relationships arising from gene-gene interactions and gene-environment interactions [3].

Another possible question with our binning procedure is that the number of observations may vary from bin to bin and thus the residual variance may also vary from bin to bin. At the first glimpse, the use of weighted least square (WLS) analysis would be a natural solution to this problem. However, our preliminary analysis shows that the WLS results are similar to those from the regression analysis of un-binned data (i.e., individual phenotypic and genetic similarities). We are not exactly sure why WLS does not work well as it should. We suspect the following reason. For the GRM constructed for a large sample, the distribution of genetic similarity between pairs of individuals is often highly skewed towards zero, that is, the majority of genetic similarity values are clustered around zero (see Figure S1 for an example of n=4000 under the AR1 model with ρ_a=0.95 and h^2=0.5). In WLS, these near-zero values would be overemphasized (i.e., they would carry much more weight), thereby having a much stronger influence on the slope of the regression line (estimated additive genetic variance or narrow-sense heritability) than the values over the rest of the genetic similarity range. This issue certainly needs to be further investigated.

Our study focuses on a comparison of LMM-REML and SDS methods for estimating additive genetic variance and thus narrow-sense heritability. Such comparison can be easily extended to include non-additive genetic variances. As shown in Su et al. [41], the dominance genomic relationship matrix (**D**) based on marker genotypes is needed to estimate the dominance genetic variance; similarly different epistatic relationship matrices based on Q and **D** matrices (i.e., Q#Q, Q#**D**, **D**#Q and **D**#**D**) are needed to estimate epistatic genetic variances. While it is quite straightforward for the SDS method to estimate all non-additive variances by extending the simple regression analysis to the multiple regression analysis, it is computationally even more challenging for the LMM-REML methods to estimate non-additive genetic variances even for a moderate-sized data set. Furthermore, far fewer software packages are available for the estimation of non-additive genetic variances. There is an ongoing debate on how much non-additive genetic variances really contribute to the 'missing heritability' [3], [42]–. According to Hill [42], such contribution would be relatively small for biologically more realistic epistatic models. Regardless, one of the key issues with the estimation of non-additive genetic variances is that unless sample size is large enough, the estimates would be highly unreliable. Thus, in the future efforts to improve the reliability of additive and non-additive variances from the analysis of large-scale genomic data sets, the LMM-REML methods will be certainly challenged but the SDS method looks very promising in terms of computational feasibility and estimation accuracy.

Our study uses all individuals, related and unrelated, in the simulated and empirical populations for the SDS and the LMM-REML analyses. This is somewhat in contrast to some recent studies, more specifically by Yang et al. [5] who focused on the estimation of additive genetic variance and narrow-sense heritability in a human population of 'unrelated' individuals with close relatives in their original population being selectively excluded from their LMM analysis. In particular, while the off-diagonal elements of the estimated GRM between 4259 individuals ranged from −0.024 to 0.585, Yang et al. [5] selectively chose individuals such that only those individuals with the genetic relatedness within the range of −0.024 to 0.024 were considered as 'unrelated' for the LMM-REML analysis. It is evident from Figure S1 that the majority of genetic similarity values for n=5000 under the AR1 model and θ_a=0.95 are clustered around zero with 95% of pairwise elements in the estimated GRM being from −0.0496 to 0.0524. This is slightly larger than but essentially similar to the range of −0.027 to 0.027 for the estimated GRM as given in Yang et al. [5]. It is also evident from Figure S1 that, of the remaining 5% elements in the estimated GRM under the AR1 model, the range from 0.0524 to 0.95 spans widely. It is this small percentage of the widespread values that largely determine the slope of the regression line in the SDS method. Similarly, we analyze the same data set using the usual LMM-REML analysis [e.g., 29] where all individuals are used regardless of their relatedness.

A distinction needs to be made between the use of simple regression analysis or its refinements for genome-wide scan of individual marker effects in a large-scale (n >100000 and m >500000) genomic studies [e.g., 23] and the use of LMM-REML analysis for the aggregate effect of all markers across the genome in our study and other studies. It should not be forgotten that a key motivation of using LMM-REML analysis is to help recover the portion of 'missing' heritability encountered in the single-marker analysis [5]. Earlier, we have already discussed some of the computational challenges with the LMM-REML analysis for a moderate-scale (n <10000 and m <100000) genomic data under our current typical computing capacity and suggested the SDS approach as a viable alternative when the data becomes larger and larger. With our binning approach to the SDS analysis, it is possible to handle genomic data sets of any size because the regression analysis is based on a fixed number of bins for phenotypic and genetic similarity indexes regardless of sample size. However, it is presently not feasible yet for the LMM-REML analysis to handle a very large data set. For example, Yang et al. [13] reported that it would take their software package GCTA ~ 4 CPU hours (AMD Opteron 2.8 GHz) to compute the GRM for a data set with 3925 individuals genotyped by 294,831 SNPs. If we use this time as a guideline, then for a simulated data set with m=500000 and n=4000, it would have taken GCTA more than 400 (=100×4) CPU hours or more than 16 days to complete this simulation alone. Thus, in most recent LMM-REML analyses [e.g., 39], a moderate-sized data set is used.

CONCLUSIONS

The SDS method that has been overshadowed by more popular LMM or Bayesian methods in recent years appears to have a bright future because of the computational challenges that the LMM and Bayesian methods are currently facing with growing availability of larger and larger genomic data sets from the genomic studies of human and domestic plants and animals. It can serve as a viable alternative framework for quantitative genomic analyses such as GWAS and genome-wide prediction. We hope that our study stimulates and renews research interests in the use of the SDS method for the analysis of large-scale genomic data sets that will become increasingly available in the future.

METHODS

The SDS Approach

There are two versions of the SDS approach. The first and true version of the SDS approach (SDS1) was originally proposed by Koch [45] and has been subsequently used in many genetics and genomics applications [3], [5], [24], [25], [30]. The second version that is based on the phenotypic covariance between pairs of related individuals (SDS2) was first put forward by Ritland [26] and subsequently discussed particularly in the genetic analysis of natural populations [27]–[29], [46]–[48]. However, the theoretical relationship between the versions has never been clarified so that they are sometimes considered as two different approaches [29]. In fact, we view them simply as the two sides of the same coin. In addition, there is a considerable amount of confusion regarding what needs to be used in the regression equation for estimation. Here we will describe both versions in details, point out the theoretical relationship between them and clarify appropriate and correct estimation procedures that should be used in each case.

SDS2.

We describe the SDS2 first because it has a more direct connection with the LMM-REML analysis described in Text S1. Recall from the LMM analysis [equation (A2) in Text S1] that the total phenotypic covariance matrix among n individuals is partitioned into genetic and residual components, $\mathbf{V} = \mathbf{G} + \mathbf{R} = \Theta\sigma_a^2 + \mathbf{I}_n\sigma_e^2$, where $\mathbf{Q} = \{\theta_{ij}\}$ is the $n \times n$ genetic relationship matrix (GRM) with θ_{ij} being the coefficient of genetic relationship between ith and jth individuals, $\mathbf{I}_n$ is an identity matrix order n, σ_a^2 is the additive genetic variance and σ_e^2 is the residual variance. Writing the phenotypic covariance between individuals i and j (v_{ij}) in $\mathbf{V}$ in terms of the phenotypic correlation between the two individuals

(r_{ij}) and their variances $(\sigma_p^2 = \sigma_{pi}^2 = \sigma_{pj}^2)$, i.e., $v_{ij} = \rho_{ij}\sigma_p^2$, we have $\mathbf{V} = \mathbf{P}\sigma_e^2$ with $\mathbf{P}$ being the phenotypic correlation matrix and,

$$\mathbf{V} = \mathbf{P}\sigma_p^2 = \boldsymbol{\Theta}\sigma_a^2 + \mathbf{I}_n\sigma_e^2 \tag{1}$$

Multiplying both sides of equation (1) by σ_p^{-2}, we obtain,

$$\mathbf{P} = \boldsymbol{\Theta}h^2 + \mathbf{I}_n(1 - h^2) \tag{2}$$

where $h^2 = \sigma_a^2/\sigma_p^2$ is the narrow-sense heritability. We now write the matrices in equations (1) and (2) in the vector form,

$$\mathbf{v} = \boldsymbol{\theta}\sigma_a^2 \text{ and } \mathbf{p} = \boldsymbol{\theta}h^2 \tag{3}$$

where $\mathbf{v} = \text{Vech}(\mathbf{V})$, $\boldsymbol{\theta} = \text{Vech}(\mathbf{Q})$, and $\mathbf{p} = \text{Vech}(\mathbf{P})$ with $\text{Vech}(\mathbf{X})$ being the "vector-half" function[49] that creates a column vector whose elements are the stacked columns of the lower triangular elements of matrix $\mathbf{X}$. Thus, for an $n \times n$ matrix $\mathbf{X}$, there are $n(n-1)/2$ elements in $\text{Vech}(\mathbf{X})$. The residual term in equation (3) disappears because $\text{Vech}(\mathbf{I}_n)$ is a vector of zeros ($\mathbf{0}$). The results in equation (3) reinforce the well-known result in classic quantitative genetics[29], [50] that for a quantitative trait with purely additive-genetic basis but no shared environmental effects, the phenotypic covariance between pairs of relatives in a population is expected to be the covariance between the additive genetic effects for the same pairs of relatives. Thus, equation (3) provides a simple regression model that can be used to estimate the variance components or heritability.

SDS1

Above we show that the linear relationship between phenotypic and additive genetic covariances or correlations between pairs of individuals can be used for predicting additive genetic variance or heritability. Similarly, the expected value of difference squared (DS) between the phenotypic values of a pair of individuals can be partitioned into the two components due to additive genetic effect and residual deviation as done in Grimes and Harvey[25]. Since the DS is symmetric for any pair of individuals (i.e., the DS is identical regardless of the order of the two individuals), Grimes and Harvey [25] called it the symmetric difference squared (SDS). This partitioning for the ith and jth individuals can be written as,

$$
\begin{aligned}
d_{ij} &= \tfrac{1}{2}E[(y_i - y_j)^2] \\
&= \tfrac{1}{2}E[(a_i - a_j)^2] + \tfrac{1}{2}E[(e_i - e_j)^2], \\
&= (1 - \theta_{ij})\sigma_a^2 + \sigma_e^2
\end{aligned}
\tag{4}
$$

where the quantity $(1-\theta_{ij})$ measures the genetic distance between the ijth pair of individuals for $i<j$. The usual assumptions are: (i) the additive genetic effects are independent of the residual deviations and (ii) there is no correlation between the residual deviations of the ijth pair of individuals. Collecting the partitioning results for all pairs of individuals including the trivial result of zero SDS for individuals with themselves, we have in matrix form,

$$
\mathbf{D} = [(\mathbf{J}_n - \mathbf{I}_n) + (\mathbf{\Theta}_d - \mathbf{\Theta})]\sigma_a^2 + (\mathbf{J}_n - \mathbf{I}_n)\sigma_e^2
\tag{5a}
$$

or

$$
\begin{bmatrix}
0 & d_{12} & \cdots & d_{1n} \\
d_{12} & 0 & \cdots & d_{2n} \\
\vdots & \vdots & \ddots & \vdots \\
d_{1n} & d_{2n} & \cdots & 0
\end{bmatrix}
=
$$

$$
\begin{bmatrix}
0 & 1-\theta_{12} & \cdots & 1-\theta_{1n} \\
1-\theta_{12} & 0 & \cdots & 1-\theta_{2n} \\
\vdots & \vdots & \ddots & \vdots \\
1-\theta_{1n} & 1-\theta_{2n} & \cdots & 0
\end{bmatrix}\sigma_a^2 +
\begin{bmatrix}
0 & 1 & \cdots & 1 \\
1 & 0 & \cdots & 1 \\
\vdots & \vdots & \ddots & \vdots \\
1 & 1 & \cdots & 0
\end{bmatrix}\sigma_e^2
\tag{5b}
$$

where $\mathbf{J}_n$ is the $n \times n$ matrix of ones, $\mathbf{I}_n$ is the identity matrix of order n, and $\mathbf{Q}_d$ is a diagonal matrix with the diagonal elements being the same as in matrix $\mathbf{Q}$.

Relationship between SDS1 and SDS2.

The SDS1 and SDS2 (covariance) versions are obviously related to each other. This relationship between the phenotypic values of the ijth pair of individuals

$$
\begin{aligned}
d_{ij} &= \tfrac{1}{2}E[(y_i - y_j)^2] \\
&= \tfrac{1}{2}E\{[(y_i - \mu) - (y_j - \mu)]^2\} \\
&= \tfrac{1}{2}E[(y_i - \mu)^2 + (y_j - \mu)^2 - 2(y_i - \mu)(y_j - \mu)] \\
&= \sigma_p^2 - v_{ij} \\
&= (1 - \rho_{ij})\sigma_p^2
\end{aligned}
\tag{6}
$$

is derived as,

It is evident from equation (6) that a perfect inverse linear relationship between the phenotypic SDS (d_{ij}) and the covariance (v_{ij}) with the intercept being the phenotypic variance (σ_P^2) and the slope being -1. Thus, if the phenotypic SDS (d_{ij}) values are known, we can immediately obtain the corresponding covariances for all pairs of individuals by reversing the relationship inequation (6) as $v_{ij} = \sigma_P^2 - d_{ij}$.

Similar relationships can be found for additive genetic SDS and residual SDS with $\frac{1}{2}E[(a_i - a_j)^2] = (1 - \theta_{ij})\sigma_a^2$ and $\frac{1}{2}E[(e_i - e_j)^2] = \sigma_e^2$.

Equation (4) shows the partitioning of the phenotypic SDS into additive genetic and residual variance components. To show the partitioning in terms of heritability, we need the standardized phenotypic SDS (d_{ij}/σ_P^2) which is simply the phenotypic distance $(1 - \rho_{ij})$. Thus, the standardized phenotypic SDS or phenotypic distance is related to the genetic distance and the heritability as,

$$d_{ij}\sigma_p^{-2} = (1 - \rho_{ij}) = (1 - \theta_{ij})h^2 + (1 - h^2), \tag{7a}$$

or in matrix form,

$$\mathbf{D}\sigma_p^{-2} = (\mathbf{J}_n - \mathbf{P}) = [(\mathbf{J}_n - \mathbf{I}_n) + (\mathbf{\Theta}_d - \mathbf{\Theta})]h^2 + (\mathbf{J}_n - \mathbf{I}_n)(1 - h^2) \tag{7b}$$

Estimation Procedure

It is evident from the above theoretical analysis that the estimation of variance components or heritability can be carried out using either the SDS1 (difference) model or SDS2 (covariance) model and both models would lead to the same estimates of variance components or heritability. Thus we will focus on our estimation procedure under the SDS2 model as it is more directly connected to the LMM analysis.

Before describing our own procedure, we outline the estimation procedure of Ritland [26] and Lynch and Walsh [29, p. 800–803] under the SDS2 model (also commonly known as Ritland's procedure). For n individuals, there are $n(n-1)/2$ pairs of individuals and phenotypic values of each pair are used to calculate the phenotypic similarity. Thus, the sample phenotypic similarity between the ith and jth individuals for $i<j$ is given by,

$$s_{ij} = (y_i - \bar{y})(y_j - \bar{y}) \tag{8}$$

where $\bar{y}$ is the mean of n phenotypic values. Thus, the additive genetic variance can be estimated from the linear regression of the sample phenotypic similarity on the genetic relatedness as,

$$s_{ij} = \alpha + \hat{\theta}_{ij}\beta + \varepsilon_{ij} \tag{9}$$

where a is the intercept, β is the regression coefficient which estimates the additive genetic variance (σ_a^2), $\hat{\theta}_{ij}$ is the estimated value of q_{ij} corresponding to the ijth element of the GRM estimated using m markers scored and e_{ij} is the residual deviation of the sample phenotypic similarity from its expected value. The intercept should be zero because it is assumed that none of individual pairs are genetically identical, nor do they share the same environment. The estimate of additive genetic variance is simply the regression coefficient (slope),

$$\beta = \hat{\sigma}_a^2 = \frac{\text{cov}(s_{ij},\hat{\theta}_{ij})}{\text{var}(\hat{\theta}_{ij})} \tag{10}$$

The narrow-sense heritability is subsequently estimated as $\hat{h}^2 = \hat{\sigma}_a^2/\hat{\sigma}_p^2$. A more direct estimate of h^2 can be obtained by regressing the standardized phenotypic similarity index, $s'_{ij} = s_{ij}/\hat{\sigma}_p^2$, on the $\hat{\theta}_{ij}$ values,

$$\hat{h}^2 = \frac{\text{cov}(s'_{ij},\hat{\theta}_{ij})}{\text{var}(\hat{\theta}_{ij})} \tag{11}$$

Ritland's procedure has two major drawbacks particularly in the context of large-scale genomic studies. First, since it is not the sample phenotypic similarity index but rather its expected value that is proportional to the additive genetic variance or heritability without error[i.e., $E(s_{ij}) = \theta_{ij}\sigma_a^2$ or $E(s'_{ij}) = \theta_{ij}h^2$], Ritland's procedure makes the direct use of sample phenotypic similarity index and thus it would lead to a biased estimate of additive genetic variance or heritability [cf. equation (9)]. Second, while Ritland's procedure is computationally much simpler than the above LMM analysis, it can still be memory- or time-consuming. For example, for 40000 individuals, the total pairs of phenotypic similarity values and genetic relatedness estimates are 799980000 and it certainly takes a long time and a large RAM capacity for the regression analysis based on such two huge arrays of values.

Here we propose a modification of Ritland's procedure to remove the two deficiencies. Our new procedure is based on a simple idea that if we sort out the coefficients of genetic relatedness in an ascending or descending order, the neighboring values of genetic relatedness should contribute very similarly to the shape and pattern of the linear regression line in equation (9) and thus they can be bracketed into the same group or bin. Now the bin averages instead of individual phenotypic and genetic similarity indexes are used for the least squares estimation of additive genetic variance,

$$\bar{s}_{bin} = \Sigma s_{ij(bin)}/n_{bin} = \bar{\theta}_{bin}\sigma_a^2 + \Sigma \varepsilon_{ij(bin)}/n_{bin} \qquad (12)$$

Clearly, the average of residual deviations (e_{ij}'s) would tend to zero for the large number of observations (i.e., large n_{bin}) within the bins, thereby leading to the estimate of additive genetic variance with minimal bias. A similar argument can be made for binning the standardized phenotypic similarity index (s'_{ij}) for estimating the heritability. In actual implementation of our binning strategy, we divide the whole range of genetic relatedness (usually 0–1) into 1000 equally spaced bins and then distribute individual similarity index values into different bins according to their levels of genetic relatedness. In other words, our binning approach uses the regression analysis based on this fixed number of bins regardless of the number of individual pairs and thus the sample size. Thus, its computing load is practically the same for samples of any sizes.

A similar estimation procedure can be given under the SDS1 model. The regression of the sample SDS ($\hat{d}_{ij}$) on estimated genetic relatedness ($\hat{\theta}_{ij}$) is given by,

$$\hat{d}_{ij} = \alpha + \hat{\theta}_{ij}\beta + \varepsilon_{ij}$$

where the intercept α estimates the phenotypic variance ($\sigma_p^2 = \sigma_a^2 + \sigma_e^2$) and the slope β estimates minus the additive genetic variance ($-\sigma_a^2$) [cf. equation (4)]. In some studies [5], [24], twice the SDS value is used for the regression analysis and thus the intercept α estimates twice the phenotypic variance ($2\sigma_p^2$) and the slope β estimates minus twice the additive genetic variance ($-2\sigma_a^2$).

Simulation Experiments

Simulation Models and Procedures

To investigate the dependence of phenotypic similarity on genetic relatedness, we need a population of individuals whose genetic relatedness covers the full range from no genetic correlation to perfect genetic correlation between pairs of individuals. Such special population is simulated using the first-order auto-regressive model (AR1). With a nearly perfect starting genetic correlation, AR1 model guarantees a full range of genetic relatedness for each and every simulated sample of sufficient size. Such high genetic correlation may be artificial and unrealistic for many human and livestock populations to which the LMM-REML analysis is often applied. However, it does approximate some situations in crop breeding where modern cultivars are genetically highly similar because (i) strong directional selection has been practiced and (ii) they all trace back to one or a few founders, for example, Canadian wheat [51].

We assume that the population consists of n individuals which are arranged in a descending order according to the degrees of their genetic relatedness. Thus, the genetic correlation between any individual and its nearest neighbor is a constant θ_a, the genetic correlation between any individual and its second nearest neighbor (i.e., one individual apart) is θ_a^2, and so on. In general, the genetic correlation between the ith and jth individuals which are $t=|i-j|$ individuals apart is θ_a^t. The correlation between n individuals under the AR1 model is,

$$\Theta_{AR1} = \begin{bmatrix} 1 & \theta_a & \cdots & \theta_a^{n-1} \\ \theta_a & 1 & \cdots & \theta_a^{n-2} \\ \vdots & \vdots & \ddots & \vdots \\ \theta_a^{n-1} & \theta_a^{n-2} & \cdots & 1 \end{bmatrix}$$

(13)

We further assume that m independent markers are genotyped for each individual. The simulation can be done by obtaining a 3×3 two-way contingency table for nine genotypes at a pair of SNP loci through sampling from a multinomial distribution. However, this procedure is time-consuming. We have found an equivalent but more efficient simulation procedure by directly sampling an $n \times m$ $\mathbf{Z}^c$ matrix consisting of m random observations from an n-variate standard normal distribution with the mean vector of zeros and the covariance matrix being given in equation (13).

However, while this direct sampling scheme can be easily implemented using existing software packages such as the mvrnorm function in MASS/R [52] when n is not too large, it can be time-consuming for very large n. For this reason, a more efficient sampling scheme based on the definition of the AR1 model is employed and implemented in the following three steps. First, a m×1 vector, $\mathbf{z}_1^c$, of random numbers are taken from the univariate standard normal distribution N(0,1) for the first individual. Second, another m×1 vector $\mathbf{z}_2^c$ of random numbers are taken again from N(0,1) for the second individual given $\mathbf{z}_1^c$, but with the correlation between $\mathbf{z}_1^c$ and $\mathbf{z}_2^c$ being θ_a. Thus a general recursive relationship for sampling the ith vector, $\mathbf{z}_i^c$ given the $(i-1)$th vector $\mathbf{z}_{i-1}^c$ is

$$\mathbf{z}_i^c = \theta_a \mathbf{z}_{i-1}^c + \sqrt{1 - \theta_a^2}\,\mathbf{\kappa}$$

where k is a $m \times 1$ vector of random numbers that are taken from $N(0,1)$. Third, all n z vectors generated in such way are collected to form the matrix $\mathbf{Z}^c = \{ \mathbf{z}_1^c, \mathbf{z}_2^c, \cdots, \mathbf{z}_n^c \}$.

Regardless of whether the $\mathbf{Z}^c$ matrix is generated by directly sampling from a multivariate normal distribution or by the recursive relationship, it needs to be converted into an indicator genotype matrix $\mathbf{Z} = \{ \mathbf{z}_1, \mathbf{z}_2, \cdots, \mathbf{z}_n \}$ with vector $\mathbf{z}_i$ containing only three values of 0, 1 and 2 to indicate three possible genotypes at each of m independent loci for the ith individual. For simplicity, we consider the proportions of the three genotypes in each individual to be $1 : 2 : 1$ so that the ranges for converting a normally distributed variate into the three genotypes coded as 0, 1 and 2 are: $(-¥, -0.67449)$, $(-0.67449, 0.67449)$ and $(0.67449, \infty)$, respectively. The genotypes could have been simulated directly through sampling from a multinomial distribution but the genetic relatedness between individuals are more conveniently accommodated through sampling from a multivariate normal distribution as in our simulation.

To simulate random additive genetic effects and phenotypic values, we assume that a quantitative trait is controlled by the quantitative trait loci (QTL) which are randomly located at one tenth of the total simulated markers (see Program S1 for step-by-step description of simulating positions and effects of causal markers). In other words, 10% of the total simulated markers are causal (relevant) variants and the remaining 90% are irrelevant markers. Thus a vector of phenotypic values ($\mathbf{y}$) for n individuals are simulated using the following LMM model,

$$\mathbf{y} = 1\mu + \mathbf{Wu} + \mathbf{e} = 1\mu + \mathbf{a} + \mathbf{e} \qquad (14)$$

where μ is the population mean, $\mathbf{W}$ is an $n \times m$ standardized genotype matrix with the ilth element being $(z_{il} - 2\hat{p}_l) / \sqrt{2\hat{p}_l(1 - \hat{p}_l)}$ and $\hat{p}_l$ being the estimated frequency of the reference allele for the lth simulated marker, and $\mathbf{e}$ is a vector of n residual effects taken from a multivariate normal distribution, $\mathbf{u}$ is a vector of m random additive genetic effects that are taken from a standard multivariate normal distribution $\mathbf{u} \sim N(0, \mathbf{I}_m \sigma_a^2 / m)$ or equivalently the genome-wide additive genetic effects $\mathbf{a} = \mathbf{Wu}$ are taken from a multivariate normal distribution, $\mathbf{a} \sim N(0, \mathbf{\Theta}\sigma_a^2)$ with Q being estimated by $\hat{\mathbf{\Theta}} = \mathbf{WW}' / m$, and $\mathbf{e} \sim N(0, \mathbf{I}_n \sigma_e^2)$. In our simulation, we have the phenotypic variance $\sigma_p^2 = \sigma_a^2 + \sigma_e^2$ so that $\sigma_a^2 = [h^2 / (1 - h^2)]\sigma_e^2$ and $\sigma_e^2 = [(1 - h^2) / h^2]\sigma_a^2$ with h^2 being the narrow-sense heritability.

Three Simulation Scenarios

We consider three simulation scenarios. In the first scenario, we investigate the degree of agreement between the actual and theoretical genetic relatedness. This scenario simulates five populations of sizes n=200, 500, 1000, 2000 and 10000 with each population being genotyped at seven marker densities (m=200, 500, 1000, 2000, 5000, 10000 and 20000). For each simulated population, the theoretical GRM between n individuals is obtained using the AR1 model as in equation (13). Here and throughout all simulations, we choose the genetic correlation between any individual and its nearest neighbor to be θ_a=0.95. The actual GRM is estimated by $\hat{\Theta} = \mathbf{WW'}/m$ using each of the seven marker densities. The degree of agreement between the actual and theoretical genetic relatedness is measured by Pearson's correlation between elements of matrices $\hat{\Theta}$ and Q_{AR1}.

In the second scenario, we want to examine the effects of sample size (n) and marker density (m) on the estimation of narrow-sense heritability. This simulation scenario consists of all combinations of the following parameter values: three levels of the heritability (h^2=0.2, 0.5, and 0.8); three levels of sample size (n=500, 1,000 and 5,000); and three levels of marker density (m=200, 2,000 and 20,000). These simulation trials are replicated 100 times. The variance components and heritability are estimated using (i) the LMM approach as implemented by two software packages, rrBLUP [14] and GCTA [13], and (ii) the SDS approach as implemented by our own R package, SDS/R (http://statgen.ualberta.ca/index.html?open=software.html). In each of 100 replicated simulations, the LMM approach as in rrBLUP and GCTA estimates the additive genetic variance and residual variance and the heritability is subsequently calculated as the ratio of the estimated additive genetic variance to the sum of the estimated two variance components. On the other hand, the SDS approach directly estimates the heritability through the regression of the phenotypic correlation between pairs of individuals on the corresponding values of genetic relatedness.

In the third scenario, we want to compare and contrast the computational efficiency of the SDS method of heritability estimation to the REML-based methods as implemented in the two software packages, rrBLUP and GCTA over a wide range of sample sizes. We choose seven sample sizes for this scenario: n=500, 1000, 2000, 5000, 10000, 20000, and 40000, but we only consider one marker density (m=2000) and one heritability (h^2=0.5). Given that the REML-based methods are very time-consuming for large n, each simulation trial is replicated only five times. The time (in seconds) required by the different estimation methods for the analysis of simulated data for each of the five replicates are recorded using proc.time, a R core function [53]. Since the GRM is required by all the estimation methods, the time needed for constructing the GRM is not included in the comparison of computational efficiency.

SUPPORTING INFORMATION

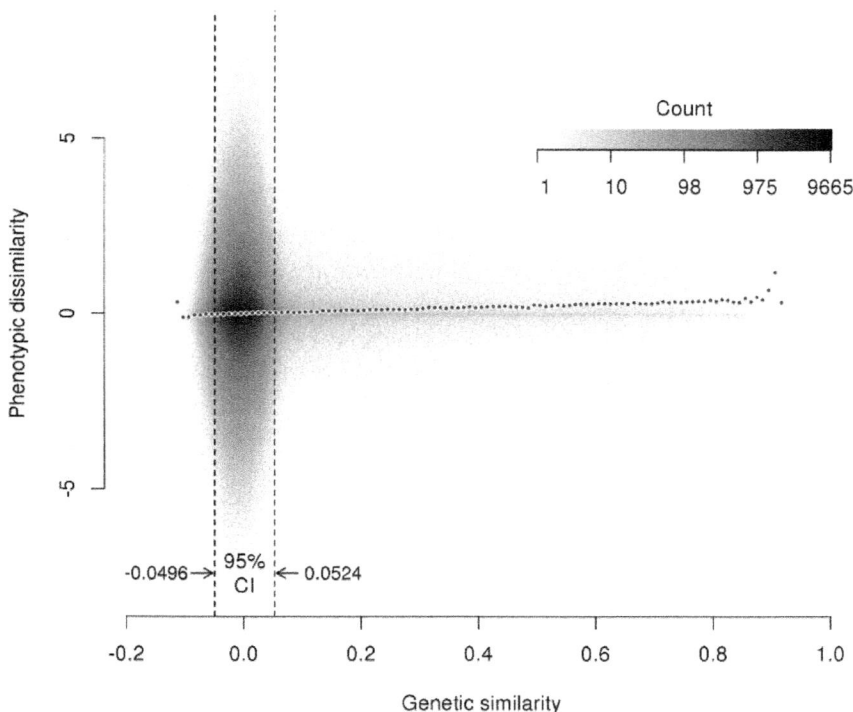

Figure S1: The genetic similarity and phenotypic similarity in a simulated population under an AR1 model with θ_a=0.95 and h^2=0.5, n=4000 and m=2000.

TEXT S1

Overview of Linear Mixed Models for Genomic Data

Here we will provide a brief overview of theory and procedures underlying the LMM-REML estimation of variance components and heritability. Such overview serves to demonstrate a natural connection with the SDS approach described in the Methods section.

The linear mixed model for fitting the effects of all the diallelic markers (e.g., SNPs) as random effects is,

$$\mathbf{y} = \mathbf{X}\boldsymbol{\beta} + \mathbf{Wu} + \mathbf{e} \tag{A1}$$

where y is a vector of n phenotypic values, β is a vector of k fixed effects including the overall mean, X is an $n \times k$ design matrix that relates the phenotypes to the fixed effects, u is a vector of m marker random additive effects with $\mathbf{u} \sim N(\mathbf{0}, \mathbf{I}_m \sigma_u^2)$, W is an $n \times m$ mean-corrected or standardized genotype matrix with the ijth element being $w_{il} = z_{il} - 2p_l$ in the mean-corrected genotype matrix or $w_{il} = (z_{il} - 2p_l)/\sqrt{2p_l(1-p_l)}$ in the standardized genotype matrix, where z_{il} is the number of copies (0, 1 and 2) of the reference allele for the lth marker of the ith individual and p_l is the frequency of the reference allele for the lth marker, and e is a vector of n residual effects with $\mathbf{e} \sim N(\mathbf{0}, \mathbf{I}_n \sigma_e^2)$. It is well known that the expected value and variance of vector y are $E(\mathbf{y}) = \mathbf{X}\boldsymbol{\beta}$ and $\mathrm{Var}(\mathbf{y}) = \mathbf{V} = \mathbf{G} + \mathbf{R}$, where the additive genetic covariance matrix can be $\mathbf{G} = [\mathbf{WW}'/2\Sigma_l^m p_l(1-p_l)]\sigma_u^2$ if W is the mean-corrected genotype matrix or $\mathbf{G} = \mathbf{WW}'\sigma_u^2$ if W is the standardized genotype matrix, and the residual covariance matrix is $\mathbf{R} = \mathbf{I}_n \sigma_e^2$. Since the standardized genotype matrix is often used in the literature for constructing the G matrix, we will only use $\mathbf{G} = \mathbf{WW}'\sigma_u^2$ in our subsequent development and discussion.

Model (A1) is equivalent to the conventional mixed model with single record per individual.

$$\mathbf{y} = \mathbf{X}\boldsymbol{\beta} + \mathbf{a} + \mathbf{e} \tag{A2}$$

if $\mathrm{Var}(\mathbf{a}) = \mathbf{G} = \boldsymbol{\Theta}\sigma_a^2 = \mathbf{WW}'\sigma_u^2$, where $\mathbf{a} = \mathbf{Wu}$ is an $n \times 1$ vector of the total additive genetic effects for the n individuals with $\mathbf{a} \sim N(\mathbf{0}, \boldsymbol{\Theta}\sigma_a^2)$ and $\boldsymbol{\Theta}$ being the $n \times n$ genetic relationship matrix (GRM) between individuals. In the past, the GRM has been estimated using known pedigrees among individuals (i.e., A matrix), with the ijth element of the GRM being $\theta_{ij} = 2\phi_{ij}$ for diploids or $\theta_{ij} = \phi_{ij}$ for haploids, where ϕ_{ij} is the kinship coefficient between the ith and jth individuals. Now the routine use of marker genotypes for estimating the GRM allows for capturing additional genetic variation due to Mendelian sampling. However, a marker-based estimate of the GRM is unbiased only if it is based on the QTL or causal variants.

The $(k + m)$ mixed-model equations (MMEs) for β and u in model (A1) can be solved to obtain the best linear unbiased estimation (BLUE) of fixed effects β and the best linear unbiased prediction (BLUP) of random effects u,

$$\begin{bmatrix} \hat{\hat{a}} \\ \hat{u} \end{bmatrix} = \begin{bmatrix} X'R^{-1}X & X'R^{-1}W \\ W R^{-1}X & W R^{-1}W + G^{-1} \end{bmatrix}^{-} \begin{bmatrix} X'R^{-1}y \\ W R^{-1}y \end{bmatrix}$$

(A3)

with superscript minus one ($^{-1}$) and superscript minus ($^{-}$) representing matrix and generalized inverses, respectively, and $G^{-1} = I_m\sigma_u^{-2}$. The coefficient matrix in equation (A3) is also known as the C matrix,

$$C = \begin{bmatrix} C_{\hat{a}} & C' \\ C_{\hat{a}u} & C \end{bmatrix} = \begin{bmatrix} X'R^{-1}X & X'R^{-1}W \\ W'R^{-1}X & W'R^{-1}W + G^{-1} \end{bmatrix}^{-}$$

(A4)

where

$$C_{\hat{a}} = (X'V^{-1}X)^{-}$$
$$C_{\hat{a}} = -GW'V^{-1}XC$$
$$C_{\hat{a}u} = (W'R^{-1}W + G^{-1})^{-1} - C_u X'V^{-1}WG$$

Similarly, the $(k + n)$ MMEs for β and a in model (A2) is given by,

$$\begin{bmatrix} \hat{\hat{a}} \\ \hat{a} \end{bmatrix} = \begin{bmatrix} X'R^{-1}X & X'R^{-1} \\ R^{-1}X & R^{-1} + G^{-1} \end{bmatrix}^{-} \begin{bmatrix} X'R^{-1}y \\ R^{-1}y \end{bmatrix}$$

(A5)

and the C matrix can be written as,

$$C = \begin{bmatrix} C_{\hat{a}} & C' \\ C_{\hat{a}u} & C \end{bmatrix} = \begin{bmatrix} X'R^{-1}X & X'R^{-1} \\ R^{-1}X & R^{-1} + G^{-1} \end{bmatrix}^{-}$$

(A6)

where

$$C_{\hat{a}} = (X'V^{-1}X)^{-}$$
$$C_{\hat{a}} = -GV^{-1}XC$$
$$C_{\hat{a}u} = (R^{-1} + G^{-1})^{-1} - C_u X'V^{-1}G$$

It should be noted that $G^{-1} = \Theta^{-1}\sigma_a^{-2}$ exists only if Θ is positive definite and thus investable. With high marker densities, the direct use of MMEs for obtaining the BLUP of marker effects u under model (A1) may become computationally challenging. In this case, model (A2) is used to predict the genetic effects a first and then the marker effects u is obtainable as,

$$\hat{u} = W'\Theta^{-1}\hat{a}$$.

(A7)

REML estimators of variance components can be obtained by several algorithms including derivative-based methods such as the Newton-Raphson algorithm and Fisher's scoring method, EM (expectation-maximization) methods and Average Information (AI) algorithm. All these methods are computationally intensive. For example, the REML estimators of genetic variance (σ_a^2) and residual variance (σ_e^2) through the EM algorithm requires iterating on,

$$\hat{\sigma}_e^2 = (\mathbf{y'R^{-1}y} - \hat{\boldsymbol{\beta}}'\mathbf{X'R^{-1}y} - \hat{\mathbf{a}}'\mathbf{R^{-1}y}) / [n - r(\mathbf{X})]$$
$$\hat{\sigma}_a^2 = [\hat{\mathbf{a}}'\boldsymbol{\Theta}^{-1}\hat{\mathbf{a}} + \hat{\sigma}_e^2 \mathrm{tr}(\boldsymbol{\Theta}^{-1}\mathbf{C}_{uu})] / n \tag{A8}$$

where $r(X)$ is the rank of matrix X and tr() stands for the trace of the matrix. The solutions require computing $\boldsymbol{\Theta}^{-1}$ and C_{uu} . Both of these matrices are difficult to compute when n is large. In implementing the AI algorithm, some of the mixed model analysis packages, such as ASREML, have avoided the inversion of the $n \times n$ **V** matrix using the Gaussian elimination of the MME to obtain the AI matrix based on sparse matrix techniques. However, as pointed out, since the marker-based GRM matrix is usually dense, the use of the sparse matrix technique for the GRM matrix will actually lead to an extra cost of memory and CPU time. In general, the relative performance of different computing strategies is dependent on the number of individuals, the number of marker loci per individual, and the number of iterations required to solve the MMEs.

AUTHOR CONTRIBUTIONS

Conceived and designed the experiments: R-CY ZH. Performed the experiments: ZH R-CY. Analyzed the data: ZH R-CY. Contributed reagents/materials/analysis tools: ZH R-CY. Wrote the paper: R-CY.

REFERENCES

1. Manolio TA, Collins FS, Cox NJ, Goldstein DB, Hindorff LA, et al. (2009) Finding the missing heritability of complex diseases. Nature 461: 747–753. doi: 10.1038/nature08494

2. Maher B (2008) Personal genomes: The case of the missing heritability. Nature 456: 18–21. doi: 10.1038/456018a

3. Zuk O, Hechter E, Sunyaev SR, Lander ES (2012) The mystery of missing heritability: Genetic interactions create phantom heritability. Proceedings of the National Academy of Sciences of the United States of America 109: 1193–1198. doi: 10.1073/pnas.1119675109

4. Zaitlen N, Kraft P (2012) Heritability in the genome-wide association era. Human Genetics 131: 1655–1664. doi: 10.1007/s00439-012-1199-6

5. Yang J, Benyamin B, McEvoy BP, Gordon S, Henders AK, et al. (2010) Common SNPs explain a large proportion of the heritability for human height. Nature Genetics 42: 565–569. doi: 10.1038/ng.608

6. Kang HM, Sul JH (2010) Service SK, Zaitlen NA, Kong SY, et al (2010)

Variance component model to account for sample structure in genome-wide association studies. Nature Genetics 42: 348–354. doi: 10.1038/ng.548

7. Zhang ZW, Ersoz E, Lai CQ, Todhunter RJ, Tiwari HK, et al. (2010) Mixed linear model approach adapted for genome-wide association studies. Nature Genetics 42: 355–360. doi: 10.1038/ng.546

8. Zhou X, Stephens M (2012) Genome-wide efficient mixed-model analysis for association studies. Nature Genetics 44: 821–824. doi: 10.1038/ng.2310

9. Henderson CR (1984) Applications of linear models in animal breeding: University of Guelph.

10. Gilmour AR, Gogel B, Cullis B, Thompson R (2009) ASReml user guide release 3.0. VSN International Ltd, Hemel Hempstead, UK.

11. Kang HM, Zaitlen NA, Wade CM, Kirby A, Heckerman D, et al. (2008) Efficient control of population structure in model organism association mapping. Genetics 178: 1709–1723. doi: 10.1534/genetics.107.080101

12. Lippert C, Listgarten J, Liu Y, Kadie CM, Davidson RI, et al. (2011) FaST linear mixed models for genome-wide association studies. Nature Methods 8: 833–835. doi: 10.1038/nmeth.1681

13. Yang JA, Lee SH, Goddard ME, Visscher PM (2011) GCTA: A tool for genome-wide complex trait analysis. American Journal of Human Genetics 88: 76–82. doi: 10.1016/j.ajhg.2010.11.011

14. Endelman JB (2011) Ridge regression and other kernels for genomic selection with R package rrBLUP. Plant Genome 4: 250–255. doi: 10.3835/plantgenome2011.08.0024

15. Bradbury PJ, Zhang Z, Kroon DE, Casstevens TM, Ramdoss Y, et al. (2007) TASSEL: software for association mapping of complex traits in diverse samples. Bioinformatics 23: 2633–2635. doi: 10.1093/bioinformatics/btm308

16. Meuwissen THE, Hayes BJ, Goddard ME (2001) Prediction of total genetic value using genome-wide dense marker maps. Genetics 157: 1819–1829.

17. de los Campos G, Hickey JM, Pong-Wong R, Daetwyler HD, Calus MPL (2013) Whole-genome regression and prediction methods applied to plant and animal breeding. Genetics 193: 327–345. doi: 10.1534/genetics.112.143313

18. Daetwyler HD, Calus MPL, Pong-Wong R, de los Campos G, Hickey JM (2013) Genomic prediction in animals and plants: Simulation of data,

validation, reporting, and benchmarking. Genetics 193: 347–365. doi: 10.1534/genetics.112.147983

19. Randall JC, Winkler TW, Kutalik Z, Berndt SI, Jackson AU, et al. (2013) Sex-stratified Genome-wide Association Studies Including 270,000 Individuals Show Sexual Dimorphism in Genetic Loci for Anthropometric Traits. PLoS Genetics 9: e1003500.

20. Berndt SI, Gustafsson S, Magi R, Ganna A, Wheeler E, et al. (2013) Genome-wide meta-analysis identifies 11 new loci for anthropometric traits and provides insights into genetic architecture. Nature Genetics 45: 501–U569.

21. Speliotes EK, Willer CJ, Berndt SI, Monda KL, Thorleifsson G, et al. (2010) Association analyses of 249,796 individuals reveal 18 new loci associated with body mass index. Nature Genetics 42: 937–948.

22. Lango Allen H, Estrada K, Lettre G, Berndt SI, Weedon MN, et al. (2010) Hundreds of variants clustered in genomic loci and biological pathways affect human height. Nature 467: 832–838.

23. Rietveld CA, Medland SE, Derringer J, Yang J, Esko T, et al. (2013) GWAS of 126,559 individuals identifies genetic variants associated with educational attainment. Science 340: 1467–1471.

24. Haseman JK, Elston RC (1972) The investigation of linkage between a quantitative trait and a marker locus. Behavior Genetics 2: 3–19. doi: 10.1007/bf01066731

25. Grimes LW, Harvey WR (1980) Estimation of genetic variances and covariances using symmetric differences squared. Journal of Animal Science 50: 632–644.

26. Ritland K (1996) A marker-based method for inferences about quantitative inheritance in natural populations. Evolution 50: 1062–1073. doi: 10.2307/2410647

27. Lynch M, Ritland K (1999) Estimation of pairwise relatedness with molecular markers. Genetics 152: 1753–1766.

28. Thomas SC (2005) The estimation of genetic relationships using molecular markers and their efficiency in estimating heritability in natural populations. Philosophical Transactions of the Royal Society B: Biological Sciences 360: 1457–1467. doi: 10.1098/rstb.2005.1675

29. Lynch M, Walsh B (1998) Genetics and analysis of quantitative traits. Sunderland, MA, USA: Sinauer Associates.

30. Visscher PM, Medland SE, Ferreira MA, Morley KI, Zhu G, et al. (2006) Assumption-free estimation of heritability from genome-wide identity-

by-descent sharing between full siblings. PLoS Genetics 2: e41. doi: 10.1371/journal.pgen.0020041

31. Zhang Y-M, Lü H-Y, Yao L-L (2008) Multiple quantitative trait loci Haseman–Elston regression using all markers on the entire genome. Theoretical and Applied Genetics 117: 683–690. doi: 10.1007/s00122-008-0809-0

32. Zhang Y-M, Gai J (2009) Methodologies for segregation analysis and QTL mapping in plants. Genetica 136: 311–318. doi: 10.1007/s10709-008-9313-3

33. Crossa J, de los Campos G, Perez P, Gianola D, Burgueno J, et al. (2010) Prediction of genetic values of quantitative traits in plant breeding using pedigree and molecular markers. Genetics 186: 713–724. doi: 10.1534/genetics.110.118521

34. Macdonald SJ, Pastinen T, Long AD (2005) The effect of polymorphisms in the enhancer of split gene complex on bristle number variation in a large wild-caught cohort of Drosophila melanogaster. Genetics 171: 1741–1756. doi: 10.1534/genetics.105.045344

35. Yang J, Zaitlen NA, Goddard ME, Visscher PM, Price AL (2014) Advantages and pitfalls in the application of mixed-model association methods. Nature Genetics 46: 100–106. doi: 10.1038/ng.2876

36. Yung LS, Yang C, Wan X, Yu W (2011) GBOOST: a GPU-based tool for detecting gene–gene interactions in genome–wide case control studies. Bioinformatics 27: 1309–1310. doi: 10.1093/bioinformatics/btr114

37. Jiang R, Zeng F, Zhang W, Wu X, Yu Z (2009) Accelerating genome-wide association studies using CUDA compatible graphics processing units; 3–5 Aug. 2009; Shanghai. IEEE. 70–76.

38. Ma L, Runesha HB, Dvorkin D, Garbe J, Da Y (2008) Parallel and serial computing tools for testing single-locus and epistatic SNP effects of quantitative traits in genome-wide association studies. BMC Bioinformatics 9: 315. doi: 10.1186/1471-2105-9-315

39. Lippert C, Quon G, Kang EY, Kadie CM, Listgarten J, et al. (2013) The benefits of selecting phenotype-specific variants for applications of mixed models in genomics. Scientific Reports 3: 1815. doi: 10.1038/srep01815

40. Goddard ME, Hayes BJ, Meuwissen THE (2011) Using the genomic relationship matrix to predict the accuracy of genomic selection. Journal of Animal Breeding and Genetics 128: 409–421. doi: 10.1111/j.1439-0388.2011.00964.x

41. Su GS, Christensen OF, Ostersen T, Henryon M, Lund MS (2012)

Estimating additive and non-additive genetic variances and predicting genetic merits using genome-wide dense single nucleotide Polymorphism Markers. PLoS ONE 7: e45293. doi: 10.1371/journal.pone.0045293

42. Hill WG (2014) Applications of population genetics to animal breeding, from Wright, Fisher and Lush to genomic prediction. Genetics 196: 1–16. doi: 10.1534/genetics.112.147850

43. Hill WG, Goddard ME, Visscher PM (2008) Data and theory point to mainly additive genetic variance for complex traits. PLoS Genetics 4: e1000008. doi: 10.1371/journal.pgen.1000008

44. Stringer S, Derks EM, Kahn RS, Hill WG, Wray NR (2013) Assumptions and properties of limiting pathway models for analysis of epistasis in complex traits. PLoS ONE 8: e68913. doi: 10.1371/journal.pone.0068913

45. Koch GG (1968) Some further remarks concerning "A General Approach to the Estimation of Variance Components". Technometrics 10: 551–558. doi: 10.1080/00401706.1968.10490601

46. Ritland K, Ritland C (1996) Inferences about quantitative inheritance based on natural population structure in the yellow monkeyflower, *Mimulus guttatus*. Evolution 50: 1074–1082. doi: 10.2307/2410648

47. Mousseau TA, Ritland K, Heath DD (1998) A novel method for estimating heritability using molecular markers. Heredity 80: 218–224. doi: 10.1046/j.1365-2540.1998.00269.x

48. Ritland K (2000) Marker-inferred relatedness as a tool for detecting heritability in nature. Molecular Ecology 9: 1195–1204. doi: 10.1046/j.1365-294x.2000.00971.x

49. Harville DA (1998) Matrix Algebra from a Statistician's Perspective. Technometrics 40: 164–164. doi: 10.1080/00401706.1998.10485214

50. Falconer DS, Mackay TFC (1996) Introduction to quantitative genetics: Longman New York. 464 p.

51. McCallum BD, DePauw RM (2008) A review of wheat cultivars grown in the Canadian prairies. Canadian Journal of Plant Science 88: 649–677. doi: 10.4141/cjps07159

52. Venables WN, Ripley BD (2002) Modern Applied Statistics With S; Chambers J, Eddy W, Härdle W, Sheather S, Tierney L, editors. New York, NY: Springer.

53. R Core Team (2012) R: A Language and Environment for Statistical Computing. Vienna, Austria: R Foundation for Statistical Computing.

54. Lynch M, Walsh B: Genetics and analysis of quantitative traits. Sunderland, MA, USA: Sinauer Associates; 1998.

55. Henderson CR: Applications of Linear Models in Animal Breeding. University of Guelph; 1984.

56. Falconer DS, Mackay TFC: Introduction to quantitative genetics. 4th edn: Longman New York; 1996.

57. VanRaden PM: Efficient methods to compute genomic predictions. J Dairy Sci 2008, 91:4414-4423.

58. Yang J, Benyamin B, McEvoy BP, Gordon S, Henders AK, Nyholt DR, Madden PA, Heath AC, Martin NG, Montgomery GW, Goddard ME, Visscher PM: Common SNPs explain a large proportion of the heritability for human height. Nat Genet 2010, 42:565-569.

59. Goddard ME, Hayes BJ, Meuwissen THE: Using the genomic relationship matrix to predict the accuracy of genomic selection. J Anim Breed Genet 2011, 128:409-421.

60. Lippert C, Quon G, Kang EY, Kadie CM, Listgarten J, Heckerman D: The benefits of selecting phenotype-specific variants for applications of mixed models in genomics. Scientific Reports 2013, 3:1815.

61. Zaitlen N, Kraft P: Heritability in the genome-wide association era. Hum Genet 2012, 131:1655-1664.

62. Mclean RA, Sanders WL, Stroup WW: A unified approach to mixed linear-models. Am Stat 1991, 45:54-64.

63. Yang R-C: Towards understanding and use of mixed-model analysis of agricultural experiments. Can J Plant Sci 2010, 90:605-627.

64. Henderson CR: Mivque and Reml Estimation of Additive and Nonadditive Genetic Variances. J Anim Sci 1985, 61:113-121.

65. Piepho HP, Ogutu JO, Schulz-Streeck T, Estaghvirou B, Gordillo A, Technow F: Efficient Computation of Ridge-Regression Best Linear Unbiased Prediction in Genomic Selection in Plant Breeding. Crop Sci 2012, 52:1093-1104.

66. Yang JA, Lee SH, Goddard ME, Visscher PM: GCTA: A tool for genome-wide complex trait analysis. Am J Hum Genet 2011, 88:76-82.

67. Gilmour AR, Gogel B, Cullis B, Thompson R: ASReml user guide release 3.0. VSN International Ltd, Hemel Hempstead, UK 2009.

68. Stranden I, Garrick DJ: Derivation of equivalent computing algorithms for genomic predictions and reliabilities of animal merit. J Dairy Sci 2009, 92:2971-2975.

Chapter 4

A VERSATILE STRATEGY FOR RAPID CONDITIONAL GENOME ENGINEERING USING LOXP SITES IN A SMALL SYNTHETIC INTRON IN PLASMODIUM FALCIPARUM

Matthew L. Jones[1], Sujaan Das[2], Hugo Belda[1], Christine R. Collins[2], Michael J. Blackman[2] and MoritzTreeck[1]

[1] Moritz Treeck Laboratory, The Francis Crick Institute, Mill Hill Laboratory, The Ridgeway, London NW71AA, United Kingdom

[2] Michael J. Blackman Laboratory, The Francis Crick Institute, Mill Hill Laboratory, The Ridgeway, London NW71AA, United Kingdom

ABSTRACT

Conditional genome engineering in the human malaria pathogen *Plasmodium falciparum* remains highly challenging. Here we describe a strategy for facile and rapid functional analysis of genes using an approach based on the Cre/lox system and tailored for organisms with short and few introns. Our method allows the conditional, site-specific removal of genomic sequences of essential and non-essential genes by placing *loxP* sites into a short synthetic intron to produce a module (loxPint) can be placed anywhere in open reading frames without compromising protein expression. When duplicated, the loxPint module serves as an intragenic recombineering point that can be used for the fusion of gene elements to reporters or the conditional introduction of point mutations. We demonstrate the robustness and versatility of the system by targeting the *P. falciparum* merozoite surface protein 1 gene (*msp1*), which has previously proven refractory to genetic interrogation, and the parasite exported kinase FIKK10.1.

INTRODUCTION

Protein function can be conditionally interrogated by systems that allow exogenous control of mRNA abundance, stability, protein turnover or localization. While these are useful tools, they often deliver only partial depletion of a gene product, leading to outcomes that can be difficult to interpret. Of the tools available to conditionally silence a gene, methods that allow conditional DNA deletion are often preferred as these completely remove the targeted gene or important parts thereof. One of the most widely used conditional DNA deletion methods relies on Cre recombinase-mediated recombination between two 34-nucleotide *loxP* sequences. To examine gene function using the Cre/lox method, *loxP* sites are introduced at positions that either flank an entire gene of interest or that lie within introns, allowing Cre-driven deletion of an entire gene or key exon1. Both of these options are however not suitable for species such as *P. falciparum* where few genes have introns and where poor transfection efficiency precludes facile introduction of multiple *loxP* sites. In addition, the *P. falciparum* genome is highly AT-rich (>80% in intergenic regions) making targeted genome modifications in these regions complicated. Finally, currently used conditional genome recombination methods in *Plasmodium* do not allow conditional domain rearrangements, point mutations, or directed fusion to heterologous sequences such as epitope tags or reporters. While a range of important human and animal pathogens such as *Trypanosomes*, several fungal species and model organisms like yeast and *Dictyostelium* contain few and short introns, no strategy has yet been developed to rapidly place *loxP* sites into intronless genes.

To substantially increase the ability to conditionally and rapidly modify genes with the potential for broad use in species that contain few introns or many mono-exonic genes, an ideal system would allow: 1) a single transfection step for target-directed gene manipulation; 2) the introduction of "silent" *loxP* sequences anywhere in the genome, including within exon sequences; 3) the conditional rearrangement of specific genetic elements such that one stretch of DNA can be rapidly replaced by another for use in domain swapping, introduction of point mutations, epitope tagging or to generate a reporter for successful recombination events; and 4) tight temporal control of Cre activity that allows rapid recombination rates regardless of the organism used.

Here we have devised a simple strategy that meets all of the above criteria. We validate its use in the human malaria parasite *Plasmodium falciparum*, an organism that presents many inherent difficulties in genetic modification. We use a standardized small module (loxPint) that consists of a short, *loxP* site-containing intron that can replace endogenous introns or can be placed in open reading frames of episomes as well as in chromosomal genes. If duplicated, the loxPint module is reconstituted after Cre-mediated recombination, enabling the fusion, replacement, or specific deletion of flanking genetic elements. To achieve rapid recombination rates we use parasite lines stably expressing a chemically-regulated Cre (DiCre)[1,2]. We demonstrate that this system allows the rapid, conditional control and modification of essential and non-essential genes alike.

RESULTS

Introduction of Silent *loxP* Sites using a Standardized Intron

Limitations of novel genetic tools in *P. falciparum* are mainly dictated by its low transfection efficiency. For that reason, facile application of the Cre/lox system, while introducing a tightly controllable conditional genetic system into the parasite, has been hampered by the fact that flanking target genes with *loxP* sites can take many months. Such manipulations also carry the risk that introduction of *loxP* sites into flanking regulatory sequences can interfere with gene expression levels. In other organisms, *loxP* sites are commonly placed in introns where they are unlikely to interfere with promoter function. However, introns are rare in the *P. falciparum* genome (45.5% of all protein coding genes are single-exon genes, 24.0% contain 2 exons and only 30.5% contain 3 or more exons) (www.plasmoDB.org), which limits the use of that strategy. We reasoned that placement of *loxP* sites into a small intron that naturally occurs in the *P. falciparum* genome might allow it to be used as a universal module that could be placed into exons, enabling the flanking of critical segments of genes with silent*loxP* sites in a single transfection step. Ideally this intron would be short and lack long repetitive A or T stretches because these often cause substantial problems with gene synthesis. We screened the *P. falciparum* genome and identified an intron of the *sera2* gene (PlasmoDB ID PF3D7_0207900) which meets these criteria. To test whether the *sera2* intron can accommodate a *loxP* site, we generated a proof-of-principle episomal expression construct, *pRex2:loxPint:gfp*, in which the intron of *rex2* was replaced with the *sera2* intron containing an integral *loxP* site (loxPint, sequence shown in supplemental table 2) and then fused in frame with the green fluorescent protein gene (*gfp*)(Fig. 1a). We chose REX2 because of its small size and because it has been well characterized as a GFP fusion protein that is efficiently exported from the intracellular parasite into the host erythrocyte. We reasoned this would allow its use as a reporter for correct loxPint splicing, even for exported proteins. This is an important property as other conditional methods that have been applied in *P. falciparum*, such as protein degradation systems, are not suitable for the study of genes encoding exported or secreted proteins. A second loxPint module was placed downstream of the *gfp* open reading frame, followed by a myc epitope tag sequence. Cre-mediated recombination between the two*loxP* sites was predicted to reconstitute the loxPint module, excising the C-terminal region of

REX2 and the entire GFP sequence, replacing this sequence with that encoding the myc epitope tag. Activation of Cre recombinase activity in the 1G5DC *P. falciparum* clone is induced by rapamycin (RAP), which mediates rapid heterodimerisation of the constitutively-expressed DiCre polypeptides. This was predicted to lead to recombination between the two *loxP* sites and excision of the 3' end of *rex2* and *gfp* (Fig. 1a). We transfected *pRex2:loxPint:gfp* into 1G5DC parasites and confirmed faithful expression and localization of REX2. GFP (Fig. 1b). To test whether loxPint was correctly spliced in transfected parasite lines, we analyzed plasmid DNA (pDNA) and reverse transcribed DNA (cDNA) of *rex2:loxPint:gfp* parasites. Two species were amplified from cDNA, one likely corresponding to the spliced variant, and one slightly higher running band that suggested incomplete splicing of the loxPint. Sanger sequencing confirmed that the lower band corresponded to the correctly spliced version of *rex2:loxPint:gfp* and verified correct retention of the loxPint following splicing (Fig. 1c). These results showed that the loxPint module allows functional splicing in a heterologous locus.

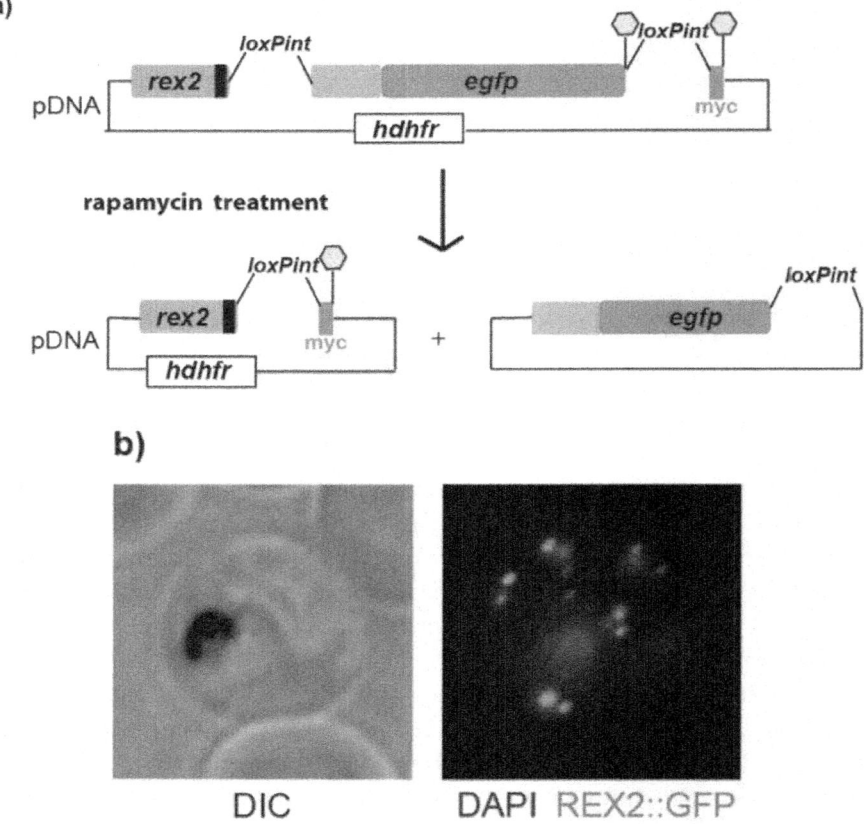

a)

rapamycin treatment

b)

DIC   DAPI REX2::GFP

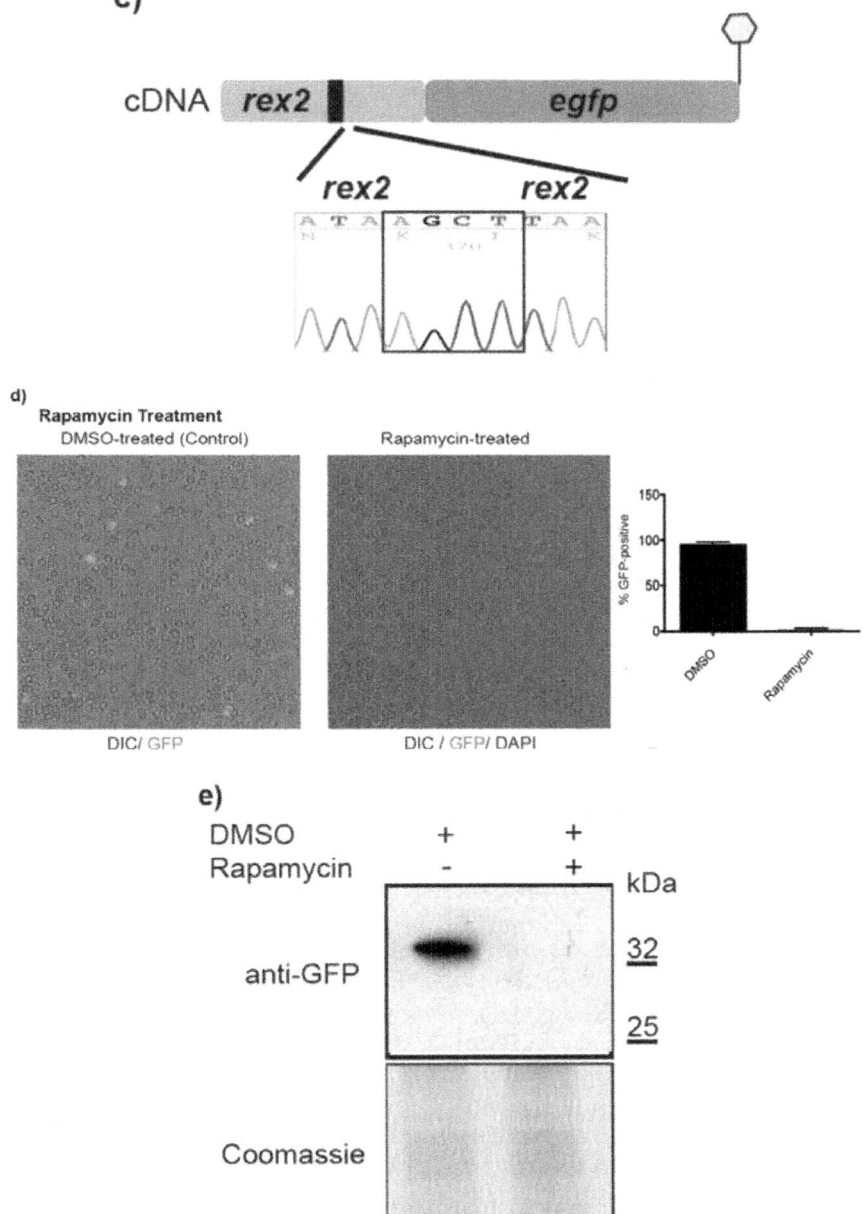

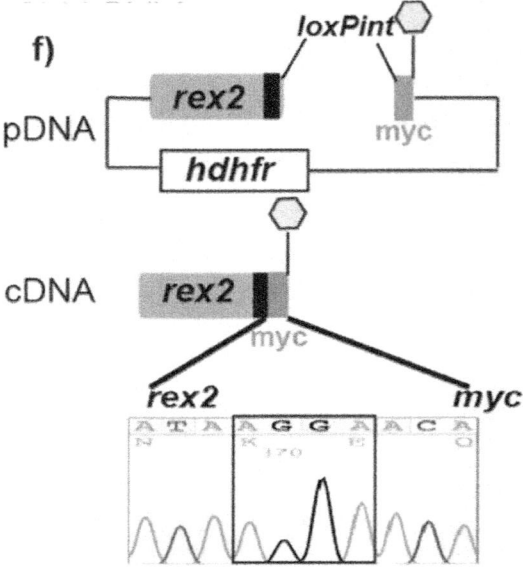

Figure 1: A *loxP* **containing synthetic intron** loxPint allows inducible, targeted DNA rearrangement. (a) Schematic shows the loxPint module in the *rex2* gene (black rectangle represents the native transmembrane domain). Stop codons are represented as yellow hexagons. A second loxPint is located downstream of the GFP coding sequence followed by a myc epitope coding sequence. Rapamycin treatment causes dimerization of two Cre-recombinase subunits in the DiCre-expressing 1G5DC *P. falciparum* line. Cre-driven LoxP-site recombination between the loxPint modules reconstitutes lox-Pint and places the myc coding sequence in frame with *rex2*. (b) Live fluorescence image showing *pRex2:loxPint:GFP* confirms Rex2.GFP expression and correct targeting to the host cell. (c) LoxPint is appropriately spliced. The chromatogram shows the nucleotides identified in sequencing reactions from cDNA from Rex2:loxPint:GFP parasites. Black box represents exon-exon boundaries. The resulting amino acid translation is shown in light grey. LoxPint is absent from sequenced cDNA and the correct *rex2* coding sequence is present, confirming correct splicing. (d) RAP-induced Cre-mediated DNA excision reduces Rex2.GFP expression. GFP positive parasites are reduced by 97.6% (+/−1.1%) upon rapamycin treatment (error bars are SD). (e) Western blot showing RAP-induced DNA excision results in the loss of Rex2.GFP expression. (f) LoxPint is reconstituted and correctly spliced after Cre-mediated recombination. Schematic shows the *pRex2:loxPint:GFP* construct after Cre-mediated excision before (pDNA) and after (cDNA) splicing. The 5' *rex2* sequence that remains after Cre-driven DNA excision is transcribed in-frame with the myc epitope coding sequence as shown by cDNA sequencing. The black box in the chromatogram highlights the exon-exon boundary.

The loxPint Module Allows Conditional Domain Fusions in an Episomal Context

To test for correct DiCre-mediated recombination between the two loxPint sequences in *pRex2:loxPint:gfp*, we treated transfected parasites with RAP for 4 h then analysed excision between the two loxPint modules by evaluating GFP expression. As expected, RAP-treated but not control parasites, showed near complete loss of GFP expression as measured by live microscopy (Fig. 1d) and Western Blot (Fig. 1e). To test whether RAP treatment had reconstituted the loxPint module in a form that was still correctly spliced, we examined the parasites by immunofluorescence analysis (IFA) with anti-myc antibodies. Unfortunately, we could not confirm a specific signal in RAP-treated parasites because the anti-myc antibodies cross-reacted strongly with other parasite antigens on IFA and the truncated REX2: myc protein was likely too small to be detected on Western blot. However, sequencing of cDNA from the RAP-treated parasites confirmed correct splicing of the reconstituted loxPint module, effectively fusing the myc tag in frame with the 3′ end of the remaining *rex2* open reading frame (Fig. 1f). This shows that duplication of the loxPint module allows directed, conditional fusion of distinct protein-coding sequences.

The loxPint Module Can Be used in an Endogenous Genomic Locus and is Quantitatively Spliced

Encouraged by these results, we decided to further validate our method by attempting to introduce a loxPint module into the open reading frame of a chromosomal parasite gene with the aim to tag the gene of interest in the same gene modification step as the introduction of two repeated loxPints. For this we chose to investigate the gene encoding an uncharacterized parasite kinase called FIKK10.1 (PF3D7_1016400) that is predicted to be exported into the host cell. The *fikk10.1* gene includes two short introns close to its 5′ and 3′ ends and encodes a protein with a C-terminal kinase domain, allowing targeting by 3′ replacement. We generated a targeting construct, *pfikk10.1:loxPint:HA*, designed to introduce a recodonised *fikk10.1* kinase domain into the endogenous locus flanked by loxPint modules (Fig. 2a) and transfected this construct into 1G5DC parasites. Integration of *pfikk10.1:loxPint:HA* was expected to result in expression of an HA-tagged chimeric FIKK10.1 under the control of its endogenous promotor (Fig. 2b). Correct integration of the construct and expression of the modified gene was confirmed by PCR, IFA, and Western Blot (Fig. 2b–d). As predicted, treatment of this engineered parasite line with RAP led to recombination between the loxPint modules flanking the recodonised *fikk10.1* kinase domain and subsequent loss of the entire kinase domain and HA tag (amino acid residues 251–641 of FIKK10.1.HA (Fig. 2c,d)). Sequencing of cDNA from DMSO-treated control and RAP-treated parasites confirmed correct splicing of the loxPint both before and after DiCre-driven DNA rearrangement (Fig. 2e). Because we had observed incomplete splicing from the episomally expressed *rex2:loxPint:gfp* we tested splicing efficiency of the loxPint in the context of the *fikk10.1* locus. A single PCR product was obtained from gDNA and cDNA, corresponding to the unspliced and spliced versions respectively, indicating complete splicing of the loxPint (Fig. 2e).

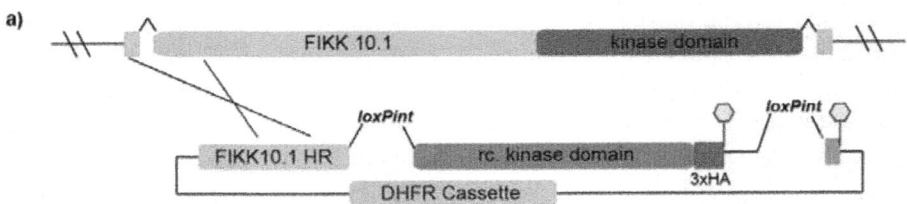

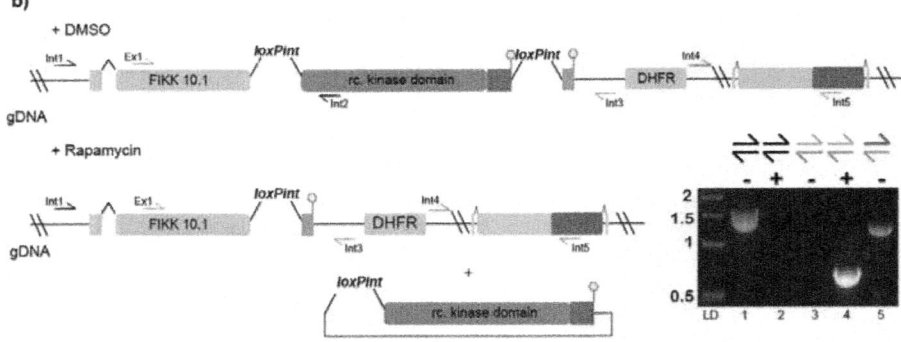

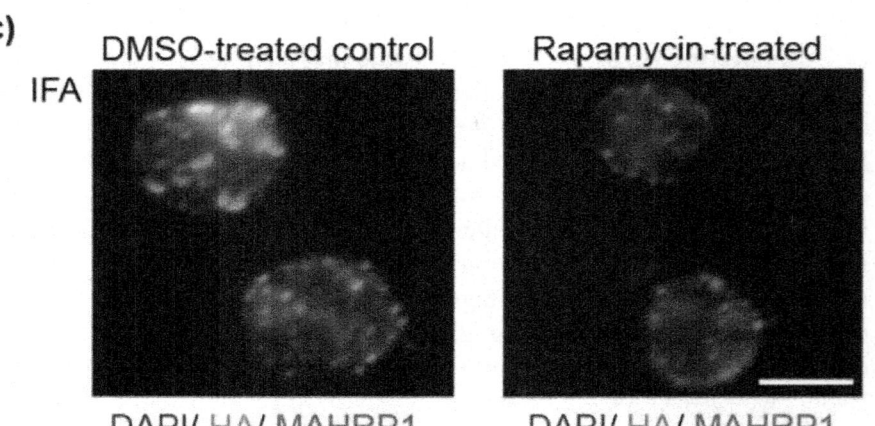

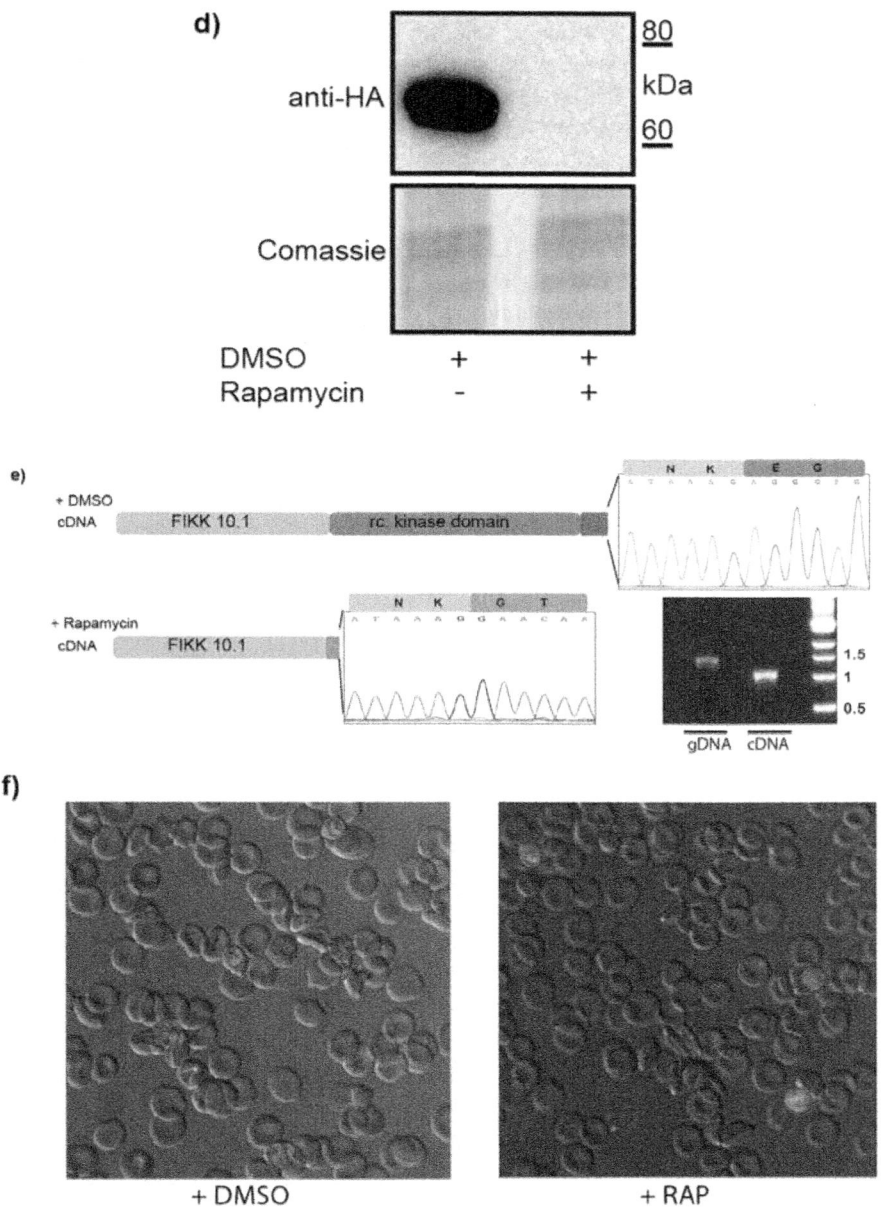

Figure 2: Use of the loxPint strategy allows the silent targeting of the exported kinase FIKK10.1 and conditional domain fusion. (a) Overview of the strategy to introduce two loxPint modules in a single transfection step using single homologous recombination at the 3′ end of FIKK10.1. (b) Schematic of the structure of the *FIKK10.1* locus

after integration of construct *pfikk10.1:loxPint:HA,* which introduces loxPint upstream of a recodonized FIKK10.1 kinase domain. The loxPint is duplicated downstream of an HA3 epitope tag sequence to allow removal of the HA3-tagged FIKK10.1 kinase domain by Cre-driven DNA excision. (**c,d**) The introduced loxPint does not interfere with FIKK10.1 expression or export to the host erythrocyte as shown by IFA and Western Blot. Cre-driven DNA excision induced by treatment with RAP results in loss of the FIKK10.1.HA3 signal in IFA and Western Blot. (**e**) cDNA sequencing results from DMSO and RAP treated FIKK10.1 loxPint parasites shows that the loxPint is correctly spliced and RAP treatment leads to reconstitution of the loxPint module, allowing conditional domain re-arrangements. PCR results show complete splicing of the loxPint in FIKK10.1 in cDNA but not gDNA. (f) Live fluorescence imaging showing GFP expression in the DMSO vs. RAP treated *pfikk10.1:loxPint:HA:gfp* parasite line.

To further verify the use of the loxPint module for conditional domain fusions we generated the parasite line FIKK10.1:loxPint:HA:GFP which shares all the features of FIKK10.1:loxPint:HA except that RAP treatment leads to fusion of the first 250 amino acids of FIKK10.1 with GFP. As expected, treatment of this parasite line with RAP led to replacement of the kinase domain of FIKK10.1 with GFP (Fig. 2f). No effects on parasite growth were observed after RAP treatment, indicating that FIKK10.1 is likely a non-essential kinase under the conditions tested here and further work will be required to identify the biological function of FIKK10.1. However, collectively these data show that the loxPint module can serve as a recombination point for rapid, efficient conditional deletion or fusion of DNA elements.

THE LOXPINT MODULE CAN BE PLACED IN ESSENTIAL GENES AND DOES NOT COMPROMISE EXPRESSION LEVELS

To test the effectiveness of the loxPint in the modification of a proven essential genomic gene we tested its utility in modification of the single exon gene encoding merozoite surface protein-17 (*msp1,* PlasmoDB ID PF3D7_0930300). Using single-crossover homologous recombination, a loxPint module was incorporated into the *msp1* coding sequence upstream of the glycosyl phosphatidyl inositol (GPI) anchoring signal that tethers this protein to the surface of the merozoite, the invasive blood-stage form of the parasite. In the same gene modification step, a second *loxP* site was introduced downstream of the *msp1* stop codon (Fig. 3a). Excision of the *loxP*-flanked sequence was predicted to produce a truncated form of MSP1 lacking its C-terminus and GPI anchor (amino acid residues 1277–1720). The construct readily integrated into the endogenous *msp1* locus of transfected 1G5DC parasites (Fig. 3a). This was confirmed by IFA and Western blot analysis (Fig. 3c,d). RAP-treatment of the engineered parasites resulted in the expected deletion of the MSP1 C-terminal domain and GPI anchor (Fig. 3b), as confirmed using monoclonal antibodies specific for the deleted C-terminal region of the protein (Fig. 3c,d). As we recently reported, RAP treated parasites showed a severe defect in egress from the host erythrocyte. In tests designed to assess the splicing efficiency of the*msp1* gene containing the loxPint, a single ~600 bp band corresponding to completely spliced transcript was amplified from cDNA, showing complete splicing of the transcript (Fig. 3e). This was confirmed by Sanger sequencing (data not shown). To test that overall expression levels of MSP1 were not compromised by the introduction of the loxPint module, we analysed parental 1G5DC and MSP1_loxPint parasites by Western blot using monoclonal antibodies specific for MSP1. This detected no significant alterations in MSP1 expression levels in msp1:loxPint parasites compared to the parental 1G5DC line (Fig. 3f). These results confirmed that the loxPint can be readily introduced into open reading frames of essential, mono-exonic endogenous genes and is efficiently spliced without impacting on gene expression levels.

a)

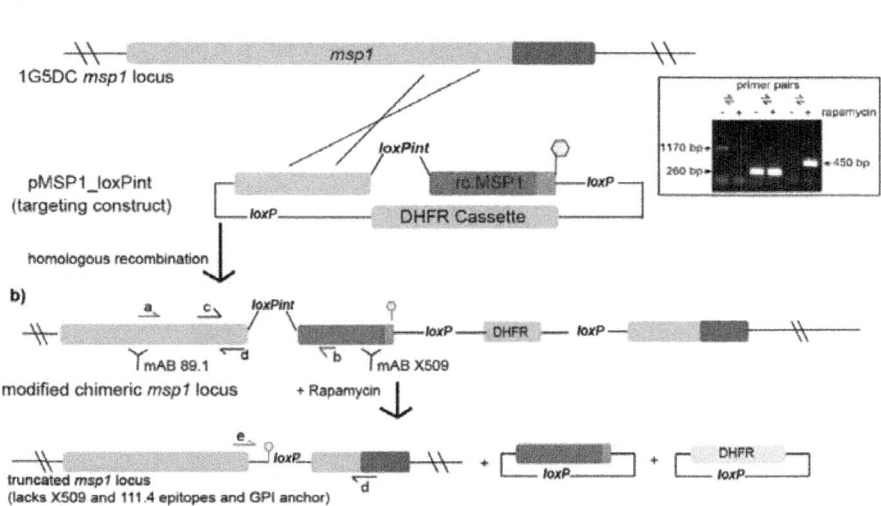

c)

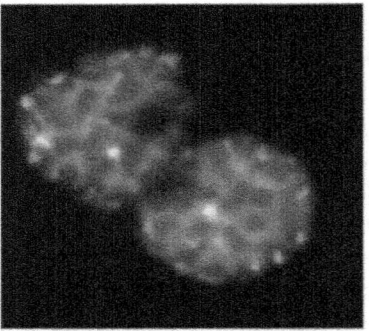

DMSO-treated control

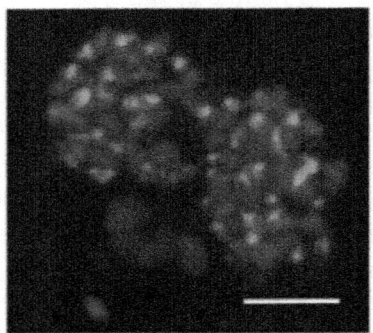

Rapamycin-treated

DAPI/ RhopH2/ MSP1   DAPI/ RhopH2/ MSP1

d)

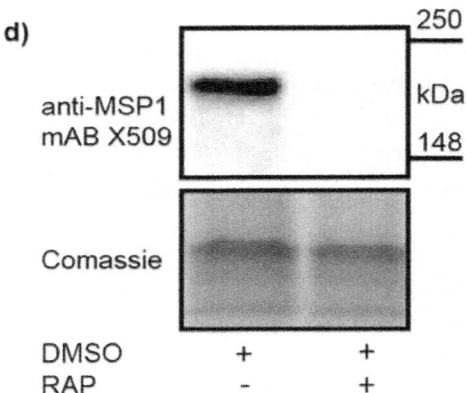

anti-MSP1
mAB X509

Comassie

DMSO	+	+
RAP	-	+

e)

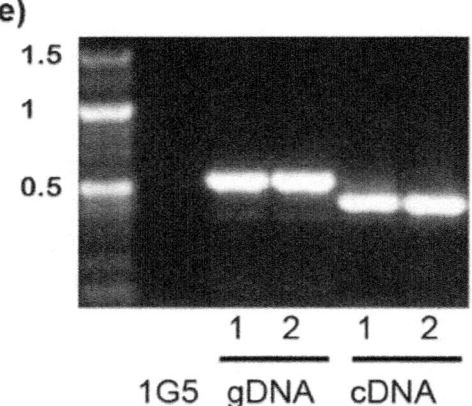

f)

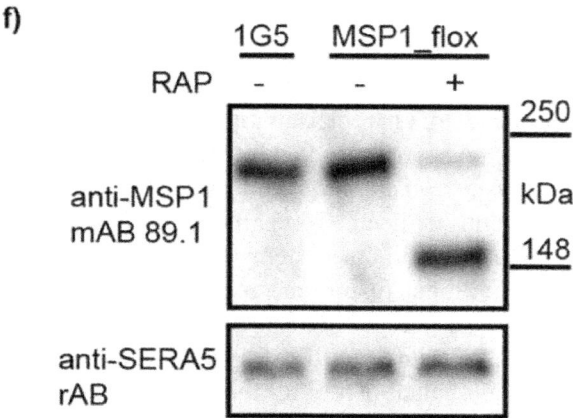

Figure 3: Use of the loxPint strategy allows the silent targeting of *msp1* in *P. falciparum*. (**a**) Schematic showing the *P. falciparum* 1G5DC *msp1* locus following the introduction of loxPint by homologous recombination via a 3' replacement. To force integration downstream of the loxPint module, recodonized MSP1 sequence is used downstream of that module (red). The endogenous C-terminal end is exchanged with a polymorphic MSP1 variant specifically recognized by mAb 111.4, thus effectively epitope tagging the gene. The modified *msp1:loxPint* locus is followed by a second *loxP* site that allows removal of the*msp1 3'* coding sequence. The *msp1* GPI anchor coding sequence is shown in blue. Yellow hexagon represents a stop codon. Correct integration of the construct into the endogenous locus is verified by PCR (insert and primer binding sites shown in (**b**)). (**b**) Cre-driven DNA excision results in the truncation of MSP1 (loss of amino acid residues 1240–1682, which includes the mAb X509 and mAb 111.4 epitopes). (**c**) The introduced loxPint is efficiently spliced and does not interfere with MSP1 expression or localisation, as shown by normal expression of MSP1 in control (DMSO-treated) schizonts by IFA. MSP1 was detected by IFA with mAb X509. Cre-driven DNA excision induced by treatment with RAP results in loss of the mAb X509 epitope. The rhoptry marker anti-RhopH2 mAb 61.3 (green) was used as a control. Schizont nuclei are stained with DAPI (blue). Scale bar, 5 µM. (**d**) Western blot, showing that reactivity with mAb X509 is lost upon RAP-treatment. (**e**) PCR showing complete splicing of the loxPint module in *msp1:loxPint* in two independent preparations of cDNA. (**f**) Western blot showing that introduction of the loxPint module into the *msp1* locus does not alter MSP1 expression levels compared to the parent 1G5 parasite line. Excision results in expression of a truncated MSP1 that is still recognised by mAb 89.1, which binds an epitope within the N-terminal part of MSP1 (see (**b**)).

DISCUSSION

In this study we have demonstrated that a small, readily synthesised genetic module comprising a *loxP* site integrated into an intron can be introduced into a range of genomic loci to: (1) replace native introns; (2) inserted into genomic exons; and (3) can even be incorporated into non-native recodonised sequences in the malaria parasite *Plasmodium falciparum*. In all cases the module is correctly and efficiently spliced. Compared to currently employed strategies for conditional gene deletion, our system represents a substantial step forward as it allows the conditional fusion and removal of DNA, irrespective of the presence of endogenous introns, in a single transfection step. Simultaneous 3′ tagging of a gene in the same modification step as the introduction of loxPint modules provides an invaluable tool for the study of uncharacterized genes, as shown for the parasite kinase FIKK10.1. In that case loss of the C-terminal HA tag served as a reporter for efficient recombination between the two loxPint modules and GFP was used as a positive reporter for efficient recombination. Importantly, the ability to exchange one stretch of DNA with another will allow conditional introduction of point mutations in the future (see Supplementary Fig. 1). Here we used single cross-over homologous recombination to introduce the loxPint module into the *P. falciparum* genome, but it is worth noting that successful introduction of an artificial exon flanked by two loxPint sequences for internal domain deletion has recently been achieved in *P. falciparum* using CRISPR/Cas9 technology (Emma Sherling, Michael Blackman and Christiaan van Ooij unpublished results (see also Supplementary Fig. S1). It is important to stress that while CRIPSR/Cas9 technology allows precise insertion of loxPint modules into genomic loci, it does not on its own allow the functional interrogation of essential genes.

The ability to conditionally disrupt genes such as FIKK10.1 that are non-essential in cell culture is just as important as it is for essential genes. Many genes that are not required during the blood stages in cell culture are readily lost. As such, phenotypes associated with gene knock-out studies that aim to interrogate for example sexual stage formation, bear a substantial risk of gene loss during the gene modification step. Our strategy, which allows the incorporation of 2 *loxP* sites in a single transfection step and delete a genetic element within a single parasite generation (~48 h) in a RAP controlled manner significantly reduces the time that it takes to generate such mutants by conventional methods, including CRIPSR/Cas9. The observation that the loxPint introduced into the non-essential FIKK10.1 parasite line is quantitatively spliced is important since it shows that selective pressure is not required for correct splicing of the loxPint and so it can be used for essential and non-essential genes alike. While we did not detect unspliced loxPint in the two genomic loci tested here, the presence of unspliced product from the REX2:GFP episome indicates that under some circumstances splicing may not be 100% efficient. Whether this is due in this case to the mRNA originating from an episome with potentially incomplete 5' and or 3' regulatory mRNA elements is unknown. However, given that we have shown that the loxPint works faithfully in two distinct endogenous loci, we are confident that it will prove a versatile novel tool to interrogate the many unknown and known essential and non-essential genes in the human malaria parasite.

The combination of a standardized small intron bearing a *loxP* site in conjunction with DiCre will open up its use to species beyond the malaria parasite where splicing machinery is present. While a strategy has been previously developed in mice where a synthetic intron was used to introduce a cassette (conditionals by inversion (COIN- module)) this is unlikely to be functional in organisms with fewer and shorter introns because of its large size. In conclusion, the method we present here is a major breakthrough in our ability to study the function of essential and non-essential genes in the malaria parasite *P. falciparum*. Because of the simplicity of its design we predict it will prove widely useful.

MATERIALS AND METHODS

P. falciparum Culture and Transfection

P. falciparum clone 1G5DC and the transfected parasite lines established here were cultured as described. Routine synchronization was by Percoll enrichment and sorbitol treatment. For transfection, either purified schizont-stage or ring-stage parasites (>5% parasitemia) were electroporated with 50–150 ug of plasmid DNA as described and selection was performed with 5 nM WR99210 (Jacobus Pharmaceuticals). To select for plasmid integration into the *fikk10.1* or*msp1* locus, transfectants were grown in the absence and then presence of 5 nM WR99210 (21 days in the absence of drug followed by growth in the presence of drug until a viable parasite population was re-established). Cycling between growth in the absence and then presence of WR99210 was performed at least twice. Plasmid integration into the *fikk10.1 or msp1* locus was verified by PCR using the primers described in Supplementary Table 1. For *fikk10.1* modification, integration was confirmed after each drug cycle.

To induce DiCre-driven *loxP* site recombination, synchronized ring-stage parasites were treated with 100 nM RAP (Sigma) or DMSO only (final concentration 1% v/v) for 4 h. Parasites were subsequently washed twice with warm RPMI and returned to culture. Samples used for nucleic acid extraction were taken at least 24 h after rapamycin treatment, and samples used for immunofluorescence (IFA) or protein extraction were taken at the end of the same asexual cycle (~44 h following RAP treatment) or in subsequent cycles.

loxPint Design and Construct Assembly

To identify an intron that could be used to silently introduce *loxP* sequences, we used a combination of manual searching and motif screening (using the search term "AGGTAA.{30,120}AGAT" with the PlasmoDB motif search feature, which incorporates bioinformatics analysis of *P. falciparum* 3D7 intron structure performed by Zhang *et al.*. This allowed the identification of several short intron sequences that could potentially be used to introduce *loxP* sites. We chose the short intron of *sera2* because it contained few extended mononucleotide tracts and could readily be synthesized by gene synthesis services. The 34 nt *loxP* sequence was inserted into the *sera2* intron after identification of its probable branch point so that splicing would likely proceed correctly even in the presence of these additional nucleotides. See Supplemental Table 2 for the complete *Sera2Intron:LoxP* sequence.

To generate the loxP:intron proof-of-principle construct *pRex2:loxPint:gfp*, we replaced the Rex2.GFP cassette of a pARL-based Rex2.GFP-encoding plasmid4 (a kind gift of Tobias Spielmann, Bernhard Nocht Institute for Tropical Medicine, Hamburg, Germany) with a *rex2.GFP* sequence in which the endogenous *rex2* intron was replaced with the loxPint sequence. The GFP-coding sequence was followed by a stop codon and a second loxPint, all followed by a myc-Tag and stop codon. This sequence was synthesized (Geneart®, see also Supplementary Table 2) then amplified using primers Rex2.POP.F and Rex2.POP.R and cloned into KpnI and AvrII-digested pARL:rex2GFP using Gibson assembly.

To generate pMSP1_loxPint, the 998 bp *msp1* targeting fragment was amplified from *P. falciparum* 3D7 genomic DNA using primers endo3D7-MSP1-BglII-targ-F and endo3D7-MSP1-R. The recodonized fragment was created by amplifying recodonized sequence from plasmid pZ-3D7-MSP138/42 (a kind gift of Dr Christian Epp, University of Heidelberg, Germany) using primers syn3D7-MSP1-F and syn3D7-MSP1+PstI-R. The fragments were then joined together by overlapping PCR (using primers endo3D7-MSP1-BglII-targ-F and syn3D7-MSP1+PstI-R) to create a fragment with 5′ BglII and 3′ PstI restriction site overhangs. This was ligated into the pHH1-MSP1$_{19}$ backbone, which has the wMSP1$_{19}$ in the 3′ end (created by digesting the plasmid pMSP1chimWT with BglII and PstI). The resultant plasmid construct was called pHH1-3D7wt. To introduce the loxPint, a ~400 bp sequence corresponding to the loxPint fragment flanked by *msp1* targeting sequence at the 5′ and 3′ ends (-CCTCAACC**AG**-loxPint-**AT**GTAACTCC-; bold letters indicate a naturally occurring AGAT motif, which effectively serves as the intron-exon boundary) was synthesised by Geneart® (see also Supplementary Table 2) and introduced into pHH1-3D7wt using restriction sites HpaI and BstEII. This generated plasmid pMSP1_loxPint. The plasmid structure was confirmed by nucleotide sequencing on both strands. See Supplementary Table 2 for the full recodonized *msp1* fragment sequence.

The FIKK10.1 targeting construct was generated by removal of the loxPint in the rex2:loxPint:gfp cassette of *pRex2:loxPint:gfp* using inverse PCR. This also introduced a point mutation into *rex2:GFP* that prevents REX2.GFP expression. The resulting plasmid was digested with NotI to introduce the *fikk10.1* homology region and recodonized kinase domain by Gibson assembly to generate the plasmid *pfikk10.1:loxPint:HA*. The *fikk10.1* homology region consists of *fikk10.1* nucleotides 1 to 1033 and was amplified using primer FK10.1HRF and FK10.1HRR. The recodonized kinase domain corresponds to *fikk10.1* nucleotides 1034 to 2319 lacking the 3' native intron. The recodonized kinase domain was synthesized (Geneart®, see also Supplementary Table 2) and then amplified using primers FK10.1RCF and FK10.1RCR. The final plasmid was sequenced to confirm correct assembly. See Supplementary Table 2 for the full recodonized *fikk10.1* kinase domain sequence including the added loxPint sequence.

All primers used in this study can be found in Supplementary Table 1.

Immunofluorescence

Live microscopy: Rex2:GFP and FIKK10.1loxPint:GFP expression was quantified by imaging live synchronized parasites that had been treated with RAP or DMSO for 4 h in at least the previous cycle. Parasites were taken fresh out of cell culture and treated with 4',6-diamidino-2-phenylindole (DAPI) to visualize nuclei. Images were taken using a Ti-E Nikon microscope using a 40x and 100x TIRF objective at room temperature equipped with an LED-illumination and an Orca-Flash4 camera. Images were processed with Nikon Elements software.

For immunofluorescence of the FIKK10.1loxPint parasite line, air-dried blood films were fixed for 4 min in ice-cold methanol and subsequently rehydrated in PBS for 5 min. Slides were blocked in 3% (w/v) bovine serum albumin (BSA) in PBS containing kanamycin (50 μg/ml) for 1 h and subsequently incubated with primary or secondary antibodies in 3% (w/v) BSA in PBS containing kanamycin (50 μg/ml) for 45 min. Antibody concentrations used were: rat anti-HA high affinity (Roche) (1:1000), rabbit anti-MAHRPI ((1:1000) a kind gift of Lindsay Parish and Julian Rayner). Images were taken using a Ti-E Nikon microscope using a 100x TIRF objective at room temperature equipped with an LED-illumination and an Orca-Flash4 camera. Images were processed with Nikon Elements software.

For MSP1 immunofluorescence, thin films of *P. falciparum* cultures were air dried, fixed in 4% paraformaldehyde (in PBS) for 30 min at room temperature and cell membranes permeabilised in 0.1% (v/v) Triton X-100 (SIGMA) for 10 min. Fixed slides were then washed three times with PBS for 10 min and blocked overnight at 4 °C in 3% (w/v) BSA in PBS. Slides were probed with anti-MSP1 mAb X509 using undiluted supernatant from culture hybridomas. Slides were also probed with anti-RhopH2 mAb 61.3 (1:250). Slides were probed with primary sera for 30 min at 37 °C and then washed for 10 min in PBS. Slides were then probed with an appropriate fluorescent secondary antibody (1:2000) and washed for 10 min with PBS. Slides were stained with DAPI and mounted in PBS/glycerol and images collected using AxioVision 3.1 software on an Axioplan 2 Imaging system (Zeiss) using a Plan-APOCHROMAT 1006/1.4 oil immersion objective.

Western Blotting

Western Blotting was performed according to standard methods. Briefly, Rex2. GFP or FIKK10.1.HA-expressing parasites were released from erythrocytes by addition of 0.1% (w/v) saponin/PBS for 5 min at room temperature. 0.1% saponin lysates were centrifuged ($>15000 \times G$) and the resulting parasite pellet was solubilized with 1X sample buffer with 5% beta-mercaptoethanol at a concentration of $2.5–5 \times 10^8$ parasites/ml. Parasite extracts were subjected to SDS PAGE and transferred onto nitrocellulose membranes. Rex2.GFP was visualized by probing the blots with rabbit anti-GFP antibodies (Abcam (1:500)) and goat anti-rabbit-HRP (Insight Biotechnology (1:4000)). FIKK10.1.HA was visualised using rat anti-HA hi affinity (Roche (1:500)) and goat anti-rat-HRP (1:4000 (Sigma)). For detection of MSP1, schizonts were Percoll-enriched according to standard methods then solubilized by addition of SDS sample buffer, subjected to SDS PAGE, and proteins transferred to nitrocellulose membranes. MSP1 was visualized by probing blots with anti-MSP1 mAb X509 or mAB 89.1 (undiluted hybridoma culture supernatant) and a goat anti-human-HRP secondary antibody (Sigma). SERA5 was visualized using polyclonal rabbit anti-SERA5 serum.

Nucleic Acid Extraction, cDNA Synthesis, and Sequencing

DNA was extracted from *P. falciparum* parasite lines using the QIAamp DNA Blood MiniKit (Qiagen). RNA was extracted as described and DNA was removed using the Ambion TURBO DNA-free kit (Applied Biosystems). cDNA was produced from isolated, DNAse-treated RNA by first-strand cDNA synthesis using random hexamers (Superscript III cDNA synthesis kit, Invitrogen). For sequencing, primers allowing PCR amplification across the LoxPint module both before and after DiCre-induced recombination were designed for *Rex2, MSP1,* and *FIKK10.1* and the resulting PCR fragment was either sequenced directly or blunt cloned into pCR.Blunt II-TOPO using a Zero Blunt PCR cloning kit (Invitrogen).

SUPPLEMENTARY INFORMATION

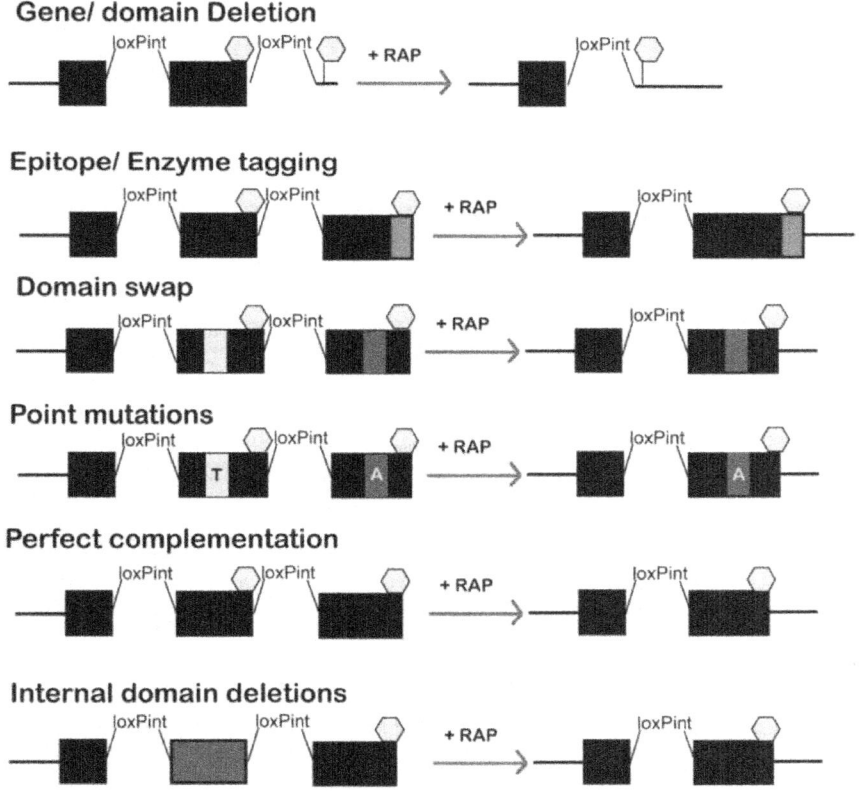

Supplemental Figure 1: Possible strategies for conditional genome engineering using the loxPint module. The reconstitution of the loxPint module after RAP treatment allows the conditional fusion or deletion of domains. This opens new avenues for conditional epitope or enzyme tagging (for example with biotin ligase A (BirA)), conditional domain swaps, introduction of point mutations or perfect complementation that is, replacing a genetic element with an identical copy. Using CRIPSR the introduction of two loxPint modules into an open reading frame to delete only parts of a gene will be feasible. Yellow hexagons represent stop codons.

Supplement Table 1: Contains all primer sequences used in this study

Used For:	Primer Name:	Primer Sequence:
pRex2:loxPint:gfp	Rex2.POP.F	GTTTTTTTTAATTTCTTACATATAACTCGACCCCGGGATGGTACCTTTATGAAAATGTATTTAGCTG
	Rex2.POP.R	GAAAAACGAACATTAAGCTGCCATATCCCCGGCGGCTGCAGTTACAGATCCTCTTCTGAG
pflkk10.1:loxPint:HA	FK10.1HRF	GCCAAGCTATTTAGGTGACACTATAGAATACTCGCGGCCGCATGACTCTTATTAATAGAAGTTATGTTTTATTTGG
	FK10.1HRR	GCTATACGGAAGTTATTGTATATTATTTTTTTTTATTTACCTTTATTATAACCATGTGTAGGTATAGAAGTTAATTTC
	FK10.1RCF	GAAATTAACTTCTATACCTACACATGGTTATAATAAAAGGTAAATAAAAAAAATAATATACAATAACTTCGTATAGC
	FK10.1RCR	CTACTAAGATCTCCTCCTAAGTCTGTTACGTTAGCGGCCGCTTAGACCGCATAATCCGGTACATCGTATGGATACG
pMSP1_loxPint	syn3D7-MSP1-F	GTGGTAGTTCAGGATCCACAAAAGAAGAAACCC
	syn3D7-MSP1+PstI-R	GCATGTCCTGCAGCTTGCCCTCTATGAGCTTTGATATGATGG
	endo3D7-MSP1-BglII-targ-F	CCAACAAAGATCTGCATCCTCTACCAATACCC
	syn3D7-MSP1+PstI-R	GCATGTCCTGCAGCTTGCCCTCTATGAGCTTTGATATGATGG
pflkk10.1:loxPint:HA integration confirmation	Int1	GTCCTTCATTAATTTGATGGTCA
	Int2	CACATAGTTTTCTCCGCACAGCACGTATTCGC
	Int3	CAACATACACATTTTTACAGTTATAAATACAATCAATTG
	Int4	CCCCAGGCTTTACACTTTATGCTTCCGGCTC
	Int5	CTTAATAAATAATCCTACTCTATCACTACCATCTC
pMSP1_loxPint integration confirmation	MSP1-UOT-FOR	GGAACATCATCTACATCCAGTCCTGG
	REV3	GTAGAGATCCTGATGTGGGGATC
	FOR1	CCATTTCTACAACAGAGATGG
	REV4	GCATTTTGTCTTGGCCAAGTTC
	P2 FOR	GTAAATAAAAAAAATAATATACAATAACTTCGTATAGCATACATTATACGAAGTTAT

Supplemental Table 2: Contains the full sequences of synthetic genes used in this study

loxPint sequence	gtaaataaaaaaaataatatacaATAACTTCGTATAGCATACATTATACGAAGTTATatatatgtatatatatatatatttatatatttttatattcttttag
Used For:	**Synthesized Sequence**
pRex2:loxPint:gfp	(full synthetic gene sequence)
pflkk10.1:loxPint:HA	(full synthetic gene sequence)
pMSP1_loxPint	(full synthetic gene sequence)

ACKNOWLEDGEMENTS

We thank Ellen Knuepfer and the members of the Treeck and the Blackman labs for critical input. We thank Lindsay Parish and Julian Rayner for the MAHRPI antibody. This work was funded by the Medical Research Council (U117532063 to M.J.B.), the Francis Crick Institute (to M.T. and M.J.B), and EC FP7 contract no. 242095 (EviMalAR). S.D. was in receipt of an EviMalAR

PhD studentship. Funding for open access charge: The Francis Crick Institute. The funders had no role in study design, data collection and interpretation, or the decision to submit the work for publication.

REFERENCES

1. Jullien, N., Sampieri, F., Enjalbert, A. & Herman, J. P. Regulation of Cre recombinase by ligand-induced complementation of inactive fragments. *Nucleic Acids Res* **31**, e131 (2003).

2. Collins, C. R. *et al.* Robust inducible Cre recombinase activity in the human malaria parasite Plasmodium falciparum enables efficient gene deletion within a single asexual erythrocytic growth cycle. *Mol Microbiol* **88**, 687–701 (2013).

3. Yap, A. *et al.* Conditional expression of apical membrane antigen 1 in Plasmodium falciparum shows it is required for erythrocyte invasion by merozoites. *Cell Microbiol* **16**, 642–656 (2014).

4. Haase, S. *et al.* Sequence requirements for the export of the Plasmodium falciparum Maurer's clefts protein REX2. *Mol Microbiol* **71**, 1003–1017 (2009).

5. Armstrong, C. M. & Goldberg, D. E. An FKBP destabilization domain modulates protein levels in Plasmodium falciparum. *Nat Methods* **4**, 1007–1009 (2007).

6. Nunes, M. C., Goldring, J. P., Doerig, C. & Scherf, A. A novel protein kinase family in Plasmodium falciparum is differentially transcribed and secreted to various cellular compartments of the host cell. *Mol Microbiol* **63**, 391–403 (2007).

7. O'Donnell, R. A., Saul, A., Cowman, A. F. & Crabb, B. S. Functional conservation of the malaria vaccine antigen MSP-119across distantly related Plasmodium species. *Nat Med* **6**, 91–95 (2000).

8. Das, S. *et al.* Processing of Plasmodium falciparum Merozoite Surface Protein MSP1 Activates a Spectrin-Binding Function Enabling Parasite Egress from RBCs. *Cell Host Microbe* **18**, 433–444 (2015).

9. Ghorbal, M. *et al.* Genome editing in the human malaria parasite Plasmodium falciparum using the CRISPR-Cas9 system. *Nat Biotechnol* **32**, 819–821 (2014).

10. Wagner, J. C., Platt, R. J., Goldfless, S. J., Zhang, F. & Niles, J. C.Efficient CRISPR-Cas9-mediated genome editing in Plasmodium falciparum. *Nat Methods* **11**, 915–918 (2014).

11. Klinz, F. J. & Gallwitz, D. Size and position of intervening sequences are critical for the splicing efficiency of pre-mRNA in the yeast Saccharomyces cerevisiae. *Nucleic Acids Res* **13**, 3791–3804 (1985).

12. Cheng, T. H., Chang, C. R., Joy, P., Yablok, S. & Gartenberg, M. R.Controlling gene expression in yeast by inducible site-specific recombination. *Nucleic Acids Res* **28**, E108 (2000).

13. Trager, W. & Jensen, J. B. Human malaria parasites in continuous culture. *Science* **193**, 673–675 (1976).

14. Moon, R. W. *et al.* Adaptation of the genetically tractable malaria pathogen Plasmodium knowlesi to continuous culture in human erythrocytes. *Proc Natl Acad Sci USA* **110**, 531–536 (2013).

15. Fidock, D. A. & Wellems, T. E. Transformation with human dihydrofolate reductase renders malaria parasites insensitive to WR99210 but does not affect the intrinsic activity of proguanil.*Proc Natl Acad Sci USA* **94**, 10931–10936 (1997).

16. Voss, T. S. *et al.* A var gene promoter controls allelic exclusion of virulence genes in Plasmodium falciparum malaria. *Nature* **439**, 1004–1008 (2006).

17. Zhang, X. *et al.* Branch point identification and sequence requirements for intron splicing in Plasmodium falciparum.*Eukaryot Cell* **10**, 1422–1428 (2011).

18. Gibson, D. G. Enzymatic assembly of overlapping DNA fragments. *Methods Enzymol* **498**, 349–361 (2011).

19. Child, M. A., Epp, C., Bujard, H. & Blackman, M. J. Regulated maturation of malaria merozoite surface protein-1 is essential for parasite growth. *Mol Microbiol* **78**, 187–202 (2010).

20. Blackman, M. J., Whittle, H. & Holder, A. A. Processing of the Plasmodium falciparum major merozoite surface protein-1: identification of a 33-kilodalton secondary processing product which is shed prior to erythrocyte invasion. *Mol Biochem Parasitol* **49**, 35–44 (1991).

21. Holder, A. A., Freeman, R. R., Uni, S. & Aikawa, M. Isolation of a Plasmodium falciparum rhoptry protein. *Mol Biochem Parasitol* **14**, 293–303 (1985).

22. Harris, P. K. *et al.* Molecular identification of a malaria merozoite surface sheddase. *PLoS Pathog* **1**, 241–251 (2005).

23. Kyes, S., Pinches, R. & Newbold, C. A simple RNA analysis method shows var and rif multigene family expression patterns in Plasmodium falciparum. *Mol Biochem Parasitol* **105**, 311–315 (2000).

Chapter 5

TREC-IN: GENE KNOCK-IN GENETIC TOOL FOR GENOMES CLONED IN YEAST

Suchismita Chandran[1] , Vladimir N Noskov[1] , Thomas H Segall-Shapiro[1] , Li Ma[1] , Caitlin Whiteis[1] , Carole Lartigue[2,3], Joerg Jores[4] , Sanjay Vashee[1] and Ray-Yuan Chuang[1]

[1]The J. Craig Venter Institute, 9704 Medical Center Drive, Rockville 20850, MD, USA

[2] INRA, UMR 1332 de Biologie du Fruit et Pathologie, F-33140, Villenave d'Ornon Bordeaux, France

[3] University Bordeaux, UMR 1332 de Biologie du Fruit et Pathologie, F-33140, Villenave d'Ornon Bordeaux, France

[4] International Livestock Research Institute (ILRI), Old Naivasha Road, 00100 Nairobi, Kenya.

ABSTRACT

Background

With the development of several new technologies using synthetic biology, it is possible to engineer genetically intractable organisms including *Mycoplasma mycoides* subspecies *capri* (*Mmc*), by cloning the intact bacterial genome in yeast, using the host yeast's genetic tools to modify the cloned genome, and subsequently transplanting the modified genome into a recipient cell to obtain mutant cells encoded by the modified genome. The recently described tandem repeat coupled with endonuclease cleavage (TREC) method has been successfully used to generate seamless deletions and point mutations in the mycoplasma genome using the yeast DNA repair machinery. But, attempts to knock-in genes in some cases have encountered a high background of transformation due to maintenance of unwanted circularization of the transforming DNA, which contains possible autonomously replicating sequence (ARS) activity. To overcome this issue, we incorporated a split marker system into the TREC method, enabling seamless gene knock-in with high efficiency. The modified method is called TREC-assisted gene knock-in (TREC-IN). Since a gene to be knocked-in is delivered by a truncated non-functional marker, the background caused by an incomplete integration is essentially eliminated.

Results

In this paper, we demonstrate applications of the TREC-IN method in gene complementation and genome minimization studies in *Mmc*. In the first example, the *Mmc dnaA* gene was seamlessly replaced by an orthologous gene, which shares a high degree of identity at the nucleotide level with the original *Mmc* gene, with high efficiency and low background. In the minimization example, we replaced an essential gene back into the genome that was present in the middle of a cluster of non-essential genes, while deleting the non-essential gene cluster, again with low backgrounds of transformation and high efficiency.

Conclusion

Although we have demonstrated the feasibility of TREC-IN in gene complementation and genome minimization studies in *Mmc*, the applicability of TREC-IN ranges widely. This method proves to be a valuable genetic tool that can be extended for genomic engineering in other genetically intractable organisms, where it may be implemented in elucidating specific metabolic pathways and in rationale vaccine design.

BACKGROUND

Mycoplasmas are the simplest and smallest living prokaryotes (0.1 µm), and although phylogenetically related to Gram-positive bacteria, lack a cell wall [1]. They also have the smallest recorded genomes (0.58 Megabases (Mb) – 1.38 Mb) for bacterial species that can replicate autonomously, and have colonized a wide range of hosts including, humans and animals [2]. However, efforts to manipulate mycoplasma genomes are fraught with difficulties owing to the lack of genetic tools available for these organisms [3]. This has made understanding the biology and elucidating the host-pathogen mechanism for any potential therapeutics, including vaccine development, challenging.

One of the early genetic tools that were developed for understanding mycoplasma biology was the generation of *OriC* plasmids that could replicate in mycoplasma cells [4–6]. Although heterologous gene expression and targeted gene disruption by single-crossover recombination were demonstrated in *Mycoplasma mycoides* subspecies *capri* (*Mmc*) and *M. capricolum* subspecies *capricolum* (*Mcc*), no recombination events were observed in the closely related *M. mycoides* subspecies *mycoides* (*Mmm*) [7, 8]. In addition, maintaining stable mutants using *OriC* plasmids turned out to be difficult and laborious [4, 7, 8]. Thus, alternate strategies were designed, including a transposon-based method to generate mutants with low passage numbers that were free of antibiotic-resistance genes [9]. Transposon-based mutagenesis has been prevalently used as a genetic tool in mycoplasmas to generate mutants of interest as well as to define essential genes required for survival [10–12]. Furthermore, a double-crossover homologous recombination method using a suicide plasmid has been described for *M. genitalium*, albeit at a very low frequency [13–15], but this method did not address seamless deletion and removal of markers [16]. Therefore, to overcome stability and marker recycling issues, we turned to yeast genetics and synthetic biology to extend the genetic toolbox of mycoplasmas. With recent advancements in synthetic genomics including, cloning of the *Mmc* genome in yeast, manipulation of the mycoplasma genome using yeast genetic tools, transplantation of the engineered mycoplasma genome from yeast to a bacterial recipient cell, and creation of the synthetic cell, expression of the engineered genome became possible [17–22]. Mycoplasma genomes including *M. genitalium* (0.6 Mb), *M. pneumoniae* (0.8 Mb), and *Mmc* (1.1 Mb) were first cloned into yeast with the idea of implementing yeast genetics tools to engineer genetically intractable organisms [17–22].

Once cloned in yeast, bacterial genomes can be theoretically manipulated by yeast genetic tools. The URA3 marker/5-FOA counter-selection is a common technique in which the marker can be recycled to create seamless gene deletions, replacements, or gene knock-ins. However, we have previously shown that this conventional two-step method was very inefficient in engineering a mycoplasma genome cloned in yeast due to instability of the genome where high background of 5-FOA resistant colonies resulted from non-specific removal of the URA3 marker [22]. Development of the tandem repeat coupled with endonuclease cleavage (TREC) method has greatly improved the efficiency of seamless gene deletions [21,22]. TREC can be also applied in gene knock-in via a single step transformation where the knock-in sequence is placed outside the cassette and immediately next to the repeated sequence (Figure 1). The removal of the cassette leaves the knock-in sequence in the target site seamlessly. Although TREC method is currently the best tool that can seamlessly engineer a genome cloned in yeast [21, 22], the process is sometimes inefficient with a high background of transformation, arising possibly due to illegitimate recombination (Figure 1). To overcome this limitation of TREC-mediated gene insertion, we developed a modified method, called TREC assisted gene knock-in (TREC-IN) that significantly improves the efficiency of gene knock-in and vastly reduces screening effort. This method relies on the split marker system whereby the gene is delivered by a non-functional truncated antibiotic resistant gene module *kanMX*, and a site-specific gene insertion is selected by functional restoration of the full length *kanMX* gene. Here, we demonstrate the feasibility of TREC-IN in the*Mmc* genome using two examples a) replacement of an endogenous gene with an orthologous one, and b) essential gene complementation in a genome reduction study.

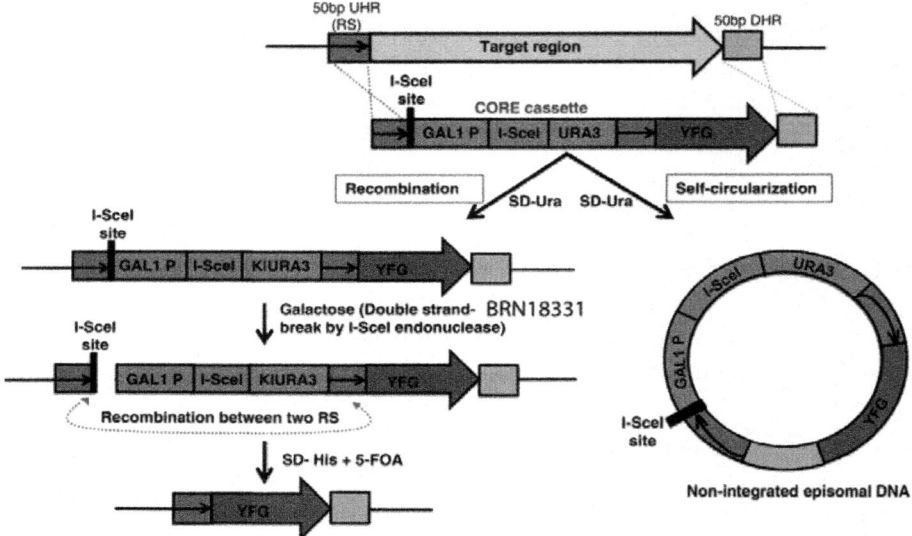

Figure 1: Brief outline of TREC and background formation. The gene to be inserted (your favorite gene, YFG), indicated by a purple arrow can be seamlessly inserted into a target site (orange arrow) via TREC. The knock-in sequence (purple arrow) is placed immediately downstream of the repeat sequence (RS) (dark blue box containing a black arrow) in the CORE cassette (gray boxes). After integration, induction of double strand break can promote homologous recombination between the two repeat sequences, leading to removal of the CORE cassette, as shown on the left. However, a fraction of transforming DNA may circularize itself through illegitimate recombination or non-homologous end joining, where broken DNA ends join. The resulting DNA may be maintained as a non-integrated plasmid if the knock-in sequence contains ARS activity, as shown on the right. UHR, indicates upstream homology region, and DHR, downstream homology region.

RESULTS

Design of the TREC-IN

The design of TREC-IN is based on the previous TREC strategy and incorporates a split marker approach with an additional step for a gene knock-in that is mediated by a functional restoration of the kanamycin resistance gene module, *kanMX*. The procedure involves three steps: first, insertion of a CORE6 cassette to the target locus; second, site-specific gene integration; and third, seamless cassette recycling (Figure 2A). In the first step, the CORE6 cassette, which consists of the 18 bp I-SceI binding site, the I-SceI endonuclease encoding gene under the control of the yeast *GAL1* promoter, the *KlURA3* gene and a 5' truncated *kanMX* gene component, is introduced to the target site. Similar to the TREC method, two sequences of about 50 bp that are homologous to the target site are added into the CORE6 cassette by PCR on 5' and 3' ends of the cassette so that they flank the CORE6 cassette (Figure 2B). Transformation of the CORE6 cassette into yeast and homologous recombination at the target site results in the replacement of the target site by the CORE6 cassette. Transformed yeast colonies are selected for uracil prototrophs, and further analyzed by PCR screening to confirm that the homologous recombination has occurred at the correct target site (Figure 2A). The second step of TREC-IN involves construction and transformation of the knock-in module containing a 3' truncated *kanMX* gene component and the knock-in sequence. The kanamycin resistance gene and the knock-in sequence are separated by a repeat sequence of about 50 bp in length, which is identical to upstream sequences of the target site in the CORE6 cassette (Figure 2C). This knock-in module is flanked at the 5' end by a region of the kanamycin resistance gene to allow for homologous recombination at the 3' end of the CORE6 cassette. On its 3' end, the knock-in module is flanked by the same homologous region that is also present on the 3' end of the CORE6 cassette to allow for recombination at the target site (Figure 2A). Upon transformation, the knock-in module integrates into the target site, resulting in an insertion containing two repeat sequences encompassing three genes (the I-SceI, *KlURA3*, and the full length *kanMX* module) and the knock-in sequence. Transformed yeast colonies are selected for resistance against the antibiotic geneticin, and then analyzed by PCR screening to confirm correct insertion. In the third step of TREC-IN, the whole cassette flanked by the two repeat sequences is removed via homologous recombination between the two repeat sequences. The efficiency of the recombination is enhanced by the double strand break (DSB) generated by the cleavage of the endonuclease I-SceI at the 18 bp recognition site in the cassette after galactose induction. The removal of the *KlURA3,* I-SceI, and the *kanMX* module counter-selected by 5-FOA would leave no scar. Only the knock-in sequence remains at the target site. Yeast cells that are resistant to 5-FOA are screened by PCR for the precise insertion of the replacement sequence (Figure 2).

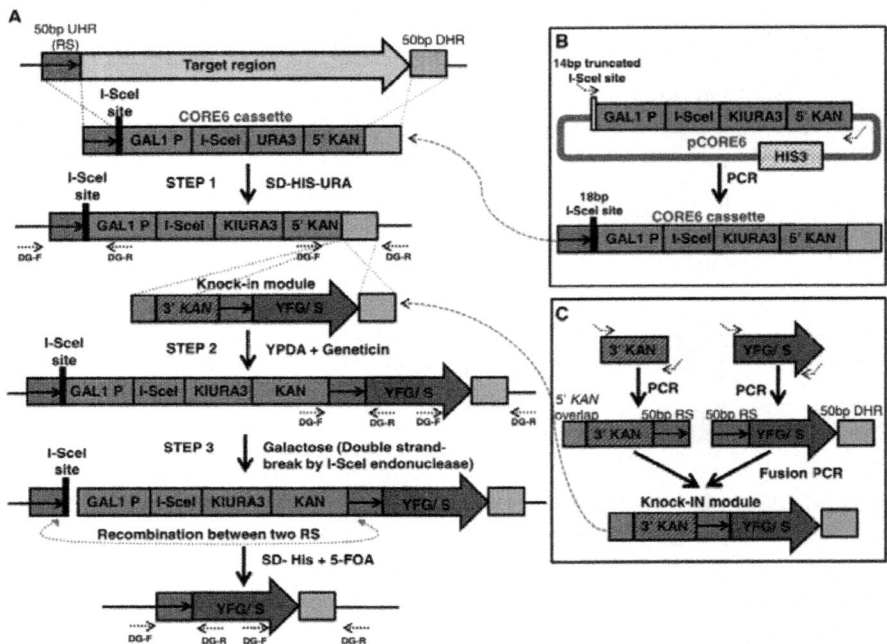

Figure 2: Schematic representation of TREC-IN. (A) Outline of the TREC-IN method. Step 1: The target region (orange arrow) is replaced with a CORE6 cassette by homologous recombination via two 50-bp homology sequences [upstream homology region (UHR) or repeat sequence (RS), indicated by dark-blue box containing a black arrow, and downstream homology region (DHR), indicated by light-blue box]. Transformed cells were selected for Uracil prototrophy. The cassette includes the reconstituted I-SceI binding site (black bar), and the I-Sce1 gene under the control of the Gal1 promoter, a Uracil marker, and 5' *kanMX* gene componenet. Step 2: The knock-in module [containing your favorite gene/ sequence (YFG/S, purple arrow) and the 3' region of the *kanMX* gene component, interspersed by the 50 bp RS, and flanked by 50 bp homology region to the 5' *kanMX* gene component in the cassette and DHR to the target site] is transformed and selected on geneticin resistance for full complementation of the *kanMX* gene component. Step 3: Yeast colonies are grown in presence of galactose to express I-SceI, which produces a double-strand break at the I-SceI site, and enhances homologous recombination (dashed double-headed arrow) between the two RS, resulting in excision of the CORE6 and knock-in modules. Colonies are grown on 5-FOA for Uracil counter selection. Primers to confirm correct insertions are shown by dashed arrows, and synthetic primers are represented by kinked arrows. **(B)** CORE6 cassette

construction. The CORE6 cassette is amplified from pCORE6 to add four nucleotides (tagg) for full reconstitution of the I-SceI binding site (black bar). 50 bp UHR and 50 bp DHR, specific to the target site, are also included in the construction. **(C)** Knock-in module construction. The knock-in module is constructed by a PCR-based fusion method. The two amplicons are the 3' *kanMX* gene component carrying the 3' region of kanamycin gene (gray striped box), the RS, and a homology region to the 5' region of the kanamycin gene in CORE6 (gray box), and the replacement gene (YFG/S) flanked by the RS and DHR.

Replacement of the *Mcc* Orthologous *dnaA* Gene in the *Mmc*genome

To demonstrate precise replacement of an orthologous gene in the *Mmc* genome, TREC-IN was applied to replace the *Mmc dnaA* gene, which is essential for chromosomal replication and viability [5, 19], with the orthologous *dnaA* gene from *Mcc*. The *Mmc* (accession no. AY277700) and *Mcc dnaA* genes (accession no. D90426) share 95% sequence identity at both the nucleotide and protein levels (analyzed using BLAST). As described in Methods, the first step of TREC-IN resulted in the precise replacement of the endogenous *dnaA* gene by the CORE6 cassette (Figure 3A, step 1). The promoter and 3' region of the *Mmc dnaA* gene were left unaltered. In the second step of TREC-IN, the orthologous *Mcc dnaA* gene was integrated downstream of the CORE6 cassette under geneticin selection (Figure 3A, step 2). In the third step, DSB at the I-SceI site promoted homologous recombination between the two repeat sequences, followed by precise and seamless insertion of the *Mcc dnaA* gene in the *Mmc* genome (Figure 3A, step 3). Each step of the deletion and replacement procedure was evaluated by PCR screening to confirm the correct insertions and junctions (Figure 3B).

Upon transformation, seven colonies were screened by PCR and all the colonies were found to be positive for CORE6 replacement [Figure 3B (a)]. In the second step, we PCR-screened 36 geneticin-resistant colonies for multiple junctions and found that 33/36 (>91%) of the colonies were positive for the precise insertion of the cassette at the targeted locus [Figure 3B (b)]. In the third step, I-SceI-mediated DSB resulted in 27/36 (75%) of the colonies showing precise removal of the cassette, resulting in a clean insertion [Figure 3B (c)]. Of note, PCR-screening [Figure 3B (b) and 3B (c)] indicated that while majority of the colonies obtained resulted in seamless replacement of the knock-in gene, the remaining 9% and 25% colonies respectively, were positive for only one or two of the junctions tested, suggesting non-specific recombination. Thus, TREC-IN proves to be a valuable genetic tool to overcome background issues and facilitate gene knock-in experiments with increased efficiency.

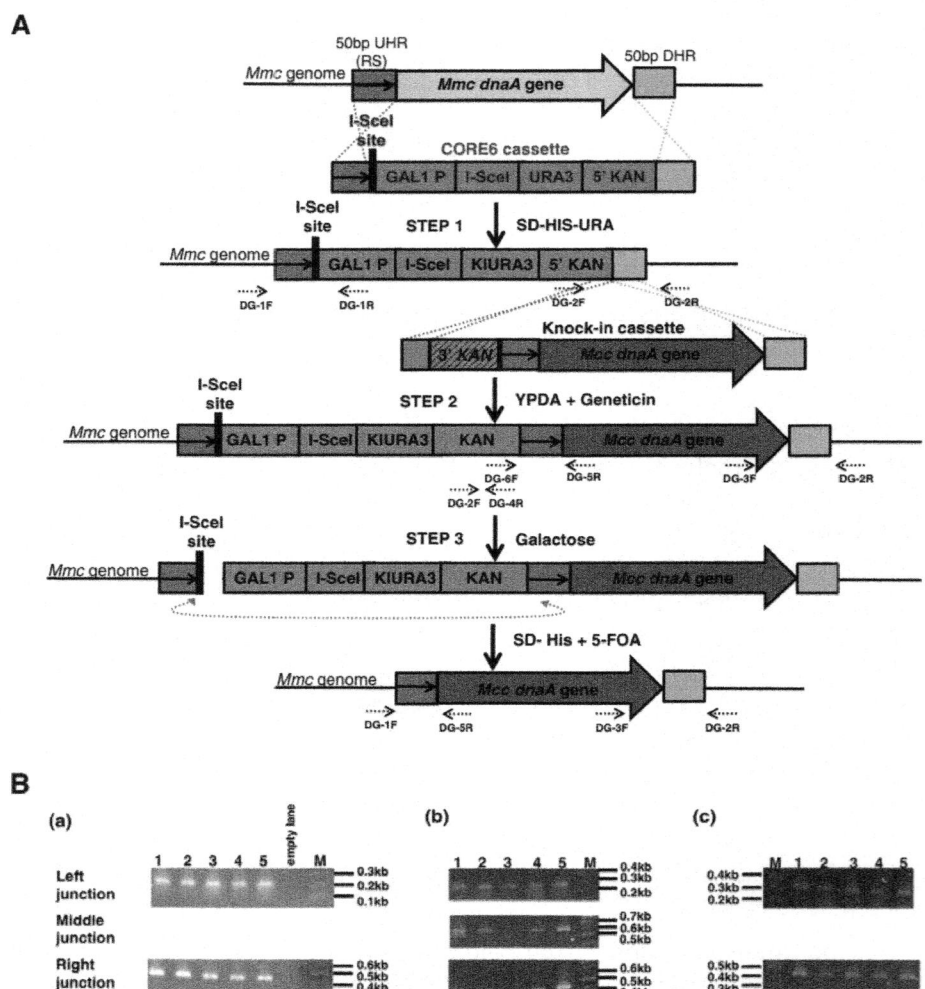

Figure 3: Replacement of *Mcc dnaA* gene in the mycoplasma genome. (A) Schematic representation of replacement of the *Mmc dnaA* gene with an orthologous gene from *Mcc*. Here, the *dnaA* gene in the *Mmc* genome (indicated by an orange arrow) was replaced by the *Mcc* orthologue (purple arrow) using TREC-IN. Diagnostic primers to confirm correct insertion of the cassettes and seamless replacement of the endogenous *dnaA* gene are indicated by dashed arrows (see Additional file 1: Figure S3 for primer information). **(B)** PCR screening to confirm replacement of the *Mmc dnaA* gene with an orthologous gene from *Mcc*. (a) DNA from yeast colonies after selection on SD-His-Ura were amplified using primers DG1F/DG1R (left junction; expected size, 222 bp) and DG2F/DG2R (right junction; expected size, 438 bp). (b) DNA from yeast colonies after selection on geneticin were amplified using primers DG2F/

DG4R (left junction; expected size, 285 bp), DG6F/DG5R (middle junction; expected size, 615 bp) and DG3F/DG2R (right junction; expected size, 446 bp). (c) DNA from yeast colonies after selection on 5-FOA were amplified using primers DG1F/DG5R (left junction; expected size, 388 bp) and DG3F/DG2R (right junction; expected size, 446 bp). A representative of 5 colonies is shown for each transformation and for each junction.

Both, the *dnaA* gene-deleted and *dnaA* gene-complemented *Mmc* genomes were transplanted to generate the mutant*Mmc* strains, as described previously [21]. As expected, genome transplantation of the *dnaA* gene-deleted genome resulted in non-viability. However, replacement of the orthologous *Mcc dnaA* gene resulted in a viable cell. The resulting colonies were of similar size to those of the control wild-type *Mmc* colonies (data not shown). Genomic DNA from the *dnaA*-replaced*Mmc* cells was isolated and analyzed by sequencing to confirm the precise and scar-less insertion of the *Mcc dnaA* gene.

Application of TREC-IN for Genome Reduction in the *Mmc* Genome

Global transposon random mutagenesis has been widely used to identify non-essential genes in minimal genome studies in prokaryotes [10–12]. Therefore, using transposons, we generated a high-resolution map of non-essential gene candidates on a synthetic *Mmc* genome (unpublished data). To carry out a top-down genome reduction strategy, consecutive non-essential genes were grouped into multigene deletion targets and were labeled non-essential gene clusters (NEGC). In some cases, several NEGCs were interspersed by single or a few Tn5-defined essential genes. To achieve genome reduction more efficiently, the TREC-IN approach was tested to remove multiple NEGCs simultaneously, and then add back the essential genes to the genome that were interspersed between them. To demonstrate this application, we chose a 16 kb region of the synthetic *Mmc* genome (*Mmc* Syn1) covering two NEGCs consisting of 10 genes, separated by a Tn5-denfined essential gene (*ssrA*) for deletion (Figure 4A and Additional file 1: Figure S2). In the first step of TREC-IN, the integration of the CORE6 cassette at the target site resulted in the deletion of the two NEGCs (*Mmc* Syn1 0152-0157 and*Mmc* Syn1 0159-0162) along with the intervening essential gene *ssrA* (*Mmc* Syn1 0158) from the *Mmc* Syn1 genome. In the second step of TREC-IN, transformation of the knock-in module resulted in the precise insertion of the *Mmc ssrA* gene back into the synthetic *Mmc* genome. The precise cluster deletion followed by insertion of the *ssrA* gene was verified by PCR screening (Figure 4B). The phenotypes of both cluster-deleted and *ssrA* gene-complemented *Mmc* strains were determined by genome transplantation. We found that the whole 16 kb deletion comprising 11 genes resulted in a non-functional genome as observed by the lack of viable cells. However, cis-complementation of the *ssrA* gene rescued the lethal phenotype. Transplantation colonies from the *ssrA* complemented synthetic *Mmc* genome were viable, and showed similar colony size to those of the control synthetic *Mmc* cells (data not shown). Genomic DNA from the *ssrA*complemented *Mmc* cells was isolated and analyzed. Sequencing of the complemented *ssrA* region in the isolated modified synthetic *Mmc* genome confirmed the precise and seamless insertion of the essential *Mmc ssrA* gene and deletion of the two NEGCs.

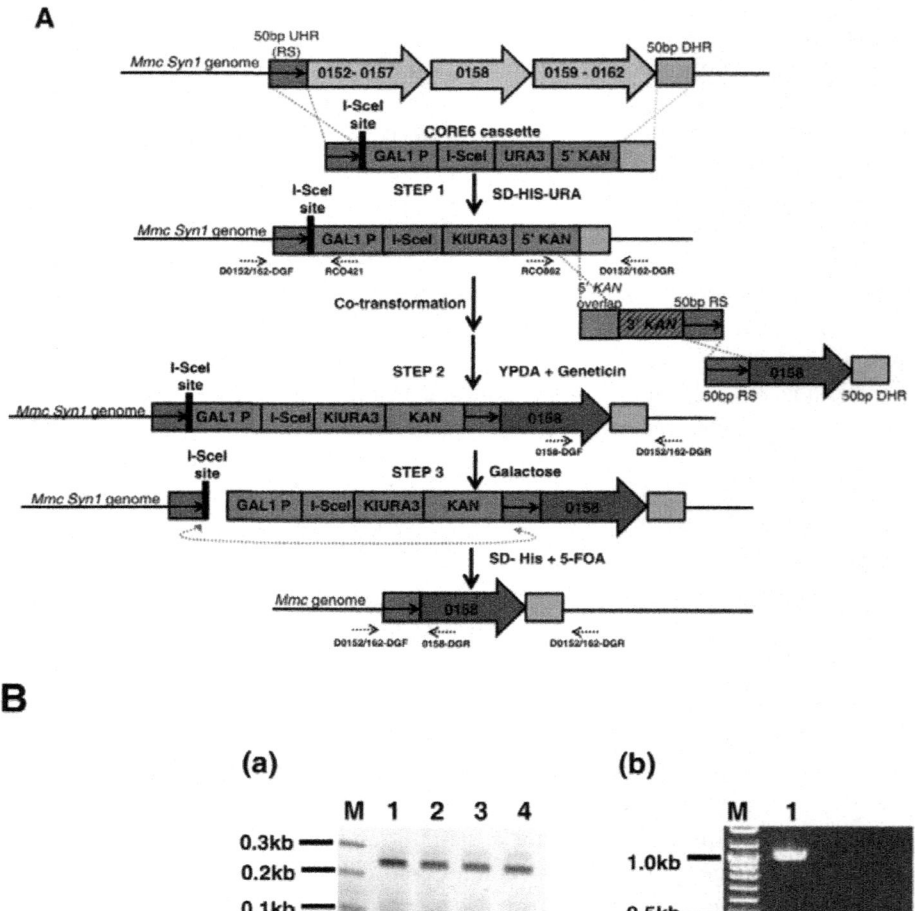

Figure 4: Cluster deletion and replacement of an essential gene using TREC-IN. (A) Schematic of replacement of the intervening *ssrA* essential gene (0158) upon cluster deletion in *Mmc* Syn1. In step 1, the target region (orange arrows) containing

the *ssrA* gene and two adjacent non-essential clusters (0152-0157 and -159-0162) were replaced with the CORE6 cassette. In step 2, a knock-in module was integrated into the target site by co-transforming two PCR products. One of the amplicons contained the *ssrA* gene, 50 bp RS, and 50 bp DHR. The other amplicon included the 3' region of the *kanMX* split marker gene component, 50 bp RS, and 50 bp homology region to the 5' *kanMX* gene component in the CORE6 cassette. Homologous recombination resulted in full complementation of the *kanMX* gene component. Yeast colonies were selected for geneticin resistance and then grown on galactose. In step 3, galactose induces I-SceI expression, which produces double-strand break at the I-SceI site and enhances intra-molecular homologous recombination (dashed double-headed blue arrow) between the two RS, resulting in excision of the CORE6 cassette. Colonies were grown on 5-FOA for Uracil counter selection. **(B)** PCR screening to confirm cluster deletion and replacement of the *ssrA* essential gene. **(a)** DNA from four yeast colonies after selection on 5-FOA were amplified using diagnostic primers (dashed arrows) D0152/162-DGF and 0158-DGR (left junction; expected size, 239bp), and **(b)** D0152/162-DGF and D0152/162-DGR (expected size of *ssrA* replacement, 1.0kb).

DISCUSSION

Mycoplasmas infect a wide range of hosts, including humans and animals, and in some cases, even contribute towards economic havoc [1, 2]. Therefore, developing better genetic tools to study and contain these pathogens has become a priority. *Mmc*, with its relatively small genome, and ease of manipulation [20, 21] is not only being probed as a model to study pathogenesis, but also as a model organism to test the concept of a minimal cell, where essential genes and functions are being determined. Additionally, *Mmc* is also being modeled as a platform to develop tools towards vaccine development that can be applied to other mycoplasma species. However, the existing genetic tool box makes it difficult to study this bacterial species.

With recent advancements in synthetic genomics, it is now possible to engineer the *Mmc* genome using yeast genetic tools, including TREC [17–22]. Development of TREC is based on a modified yeast system where generation of seamless deletions [22] and point mutations (unpublished data) in the mycoplasma genome is now made possible by using the yeast DNA repair machinery. In principle, the TREC method can be employed to insert genes of interest into the *Mmc* genome. However, several attempts to knock-in an *Mmc* gene into the *Mmc* genome were inefficient with a high background of transformation. Since yeast ARSs are A-T rich, and the mycoplasma genome is relatively A-T rich, it is reasonable to speculate that gene knock-ins containing A-T rich sequences of mycoplasma genomes likely contain ARS activity [17]. Akada and colleagues reported that a gene containing an ARS performs inefficient chromosomal integration [23]. Thus, a portion of the transforming DNA can circularize through illegitimate recombination or NHEJ [24], and be maintained as a non-integrated free plasmid in the yeast cell (unpublished results, CL). To circumvent this problem, we developed TREC-IN, which can efficiently produce gene knock-ins without leaving any scars. Since the TREC-IN method encompasses elements of the TREC method and a split marker system for seamless replacement of nucleotide sequences at any given location on the genome, background issues arising from unwanted ARS activity and A-T rich content are greatly reduced. In the example of *dnaA* gene replacement (Figure 2) TREC-IN was employed to replace the A-T rich *Mmc* gene with an orthologous gene from *Mcc*. However in this case, direct comparison with TREC was not possible because the orthologous genes share a high degree of homology (95%) and the TREC design would not be able to resolve partial recombination occurring between the two genes, as expected. Therefore by using TREC-IN, efficiency of replacement is vastly improved with frequencies of obtaining a positive clone nearing 75% (see Results), thus circumventing the cumbersome screening process of TREC which would be labor and time intensive. Of note, comparing efficiencies between TREC and TREC-IN proves to be complicated as it varies on a case-case basis where A-T content and secondary structure has to be taken into consideration.

Furthermore, TREC-IN can also been extended to delete genes with possible ARS-like activity from the *Mmc* genome, which are very difficult to achieve by the TREC method. For example, we made several attempts to delete the glycerol facilitator (*glpF*) gene from the *Mmc* genome using TREC (unpublished results). We found transformations yielded increased levels of background colonies growing on selective media, without the correct replacement. Yet, when the TREC-IN strategy was applied to delete the *Mmc glpF* gene, all colonies obtained contained the precise and seamless deletion of the *glpF* gene (manuscript in preparation, SC, LM, CL, JJ, RC, SV). In contrast to the TREC method, no background colonies were observed. While TREC-IN depends upon integration of the CORE cassette to the target site, there is some flexibility in choosing the integration sites. Depending on the case, the design can be modified to target sequences that may reside in upstream or downstream adjacent genes if the original target sites prove to be difficult. The adjacent genes can then be restored by including the deleted sequence in the knock-in fragment. The *glpF* deletion described above provides such an example. In this case, the design for the downstream homologous region (Figure 2) was modified to include part of the neighboring*glpK* gene, in order to bypass a specificity issue at the 3′ end of the *glpF* gene. The missing *Mmc glpK* region was complemented in the second step of the TREC-IN method.

TREC has been employed to delete target sites of greater than 70 kb seamlessly in *Mmc* (unpublished results), and since TREC-IN utilizes elements of the TREC, it can be speculated that TREC-IN can also be used to seamlessly delete similarly large nucleotide tracts. Although analysis of knock-in sequences larger than 3 kb has not been carried out, it is theoretically possible that TREC-IN would be able to handle larger fragments, possibly with lower efficiency in trying to complete homologous recombination. In summary, TREC-IN proved to be useful in modifying regions of the genome that tend to be difficult to engineer either due to high A-T content or ARS activity, where TREC or other conventional yeast genetic tools maybe limiting.

CONCLUSION

The TREC-IN method proves to be a powerful genetic tool for manipulating mycoplasma genome. In addition to finding applications in our top-down genome minimization of *Mmc* (Figure 4), this method can be employed to explore homologous complementation studies in other related organisms, including *M. leachii*, and *M. putrefaciens* efficiently without the cumbersome screening process that would be required by TREC alone. By using TREC-IN to manipulate metabolic pathways, pathogenic and virulence factors may be studied with relative ease; thereby facilitating better vaccine design against some of the economically devastating livestock diseases such as contagious bovine pleuropneumonia caused by mycoplasmas [25]. In our studies of mycoplasma biology, TREC and TREC-IN dramatically increased our ability to manipulate the genomes of genetically intractable bacteria. As synthetic genomics techniques are extended to other bacteria that are difficult to manipulate genetically, TREC and TREC-IN will become even more valuable as tools for engineering bacterial genomes cloned as yeast centromeric plasmids.

METHODS

Yeast Strain and Media

The yeast *Saccharomyces cerevisiae, strain* VL6-48 (*MATahis3-Δ200 trp1-Δ1 KlURA3-Δ1 lys2 ade2-101 met14*) containing either the 1.08 Mb genome of *Mycoplasma mycoides* subspecies *capri* (*Mmc*) with a yeast centromeric plasmid (YCP) [21], or the synthetic *Mmc* genome (*Mmc* Syn1) [19] were employed. Yeast cells were grown and maintained in either the synthetic minimal medium containing dextrose (SD) [21], or the standard rich medium containing glucose (YPD) or galactose (YPG) [22]. SD medium was supplemented with 5-fluoroorotic acid (5-FOA), for KlURA3 counter-selection [21, 26].

Preparation of Mutagenesis Cassettes

A. Construction of pCORE6 Plasmid

The pCORE6 plasmid (GenBank accession number KP282615) was constructed by cloning the 5' region of the kanamycin resistance gene along with its promoter (5' *KanMX*, 1-859 bp) into the previously constructed pCORE3 plasmid (unpublished) at the EcoR I site (Additional file 1: Figure S1). More precisely, 5' *kanMX* was amplified from the previously described pFA6a-KanMX plasmid [27] using primers, RCO858 (CAG*GAATTC*GACATGGAGGCCCAGAATAC) and RCO859 (ATC*GAATTC*GGCCAGCCATTACGCTCGT), containing the EcoR I restriction site (*GAATTC*) at each extremity. The pCORE3 plasmid, which includes a 14 bp incomplete I-SceI binding site (white box), a Gal1 promoter, an I-SceI gene, and yeast*KlURA3* prototrophic gene (gray boxes), was linearized with EcoR I. This plasmid also contains a *HIS3* gene and can be selected for histidine autotrophy (Additional file 1: Figure S1).

The pCORE3 plasmid and the 5' *kanMX* amplified product were then ligated to form pCORE6 (Additional file 1: Figures S1 and S4). In these constructions, the I-SceI restriction site is maintained in a truncated form (**GATAACAGGGTAAT**) (white bar) because leaky expression of the I-SceI endonuclease (if the plasmid is propagated in *Escherichia coli*), would result in cleavage of the pCORE3 and pCORE6 plasmids at the I-SceI site. Therefore, four additional nucleotide sequences (tagg) must be added to restore the complete 18 bp I-SceI site during amplification of the CORE6 knock-out cassette (black bar) (Figure 2A,B) (see below).

B. Preparation of Mutagenesis Cassette for the Replacement of Mmc dnaAgene by Mcc dnaAgene in Mmc Genome

A modified version of the CORE cassette described previously [22] was constructed as follows. Briefly, the CORE6 cassette includes an 18 bp I-SceI binding site (black bar), followed by a *GAL1* promoter, I-SceI endonuclease gene, and *KlURA3* gene (gray boxes). The CORE6 contains an additional sequence, which includes the 5' region of the kanamycin resistance gene component (5' *kanMX*) [Promoter for the Translation Elongation Factor (PTEF) followed by the 5' kanamycin resistance gene sequence (1 to 859 bp)], which forms part of the split marker system (Figure 2, Additional file 1: Figures S1 and S4). The CORE6 cassette was amplified by polymerase chain reaction (PCR) from the plasmid pCORE6 (Figure 2B) using the chimeric primers, CORE6-F (*GTT TTC CAC ATT TTT AAC AAGTGT TTA ACT ATA ATATTT TTG GAG ACA AAT* tag **GATAA CAG GGT AAT**ACG GAT TA) and CORE6-R-modWT

(*GTT AAT TTGTGG ATA ACT GTT AAT AAG TTA GGTTTA AAT AGCTAT TTT TAG* GCC AGC CAT TAC GCT CGT) (Figure 3A, step 1). The chimeric primers contained about 51 bp homology (italicized), upstream (CORE6-F) or downstream (CORE6-R-modWT) to the target site on the *Mmc* genome (Additional file 1: Figure S3).

In addition, a second cassette called the knock-in module carrying the 3′ *kanMX* gene component (3′ kanamycin resistance gene sequence (610 to 1357 bp) along with the terminator TEF), a repeat sequence (51 bp homology to the upstream target site of the modified CORE6 knock-out cassette), and the replacement orthologous *Mcc dnaA* gene was constructed in a two-step process (Figure 3A, step 2). Two overlapping PCR amplicons were produced in the first step and assembled in a second step as described below (Figures 2C and 3A). In the first step, the 3′region of the *kanMX* gene component containing a 250 bp overlapping region corresponding to the 5′ sequence of *kanMX* gene component in the CORE6 cassette was generated by PCR using the plasmid pFA6a-kanMX_AJ002680 [27] as the DNA template along with chimeric primers, 3′Kanoverlap+5′Kan-infusion-F1 (CTG ATG ATG CAT GGT TAC TCA CC) and 3′Kan-repeat-infusion-R1 (ttc *at A TTT GTC TCC AAA AAT ATT ATA GTT AAA CAC TTG TTA AAA ATG TGG AAA AC C AGT ATA GCG ACC AGC ATT C*). The chimeric primer, 3′Kan-repeat-infusion-R1 included a 51 bp repeat sequence (italicized), which is complementary to the upstream sequence from the target site on *Mmc* (*GTT TTC CAC ATT TTT AAC AAG TGT TTA ACT ATA ATA TTT TTG GAG ACA AAT*). Similarly, another PCR fragment containing the orthologous *Mcc dnaA* gene was generated using chimeric primers, repeat-MccdanA-infusion-F2 (*CTA TAC TG G TTT TCC ACA TTT TTA ACA AGT GTT TAA CTA TAA TAT TTT TGG AGA CAA A*ta tga acc taa acg ata ttt taa aag) and Mcc-R1-mod (*GTT AAT TTGTGG ATA ACT GTT AAT AAg tTA GGTTTA AAT AGCTAT TTT TA*T TAT TTT GTT AAA ATT TTATTCTTT AAA ATATCA ACA GTC), and the *Mcc* genomic DNA as template (Figures 2C and 3A). The chimeric primer, repeat-MccdnaA-infusion-F2 included 51 bases complementary to the primer, 3′Kan-repeat-infusion-R1 to create the overlap between the two amplicons. The chimeric primer Mcc-R1-mod included a 50 bp homology to the downstream target site on the *Mmc* genome. In the second step, the linear knock-in module was finally assembled by a PCR-based fusion technique [28] of the two individually synthesized PCR products, the 3′ kanamycin amplicon including the 51 bp repeat, and the replacement *Mcc dnaA* amplicon also carrying the 51 bp repeat sequence (Figures 2C and 3A). All primers were synthesized by Integrated DNA Technologies (Coralville, IA, USA).

C. Preparation of Mutagenesis Cassette for Cluster Deletion and Complementation in the Synthetic Mmcgenome

The CORE6 cassette was PCR-amplified using the plasmid pCORE6 as template, and with chimeric primers D0152/162-F (*AAA ATA AAA ATT CTC TAT AAA ATA TAT TTT GTA AAC TAG AAA GGA AAA GA* T AGG GAT AAC AGG GTA ATA CGG ATT AG) and D0152/162-R (*TTT TTA TTA AAA TAT TTT AAT TAA ATT CAT TAT ATT AAA AGG ATA AAT AA* G GCC AGC CAT TAC GCT CG) (Figure 4A, step 1). In order to introduce the 50 bp repeat sequence (italicized) (*AAA ATA AAA ATT CTC TAT AAA ATA TAT TTT GTA AAC TAG AAA GGA AAA GA)* to the knock-in module, the 3 '*kanMX* gene component was amplified by two rounds of PCR. In the first round, PCR was performed for 18 cycles using the plasmid pFA6a-kanMX_AJ002680 [27] (Figure 4A, step 2) as the DNA template along with primers, 3'Kan-F (CTG ATG ATG CAT GGT TAC TC) and 3' Kan-0158-R1 (TCT AGT TTA CAA AAT ATA TTT TAT AGA GAA TTT TTA TTT TCA GTA TAG CGA CCA GCA TT) to generate a 788 bp amplicon. The second round of PCR was conducted for 22 cycles using the 788 bp PCR product as the DNA template along with primers, 3'Kan-F and 3' Kan-0158-R2 (TTA TTA ATT AAT AAG GAG TAA ATC TTT TCC TTT CTA GTT TAC AAA ATA TAT TTT ATA GA) to generate a 820 bp PCR product where the 50 bp homology (underlined) to the upstream target site was incorporated right after the 3' *kanMX*gene component. The knock-in gene *Mmc ssrA* gene (679 bp) was amplified by PCR using the synthetic *Mmc* genome (*Mmc*Syn1) [19] as DNA template along with primers, 0158-F (TAT ATT TTG TAA ACT AGA AAG GAA AAG ATT TAC TCC TTA TTA ATT AAT AAT AAC AA) and 0158-R (TTT TTA TTA AAA TAT TTT AAT TAA ATT CAT TAT ATT AAA AGG ATA AAT AAA CTA ATC AAT CCT AAT AAA TAC TTA G). A final knock-in module (1,527 bp) consisting of the 3' *kanMX* gene component, the 50 bp repeat sequence, and the *ssrA* gene was assembled by Gibson Assembly method [29]. All primers were synthesized by Integrated DNA Technologies (Coralville, IA, USA).

Transformation and PCR Analysis

Transformation of the modified CORE6 cassette or the knock-in module was performed with lithium acetate as described previously [30]. In all experiments, about 1μg of DNA construct and 25μg of salmon sperm carrier DNA (Sigma, Saint Louis, MO) were used. Transformed yeast were plated on appropriate selection media and incubated at 30°C for 48 hours. Based on the markers present in the DNA cassette and the mycoplasma genome, transformed yeast cells were selected on SD medium minus His (Teknova, CA), SD medium minus His and minus Ura, or YPD containing 0.2 mg/ml geneticin after a period of recovery in YPD (Figure 2).

Yeast colonies growing on selective media were re-streaked and total DNA was isolated for PCR screening [31]. The correct integration of each mutagenesis cassette was verified by PCR screening using diagnostic primers located upstream and downstream of the target sites (Figures 3 and 4 and Additional file 1: Figure S3). All primers were synthesized by Integrated DNA Technologies (Coralville, IA, USA).

Transplantation

Total DNA, including the intact donor genomic DNA from yeast colonies were isolated using a CHEF Mammalian Genomic DNA Plug Kit as per the manufacturer's instructions (Bio-Rad, Hercules, CA). DNA isolated from yeast cells carrying the *Mmc* modified genome was transplanted into *Mcc* recipient cells with polyethylene glycol as described previously [21, 31]. The transplanted cells were selected for tetracycline resistance (the *tetM* gene and the β-galactosidase genes (*lacZ*) being present on the *Mmc* chromosome). *Mmc* genomic DNA containing the *Mcc dnaA* gene was isolated from the transplants using the BioRobot M48 workstation (Qiagen, Valencia, CA) as per the manufacturer's instructions. The isolated *Mmc* genomic DNA from the bacteria transplants was sequenced to confirm the precise, seamless insertion of the *Mcc dnaA* gene (JCVI Sequencing Facility, MD).

ABBREVIATIONS

ARS: Autonomously replicating sequence

DSB: Double strand break

Gal: Galactose promoter

His: Histidine

KanMX : Kanamycin resistance gene module

lacZ : β-galactosidase genes

Mb: Mega base

Mcc : *Mycoplasma capricolum* subspecies *capricolum*

Mmc : *Mycoplasma mycoides* subspecies *capri*

Mmc Syn1 : Synthetic *Mmc* genome

Mmm : *Mycoplasma mycoides* subspecies *mycoides*

NEGC: Non-essential gene clusters

NHEJ: Non-homologous end Joining

PCR: Polymerase chain reaction

PTEF: Promoter for the translation elongation factor

SD: Synthetic minimal medium containing Dextrose

TEF: Terminator

TetM : Tetracycline resistance

TREC: Tandem repeat coupled with Endonuclease cleavage

TREC-IN: TREC-assisted gene knock-in

Ura: Uracil

YCP: Yeast Centromeric Plasmid

YFD: Rich medium containing glucose

YPG: Rich medium containing Galactose.

ACKNOWLEDGEMENTS

We thank Eva Albalghiti for her contribution towards screening of yeast colonies, and Nacyra Asad-Garcia for the transplantation work. We also thank Dr. Alain Blanchard and Dr. John Glass for useful discussion and comments on the paper. Additional support for Dr. Joerg Jores was received from the CGIAR research program on Livestockand Fish.

FUNDING

This work was supported in part by the National Science Foundation [grant number IOS-1110151 (to S.V., C.L., and J.J.)], DARPA Contract # HR0011-12-C-0063, and Synthetic Genomics, Inc. Funding for open access charge: Synthetic Genomics, Inc.

ELECTRONIC SUPPLEMENTARY MATERIAL

Additional File 1

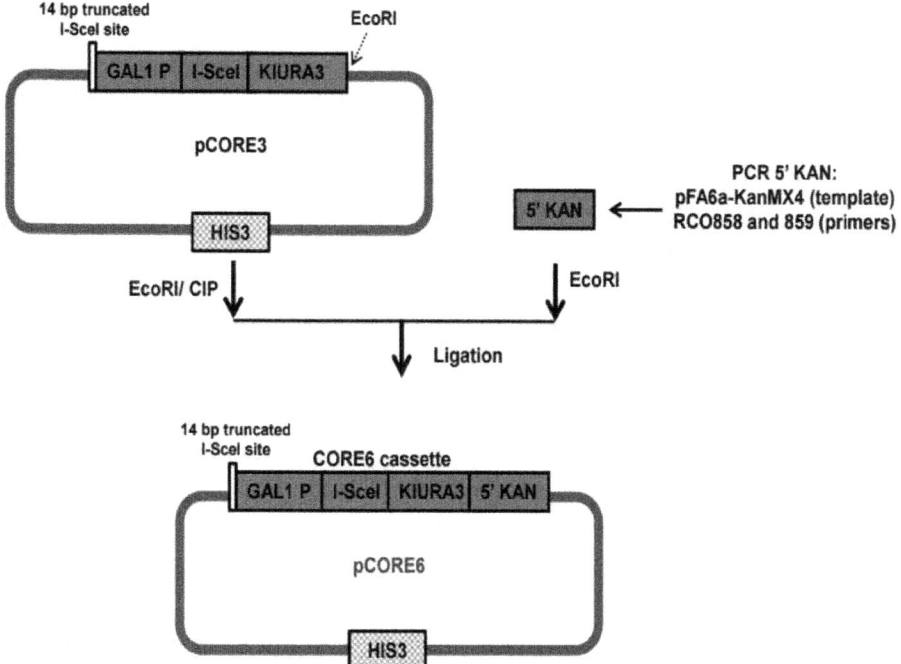

Figure S1: Construction of pCORE6 plasmid. The pCORE6 plasmid was constructed from the previously constructed pCORE3 plasmid (unpublished), and the 5' region of the kanamycin resistance gene (5' *KanMX* gene component) along with its promoter, PTEF, which was amplified from the previously described pFA6a-KanMX plasmid. The pCORE3 plasmid includes a 14 bp incomplete I-SceI binding site (white bar), a Gal1 promoter, an I-SceI restriction enzyme gene, and yeast KlURA3 prototrophic gene (gray boxes). The plasmid can be selected for HIS3 prototrophy. The pCORE6 also contains a 14 bp incomplete I-SceI site instead of the 18 bp complete sequence for stability reasons, and an additional 4 bp (TAGG) must be added on during PCR for generation of the complete CORE6 knock-out cassette.

Size (bp)	Gene No.	Annotation
1864	Mmc Syn1_0152	PTS system glucose-specific eiic ba component
2269	Mmc Syn1_0153	glycosyl hydrolase, family 31
1354	Mmc Syn1_0154	leucyl aminopeptidase
601	Mmc Syn1_0155	conserved hypothetical protein
664	Mmc Syn1_0156	tRNA (guanine-N(7)-)-methyltransferase
1402	Mmc Syn1_0157	magnesium transporter
409	Mmc Syn1_0158	ssrA
2128	Mmc Syn1_0159	putative liporotein
751	Mmc Syn1_0160	abortive infection protein AbiGII
595	Mmc Syn1_0161	abortive infection protein AbiGI
1205	Mmc Syn1_0162	putative lipoprotein

Figure S2. Genes in the two non-essential gene clusters (NEGCs) separated by the Tn5-defined essential gene, *ssrA* in the *Mmc* synthetic genome (*Mmc* Syn1). Genes 0152 – 0157 belong to the first NEGC, while genes 0159–0162 belong to the second NEGC. Gene 0158 is the essential *ssrA* gene that is present between the two NEGCs.

Primer	Primer sequence (5' → 3')	Notes
DG-1F	TAG TTA ATT GTT GAT AAG TTG AT	Diagnostic PCR product (222bp) with primer, DG1-R, diagnostic PCR product (1733bp) with primer, DG2-R, and diagnostic PCR product (338bp) with primer, DG5-R.
DG-1R	GAG TCT TCC TTC GGA GG	Diagnostic PCR product (222bp) with primer, DG1-F.
DG-2F	GCA AAA CAG CAT TCC AGG	Diagnostic PCR product (438bp) with primer, DG2-R.
DG-2R	GGT CAA TTA CTT TAG CTG CTT TTG	Diagnostic PCR product (438bp) with primer, DG2-F, and diagnostic PCR product (1733bp) with primer, DG1-F.
DG-3F	GGA AAG GCT AGA AGT AAA TCA ATT G	Diagnostic PCR product (446bp) with primer, DG2-R.
DG-4R	CAC CAT GAG TGA CGA CTG	Diagnostic PCR product (285bp) with primer, DG2-F.
DG-5R	GTT GGA CGA AAT TGT TTT ATA G	Diagnostic PCR product (615bp) with primer, DG6-F, and diagnostic PCR product (338bp) with primer, DG1-F.
DG-6F	GTT GTA TTG ATG TTG GAC G	Diagnostic PCR product (615bp) with primer, DG5-R.
DG-7F	GTAATCGGTTCAAGTAATGAAC AAG	For additional sequencing.
D0152/162-DGF	AAAATAAAAATTCTCTATAAAAT ATATTTTGTAAACTAGAAAGGA AAAGATAGGGATAACAGGGTA ATACGGATTAG	Diagnostic PCR product (230bp) with primer, RCO421, and diagnostic PCR product (1005bp) with primer, D0152/162-DGR.
D0152/162-DGR	TTTTTATTAAAATATTTTAATTA AATTCATTATATTAAAAGGATA AATAAGGCCAGCCATTACGCT CG	Diagnostic PCR product (286bp) with primer, RCO862, and diagnostic PCR product (1005bp) with primer, D0152/162-DGF.
0158-DGF	TATATTTTGTAAACTAGAAAGG AAAAGATTTACTCCTTATTAATT AATAATAACAA	Diagnostic PCR product (283bp) with primer, D0152/162-DGR.
0158-DGR	TTTTTATTAAAATATTTTAATTA AATTCAT	Diagnostic PCR product (239bp) with primer, D0152/162-DGF.
RCO421	CTTCGGAGGGCTGTCACC	Diagnostic PCR product (230bp) with primer, D0152/162-DGF.
RCO862	GTTGCATTCGATTCCTGTTTG	Diagnostic PCR product (286bp) with primer, D0152/162-DGR.

Figure S3. Diagnostic primers to confirm the correct insertion of the CORE6 knock-out cassette and knock-in cassette by TREC-IN in the *Mmc* genome. Diagnostic primers to assess for the correct junctions and precise insertion of the replaced *Mcc* ort hologous *dnaA* gene, and the essential *ssrA* gene in the *Mmc* genome are listed.

```
TCGCGCGTTTCGGTGATGACGGTGAAAACCTCTGACACATGCAGCTCCCGGAGACGGTCACAGCTTGTCTGTAAGCGGATGCCGGGAGCAGACAAGCCCGTCAGGGCGCGTCAGCGGGTGTTGGCGGGTGTCGGGG
CTGGCTTAACTATGCGGCATCACAAGCACGATTGTACTGAAGAGTGCACCATAATTCCGTTTTAAGAGCTTGGTGAGCGCTAGGAGTCACTGCCAGGTATCGTTTGAACACGGAGAGTTAGTCAGGGAAGTCATAACACAGTCCT
TTCCGGCAATTTTCTTTTTCTATTACTCTTGGCGTCCTCTAGTACAGCTCTATATTTTTTTATGCGTCGGTAATGATTTTCATTTTTTTTTTTTCCAGCTACGCGGATGACTCTTTTTTTTCTTAGCGATTGGCATTATCAGATAATGA
ATTATACATTATATAAAGTAATGTGATTTCTTCGAAGAATATACTAAAAATGAGCAGGCAAGATAAACGAAGGCAAAGATGACAGAAGCCCTAGTAAAGCGTATTACAAATGAAACCAAGATTCAGATTGCGATCT
CTTTAAAGGGTGGTCCCCTAGAGATAGAGGACTCGATCTTCCCAGAAAAAGAGGCAGAAAGCTAGCGAAGCCCACACAATCGCAAGTGATTAAACGTCCACACAGGTATAGGGTTTCTGGACCATATGATACATGCT
CTGGCCAAGCATTCCGGGCTGGTCGCTAATCGTTGAGTGCATTGGTGACTTACACATAGAACGACCATCACACCACTGAAGACTGCGGGATTGCTCTCGGTCAAGCTTTTAAAGAGGCCCTACTGGCGCGTGGAGTAAAAA
GGTTTGGATCAGGATTTGCGGCTTTGGATGAGGCACTTTCCAGAACGCGTGGTAGATCTTTCGAACAGGCGTGTCGAACTTGGTTTGCAAAGGGACAAAAGTACGAGATCTCTCTTGCCAGATGATCCGCGCA
TTTTCTTGAAAGCTTTGCAGAGGCTAGCAGAATTACCCTCCACGTTGATTGTCTGCGAGGCAAGAATGATCATCACCGTAGTGAGAGTGCGTTCAACGCTCTTGCGGTTGCCATAAGAGAGCCACCTCGCCCAATGGTA
CCAACGATGTTCCCTCCACCAAAGGTGTTCTTATGTAGTGACACGCGATTATTTAAAGCTGCAGCATACGATATATATCAGTGTGTATATATGTATACCTATGAATGTCAGTAAGTATGTATACGAACAGTATGATACTGAAGAT
GACAAGGTAATGCATCATCTCTATACGTGTCATTCTGAACGAAGGCGCGCTTTCCTTTTTCTCTTTTTTGCTTTTTCTTTTTTTTTCTCTTGAAACTCGATCGATCATATGCGGTGTGAAATACCGCACAGATGCGTAAGGAAGAAAAT
```

Figure S4. pCORE6 sequence. The CORE6 knock-out cassette (GenBank accession number KP282615) is color-coded as follows: the 14 bp incomplete I-SceI binding site (red), Gal1 promoter (dark green), I-SceI endonuclease (orange), KlURA3 gene along with its promoter and terminator (blue), and the promoter for the translation elongation factor (PTEF) (yellow) followed by the 5′ region of the kanamycin resistance gene (purple). (PDF 251 KB)

AUTHORS' CONTRIBUTIONS

SC performed the experiments related to the replacement of the *Mcc* orthologous *dnaA* gene in the *Mmc* genome and wrote the manuscript. VN helped with the design of the TREC-IN method and carried out preliminary experiments. TS-S initially performed the experiments. LM performed the experiments related to the application of TREC-IN for genome reduction in the *Mmc* genome. CW performed the replacement experiments. CL, JJ, and SV participated in the design of the study and helped to draft the manuscript. R-YC conceived and designed the TREC-IN method and helped to draft the manuscript. All authors read and approved the final manuscript.

REFERENCES

1. Dybvig K: Mycoplasmal genetics. *Annu Rev Microbiol* 1990, 44:81–104. 10.1146/annurev.mi.44.100190.000501

2. Razin S, Yogev D, Naot Y: Molecular biology and pathogenicity of mycoplasmas. *Microbiol Mol Biol Rev* 1998,62(4):1094–1156.

3. Halbedel S, Stulke J: Tools for the genetic analysis of Mycoplasma. *Int J Med Microbiol* 2007,297(1):37–44. 10.1016/j.ijmm.2006.11.001

4. Cordova CM, Lartigue C, Sirand-Pugnet P, Renaudin J, Cunha RA, Blanchard A: Identification of the origin of replication of the Mycoplasma pulmonis chromosome and its use in oriC replicative plasmids. *J Bacteriol* 2002,184(19):5426–5435. 10.1128/JB.184.19.5426-5435.2002

5. Lartigue C, Blanchard A, Renaudin J, Thiaucourt F, Sirand-Pugnet P: Host specificity of mollicutes oriC plasmids: functional analysis of replication origin. *Nucleic Acids Res* 2003,31(22):6610–6618. 10.1093/nar/gkg848

6. Renaudin J, Marais A, Verdin E, Duret S, Foissac X, Laigret F, Bove JM: Integrative and free Spiroplasma citri oriC plasmids: expression of the Spiroplasma phoeniceum spiralin in Spiroplasma citri. *J Bacteriol* 1995,177(10):2870–2877.

7. Janis C, Lartigue C, Frey J, Wroblewski H, Thiaucourt F, Blanchard A, Sirand-Pugnet P: Versatile use of oriC plasmids for functional genomics of Mycoplasma capricolum subsp. capricolum. *Appl Environ Microbiol* 2005,71(6):2888–2893. 10.1128/AEM.71.6.2888-2893.2005

8. Lee SW, Browning GF, Markham PF: Development of a replicable oriC plasmid for Mycoplasma gallisepticum and Mycoplasma imitans, and gene disruption through homologous recombination in M. gallisepticum. *Microbiology* 2008,154(Pt 9):2571–2580.

9. Janis C, Bischof D, Gourgues G, Frey J, Blanchard A, Sirand-Pugnet P: Unmarked insertional mutagenesis in the bovine pathogen Mycoplasma mycoides subsp. mycoides SC: characterization of a lppQ mutant. *Microbiology* 2008,154(Pt 8):2427–2436.

10. Hutchison CA, Peterson SN, Gill SR, Cline RT, White O, Fraser CM, Smith HO, Venter JC: Global transposon mutagenesis and a minimal Mycoplasma genome. *Science* 1999,286(5447):2165–2169. 10.1126/science.286.5447.2165

11. Glass JI, Assad-Garcia N, Alperovich N, Yooseph S, Lewis MR, Maruf M, Hutchison CA 3rd, Smith HO, Venter JC: Essential genes of a minimal

bacterium. *Proc Natl Acad Sci U S A* 2006,103(2):425–430. 10.1073/pnas.0510013103

12. Voelker LL, Dybvig K: Transposon mutagenesis. *Methods Mol Biol* 1998, 104:235–238.

13. Burgos R, Pich OQ, Querol E, Pinol J: Deletion of the Mycoplasma genitalium MG_217 gene modifies cell gliding behaviour by altering terminal organelle curvature. *Mol Microbiol* 2008,69(4):1029–1040. 10.1111/j.1365-2958.2008.06343.x

14. Dhandayuthapani S, Blaylock MW, Bebear CM, Rasmussen WG, Baseman JB: Peptide methionine sulfoxide reductase (MsrA) is a virulence determinant in Mycoplasma genitalium. *J Bacteriol* 2001,183(19):5645–5650. 10.1128/JB.183.19.5645-5650.2001

15. Dhandayuthapani S, Rasmussen WG, Baseman JB: Disruption of gene mg218 of Mycoplasma genitalium through homologous recombination leads to an adherence-deficient phenotype. *Proc Natl Acad Sci U S A* 1999,96(9):5227–5232. 10.1073/pnas.96.9.5227

16. Allam AB, Reyes L, Assad-Garcia N, Glass JI, Brown MB: Enhancement of targeted homologous recombination in Mycoplasma mycoides subsp. capri by inclusion of heterologous recA. *Appl Environ Microbiol* 2010,76(20):6951–6954. 10.1128/AEM.00056-10

17. Benders GA, Noskov VN, Denisova EA, Lartigue C, Gibson DG, Assad-Garcia N, Chuang RY, Carrera W, Moodie M, Algire MA, Phan Q, Alperovich N, Vashee S, Merryman C, Venter JC, Smith HO, Glass JI, Hutchison CA 3rd: Cloning whole bacterial genomes in yeast.*Nucleic Acids Res* 2010,38(8):2558–2569. 10.1093/nar/gkq119

18. Gibson DG, Benders GA, Andrews-Pfannkoch C, Denisova EA, Baden-Tillson H, Zaveri J, Stockwell TB, Brownley A, Thomas DW, Algire MA, Merryman C, Young L, Noskov VN, Glass JI, Venter JC, Hutchison CA 3rd, Smith HO: Complete chemical synthesis, assembly, and cloning of a Mycoplasma genitalium genome. *Science* 2008,319(5867):1215–1220. 10.1126/science.1151721

19. Gibson DG, Glass JI, Lartigue C, Noskov VN, Chuang RY, Algire MA, Benders GA, Montague MG, Ma L, Moodie MM, Merryman C, Vashee S, Krishnakumar R, Assad-Garcia N, Andrews-Pfannkoch C, Denisova EA, Young L, Qi ZQ, Segall-Shapiro TH, Calvey CH, Parmar PP, Hutchison CA 3rd, Smith HO, Venter JC: Creation of a bacterial cell controlled by a chemically synthesized genome. *Science*2010,329(5987):52–56. 10.1126/science.1190719

20. Lartigue C, Glass JI, Alperovich N, Pieper R, Parmar PP, Hutchison CA 3rd, Smith HO, Venter JC: Genome transplantation in bacteria: changing one species to another. *Science* 2007,317(5838):632–638. 10.1126/science.1144622

21. Lartigue C, Vashee S, Algire MA, Chuang RY, Benders GA, Ma L, Noskov VN, Denisova EA, Gibson DG, Assad-Garcia N, Alperovich N, Thomas DW, Merryman C, Hutchison CA 3rd, Smith HO, Venter JC, Glass JI: Creating bacterial strains from genomes that have been cloned and engineered in yeast. *Science* 2009,325(5948):1693–1696. 10.1126/science.1173759

22. Noskov VN, Segall-Shapiro TH, Chuang RY: Tandem repeat coupled with endonuclease cleavage (TREC): a seamless modification tool for genome engineering in yeast. *Nucleic Acids Res* 2010,38(8):2570–2576. 10.1093/nar/gkq099

23. Akada R, Hirosawa I, Kawahata M, Hoshida H, Nishizawa Y: Sets of integrating plasmids and gene disruption cassettes containing improved counter-selection markers designed for repeated use in budding yeast. *Yeast* 2002,19(5):393–402. 10.1002/yea.841

24. Daley JM, Palmbos PL, Wu D, Wilson TE: Nonhomologous end joining in yeast. *Annu Rev Genet* 2005, 39:431–451. 10.1146/annurev.genet.39.073003.113340

25. Jores J, Mariner JC, Naessens J: Development of an improved vaccine for contagious bovine pleuropneumonia: an African perspective on challenges and proposed actions. *Vet Res* 2013, 44:122. 10.1186/1297-9716-44-122

26. Boeke JD, LaCroute F, Fink GR: A positive selection for mutants lacking orotidine-5′-phosphate decarboxylase activity in yeast: 5-fluoro-orotic acid resistance. *Mol Endocrinol* 1984,197(2):345–346.

27. Wach A, Brachat A, Pohlmann R, Philippsen P: New heterologous modules for classical or PCR-based gene disruptions in Saccharomyces cerevisiae. *Yeast* 1994,10(13):1793–1808. 10.1002/yea.320101310

28. Shevchuk NA, Bryksin AV, Nusinovich YA, Cabello FC, Sutherland M, Ladisch S: Construction of long DNA molecules using long PCR-based fusion of several fragments simultaneously. *Nucleic Acids Res* 2004,32(2):e19. 10.1093/nar/gnh014

29. Gibson DG, Young L, Chuang RY, Venter JC, Hutchison CA 3rd, Smith HO: Enzymatic assembly of DNA molecules up to several hundred kilobases. *Nat Methods* 2009,6(5):343–345. 10.1038/nmeth.1318

30.   Gietz D, St Jean A, Woods RA, Schiestl RH: Improved method for high efficiency transformation of intact yeast cells. *Nucleic Acids Res* 1992,20(6):1425. 10.1093/nar/20.6.1425

31.   Noskov V, Kouprina N, Leem SH, Koriabine M, Barrett JC, Larionov V: A genetic system for direct selection of gene-positive clones during recombinational cloning in yeast. *Nucleic Acids Res* 2002,30(2):E8. 10.1093/nar/30.2.e8

Chapter 6

ENZYMATIC ENGINEERING OF THE PORCINE GENOME WITH TRANSPOSONS AND RECOMBINASES

Karl J Clark[1,2,3], Daniel F Carlson[1,2,3], Linda K Foster[1], Byung-Whi Kong[1], Douglas N Foster[1,3] and Scott C Fahrenkrug[1,2,3]

[1]Department of Animal Science, University of Minnesota, St. Paul, MN, USA

[2]The Arnold and Mabel Beckman Center for Transposon Research, University of Minnesota, Minneapolis, MN, USA

[3]The University of Minnesota Animal Biotechnology Center, University of Minnesota, St. Paul, MN, USA

ABSTRACT

Background

Swine is an important agricultural commodity and biomedical model. Manipulation of the pig genome provides opportunity to improve production efficiency, enhance disease resistance, and add value to swine products. Genetic engineering can also expand the utility of pigs for modeling human disease, developing clinical treatment methodologies, or donating tissues for xenotransplantation. Realizing the full potential of pig genetic engineering requires translation of the complete repertoire of genetic tools currently employed in smaller model organisms to practical use in pigs.

Results

Application of transposon and recombinase technologies for manipulation of the swine genome requires characterization of their activity in pig cells. We tested four transposon systems- *Sleeping Beauty*, *Tol2*, *piggyBac*, and *Passport* in cultured porcine cells. Transposons increased the efficiency of DNA integration up to 28-fold above background and provided for precise delivery of 1 to 15 transgenes per cell. Both Cre and Flp recombinase were functional in pig cells as measured by their ability to remove a positive-negative selection cassette from 16 independent clones and over 20 independent genomic locations. We also demonstrated a Cre-dependent genetic switch capable of eliminating an intervening positive-negative selection cassette and activating GFP expression from episomal and genome-resident transposons.

Conclusion

We have demonstrated for the first time that transposons and recombinases are capable of mobilizing DNA into and out of the porcine genome in a precise and efficient manner. This study provides the basis for developing transposon and recombinase based tools for genetic engineering of the swine genome.

BACKGROUND

Recent developments in livestock transgenesis, including somatic cell nuclear transfer (SCNT, cloning) [1], and stem cell biology [2, 3] have energized plans to engineer the pig genome for both agricultural and emerging biomedical markets. Although pronuclear injection (PNI) and SCNT are proven methods for gene supplementation and gene targeting, respectively, more sophisticated methods for manipulating the pig genome have been lacking. Tandem gene targeting and SCNT provides a method for the precise introduction of transgenes or alternate alleles, but the inherent inefficiency of homologous recombination and donor-cell senescence limits its efficiency. Transgenesis by random integration of naked DNA has proven much more efficient for gene supplementation, whether using PNI or SCNT. However, random integration of naked DNA is often accompanied by transgene instability [4, 5], transgene concatemerization [6, 7], loss of transgene expression due to methylation [8–13], and short deletions, inversions and duplications at the site of transgene integration [14–25]. In addition, the lack of precision associated with random integration of naked DNA limits transgene manipulation and control post-integration.

DNA "cut and paste" transposons have been widely used for precise and efficient delivery of DNA expression cassettes into invertebrate and plant genomes. Over the past ten years, several DNA cut and paste transposon systems have been shown to function in vertebrate cells, including *Sleeping Beauty* (SB) [26, 27], *Passport* (PP) [28, 29], *Tol2* [30, 31], and *piggyBac* (PB) [32–34]. In addition, transposons have been used for germline transgenesis of fish [35–37], frogs [38–40], and mice [32,41–43] and for transgenesis of mouse somatic and embryonic stem cells [44–46]. It is noteworthy that although transposons function in a wide array of cell types, their efficiency can differ from species to species or even within various cell types of one species. The function and efficacy of vertebrate transposons in pig cells had not previously been examined. Demonstration that one or more transposon systems functions efficiently in porcine cells would provide a rationale for investigating their use in PNI and SCNT. In addition, the precision of transpositional transgenesis (TnT) provides a segue to the development of conditional expression systems for application in pigs and porcine cells.

Many genes have roles in multiple tissues and/or at multiple times during growth and development. Due to a requirement for strict regulation, global ectopic transgene-expression or gene-knockout will be an implausible approach for many targets. To overcome these limitations, binary systems based on transcriptional transactivation or DNA recombination have been developed and applied in model organisms for conditional gene-expression or silencing [47]. Although the tetracycline transcriptional activator system [48] has been demonstrated to function in transgenic pigs [49, 50], recombinases have not. Cre and Flp recombinases catalyze a conservative DNA recombination event between two short recombinase recognition sites (RRS), *loxP* and *FRT*, respectively [51]. This results in the deletion or inversion of the DNA between two RRS- depending on their orientation. Deletion or inversion of sequences in transgenes can be used as genetic switches to activate or silence gene expression in specific cells, at particular times, or under prescribed conditions. Applications beyond conditional gene expression include the removal/recycling of selectable markers or transgenes [52] or chromosome engineering [53]. The successful application of recombinase technologies to porcine genetics requires the demonstration of Cre and/or Flp activity in porcine cells and the efficient delivery of RRS sites and recombinase-based expression vectors to the porcine genome.

In order to assess the utility of DNA transposons and recombinases for enzymatic engineering of the porcine genome, we tested four transposon systems and two recombinases. The SB, PP, *Tol2*, and PB transposon systems were able to function in cells derived from pig tissues and significantly improved the rate of transgenesis *in vitro*. Cre and Flp recombinases were capable of removing antibiotic selection cassettes in porcine cells and conditionally activating transgenes in porcine cells, demonstrating the potential for their applications to "leave no trace" and/or conditional porcine genetic engineering.

RESULTS

Sleeping Beauty Activity in Porcine Cells

To test the ability of the SB transposon systems to mediate transposition into the porcine genome, a transposon vector (pT2-FloxP-PTK) and a transposase expression vector (pKUb-SB11) were constructed (Fig 1A). The transposon vector encodes a puromycin-thymidine kinase (PuroΔTK, PTK) fusion protein [54] between the inverted repeats of the SB transposon system. The PTK cassette was flanked by both *FRT* and *loxP* sites so that it could be used as a substrate for testing both Cre and Flp recombinases (see below). Pig fetal fibroblasts (PFF) or porcine endometrial gland epithelium (PEGE) cells were transfected with the PTK transposon along with the SB expression vector, a vector encoding non-functional SB (pKUb-SBΔDDE), or a β-galactosidase expression vector (pCMV-β). After the transfection period, cells with integrations were rendered resistant to puromycin selection, and formed clonal cell colonies after 9–12 days. Clones were stained with methylene blue and quantified (Fig. 1B). The transposase catalyzed 2.5× (PFF) -10× (PEGE) more colony formation versus transfection with a non-functional transposase (ΔDDE) or β-galactosidase. This difference in the rate of clone formation corresponds to TnT versus the background rate of non-transpositional transgenesis.

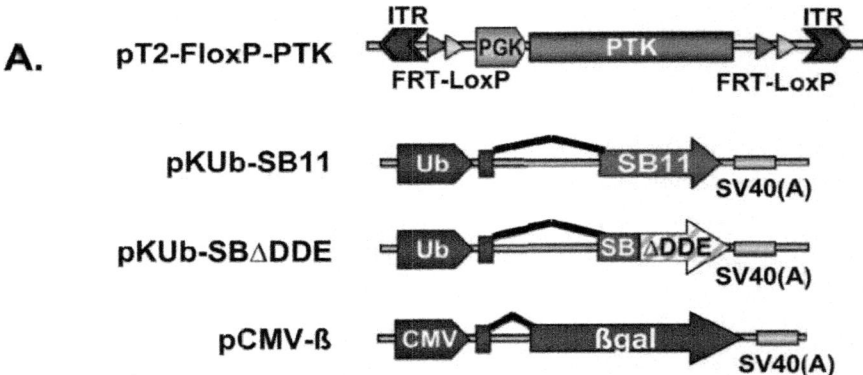

B.

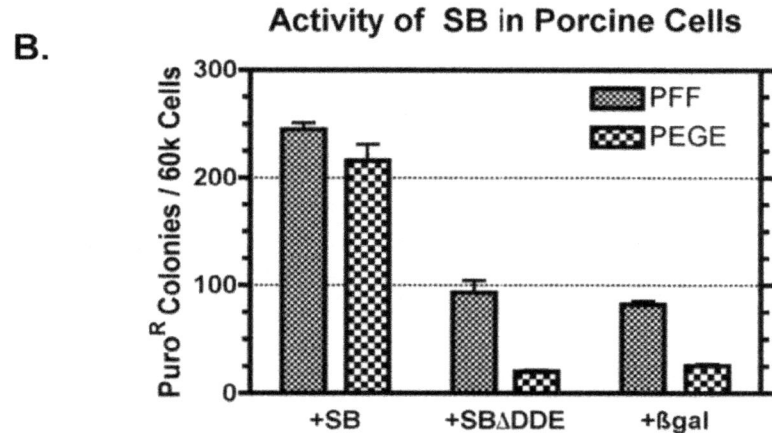

Figure 1: *Sleeping Beauty* function in pig cells. A) Diagrams of the DNA vectors transfected into pig cells. pT2-FloxP-PTK is the experimental SB transposon. The transposon is flanked by inverted terminal repeats (ITR). The puromycin phosphotransferase-thymidine kinase fusion protein (PTK) is flanked by recombinase recognition sites, *FRT* and *loxP*, for Flp and Cre, respectively. pKUb-SB11 is the source of transposase and is expressed from the ubiquitin promoter (Ub). pKUb-SBΔDDE is a non-functional version of transposase because of an internal deletion within the catalytic domain. pCMV-β functions as negative control. B) The colony forming ability of pT2-FloxP-PTK in pig fetal fibroblast (PFF) and porcine endometrial gland epithelium (PEGE) was determined by counting puromycin resistant colonies after plating 60,000 cells on 10 cm dishes when pT2-FloxP-PTK was co-transfected with pKUb-SB11 (+SB), pKUb-SBΔDDE (+SBΔDDE), or pCMV-β (+βgal). The addition of functional transposase (+SB) versus a non-functional transposase (SBΔDDE) or pCMV-β (Bgal) was determined to be significant by analysis with an unpaired t-test (p-values < 0.000002).

Multiple Transposon Systems Function in Porcine Cells

The success of the SB transposon system prompted investigation of three additional transposon systems. In addition to retesting the SB transposon system in PEGE cells, we also tested PP (an additional member of the Tc1 transposon family [55]), *Tol2* (a member of the hAT transposon family [56]), and PB, the founding member of the *piggyBac* transposon family [57]). PEGE cells are one of a few immortalized pig cell lines available, transfect consistently (8–15%), and form tight non-migrating clonal colonies- essential characteristics for the colony forming assays performed. The PTK expression cassette was placed between inverted repeats corresponding to each transposon; pKT2P-PTK, pPTnP-PTK, pGTol2P-PTK, and pPBT-PTK, respectively

(Fig. 2A). PEGE cells were co-transfected with each of these transposons along with their corresponding transposase expression construct; pKUb-SB11, pKC-PTs1, pCMV-Tol2, or pKC-PB, respectively. Each transposon vector was also co-transfected with pCMV-β to determine the background rate of non-transpositional integration. Transfected PEGE cells were placed under puromycin selection for 9–12 days, colonies fixed, stained, and enumerated. Again, transfection of PEGE cells with both components of the SB system (Fig. 2B) resulted in over 200 colonies per 60,000 plated cells, or about 3.3% of transfected cells based on an average 10% transfection efficiency. This represented a 13.5-fold increase over transfection without transposase. Similar enhancements to transgenesis were seen for all the transposon systems. PP produced an average of over 100 colonies per 60,000 cells; a 5-fold increase over transfection without transposase (Fig 2C). The inclusion of *Tol2* transposase resulted in the generation of puromycin resistant colonies at a rate 21-fold over transfections without transposase (Fig. 2D), producing on average over 240 colonies per 60,000 cells. The PB transposon system (Fig. 2E) yielded an average of over 320 colonies per 60,000 cells (about 5% of transfected cells), representing a 28-fold increase over transfection without transposase.

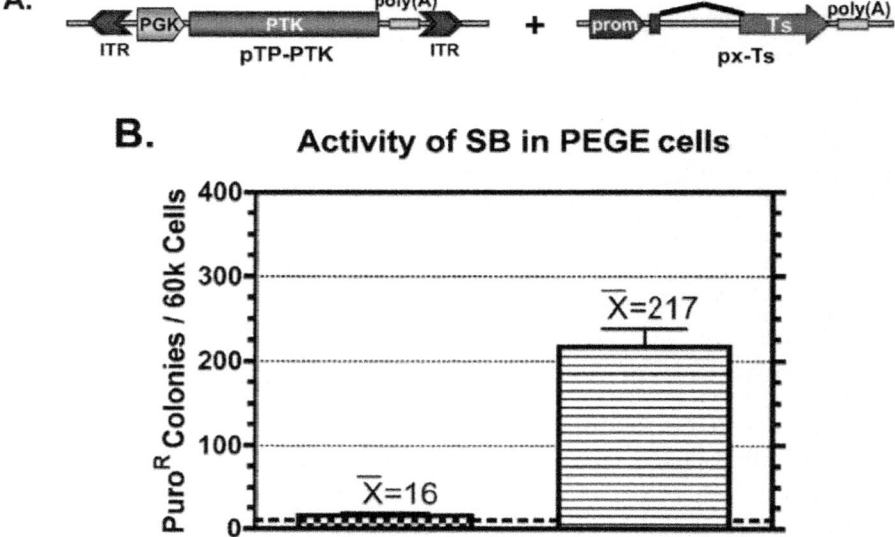

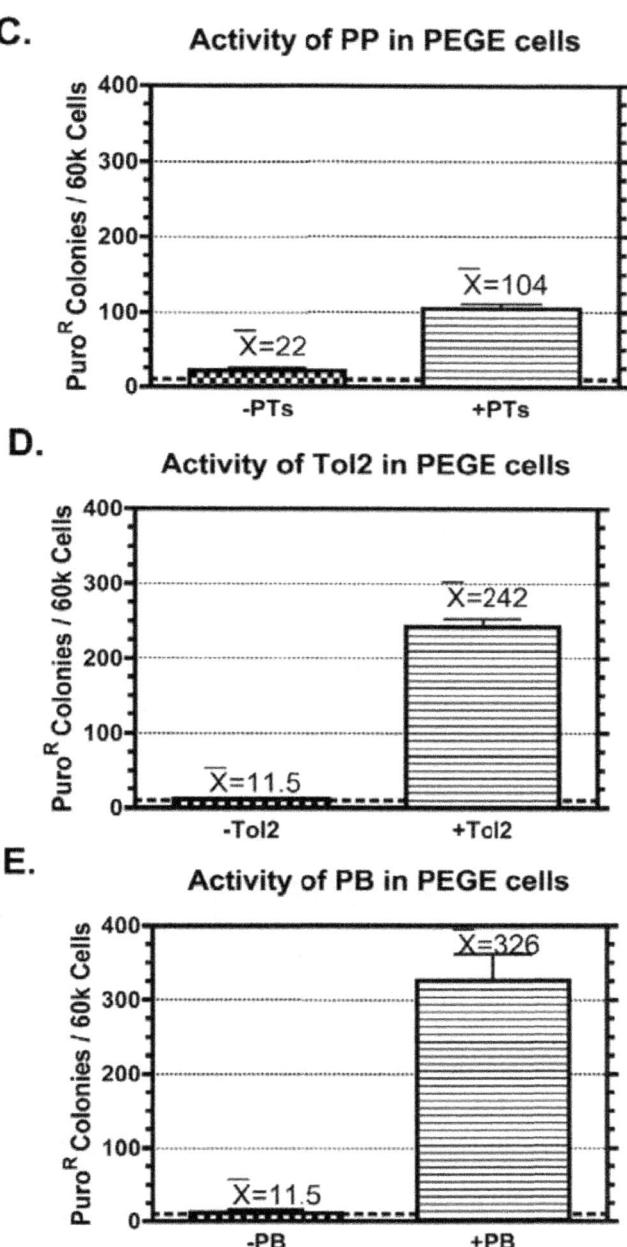

Figure 2: Activity of multiple transposon systems in PEGE cells. A) A drawing of a generic transposon (pTP-PTK) used for colony formation assays. The transposons used, except the transposon-specific inverted terminal repeats, are identical. The vector backbones of the transposons are also identical ex-

cept for pGTol2P-PTK. The pKx-Ts drawing is a generic representation of the transposase-expressing vector. The promoter choices include Ub, CMV, and mCAGs for SB, *Tol2*, and PB and PP, respectively. The vector backbones and poly(A) signals are identical except for pCMV-Tol2 B-E) The number of colonies formed with SB, PP, *Tol2*, or PB PTK transposons are shown with βgal instead of transposase (-Ts) and with transposase (+Ts), where Ts is SB, PP, *Tol2*, or PB. In each case, the significance of transposase was verified with an unpaired t-test (p-values ≤ 0.00002).

Molecular Characterization of Transposition

Integration of DNA transposons produces target-site duplications upon integration into the genome. Analogous to SB and other Tc1 type transposons, the target site preference for PP is a TA dinucleotide. Target-site preference for the PB transposon is a TTAA tetranucleotide [33]. Integration of *Tol2* results in a target-site duplication of eight bases but does not rely on specific primary sequence, instead targeting a characteristic local deformation of DNA [58]. Blocked linker-mediated PCR was used to clone junction fragments after transfection of PEGE cells with each transposon system. Characteristic integration footprints were observed for each transposon system (Fig 3). Junction sequences were compared to sequences in GenBank using BLAST [59]. Despite the small amount of contemporaneous porcine genome sequence available, some flanking DNAs of each transposon system were found to have high identity to the pig genome, in most cases in abundant repetitive elements. This demonstrates *bona fide* transposition into the porcine genome for each transposon class.

```
gctggatccagatccta TA CAGTTGAAGT
ATTGATATATAATTCACA TA CAGTTGAAGT
TCAATCATCACACTATGG TA CAGTTGAAGT    SB
ATATTACACAAGATATAT TA CAGTTGAAGT
GTAATGTTCCATTGTGTA TA CAGTTGGAGT
ACAAACAAGAACCACTAC TA CAGTTGAAGT
```

```
CAAGGCACTG TA atcggtaccatttaaatc
CAAGGCACTG TA CTTGGGCAAGATGCTTAA
CAAGGCACTG TA TATCCTAATGCCTAGAGA    PP
CAAGGCACTG TA ATCGGTACCCATGGTTGT
CAAGGCACTG TA TATTCAAGAAATCAAAAA
CAAGGCACTG TA TACGGTACCATTTGCTTG
```

```
actatagggcga ATTGGGCC CAGAGGTGTA
GAGATTAAGGTG CTAGTAGG CAGAGGTGTA
ATGCTCAAGCCC CCAGCCCC CAGAGGTGTA    Tol2
GACTTTAGCTAC CTGCCCAG CAGAGGTGTA
ACAACCAAGCCT CCAAGGTC CAGAGGTGTA
TCAAGTCAAGGA GTCTCAAC CAGAGGTGTA
```

```
CTTTCTAGGG TTAA tctaggtaccatttaa
CTTTCTAGGG TTAA GCACAAACACTGCTGC
CTTTCTAGGG TTAA GAGCCCCCTGCTCATC    PB
CTTTCTAGGG TTAA CTTGATCAGAGATATA
CTTTCTAGGG TTAA TAGTTAGCAACAGCCT
CTTTCTAGGG TTAA ACTCTAGCATGGTTGT
```

Figure 3: Examples of transposon insertion junctions. Transposon junctions amplified from PEGE cells are shown in groups of five with expected non-transposed vector sequence (lowercase) highlighted above. From top to bottom, SB (ITR-L), PP (ITR-R), *Tol2* (ITR-L), and PB (ITR-R), and. Target site duplications (bold) for each transposon are separated from genomic DNA and corresponding (ITR) by a space.

One characteristic advantage of transposase-mediated integration is the precise incorporation of one or more independently transposed gene expression cassettes, without adjacent plasmid vector. In order to observe representative integration events, DNA was isolated from 8 or 9 selected clones from each transposon and analyzed by Southern hybridization (Fig 4). Non-transposase mediated integrations, often head to tail concatemer repeats, have a predictable hybridizing fragment size following restriction enzyme digestion. However, transposon mediated events have unique DNA outside of the ITRs and therefore have unpredictable and varying fragment lengths. The enhancement of transgenesis by transposition (as detected by increased colony formation) was substantiated by the presence of inserts of varying size in cellular clones, in most cases without concatemers. The level of TnT can also be measured by counting the number of independent integrations per cellular clone. The more active transposons *Tol2* and PB, display multiple (up to 15) independent integration events. The wild-type PP transposon system mediated a single integration event per cellular clone, reflecting its lower activity in PEGE cells, whereas the engineered SB system displayed an intermediate number of insertions.

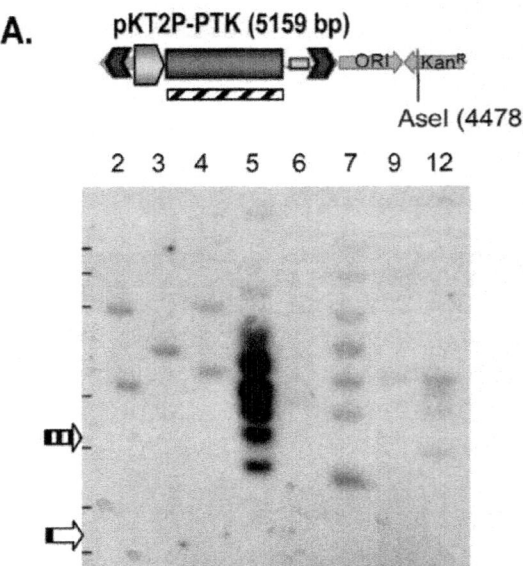

A. pKT2P-PTK (5159 bp)

Asel (4478)

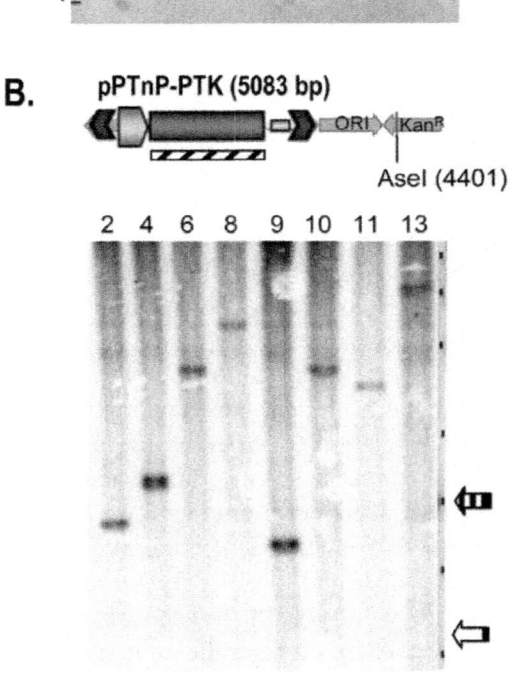

B. pPTnP-PTK (5083 bp)

Asel (4401)

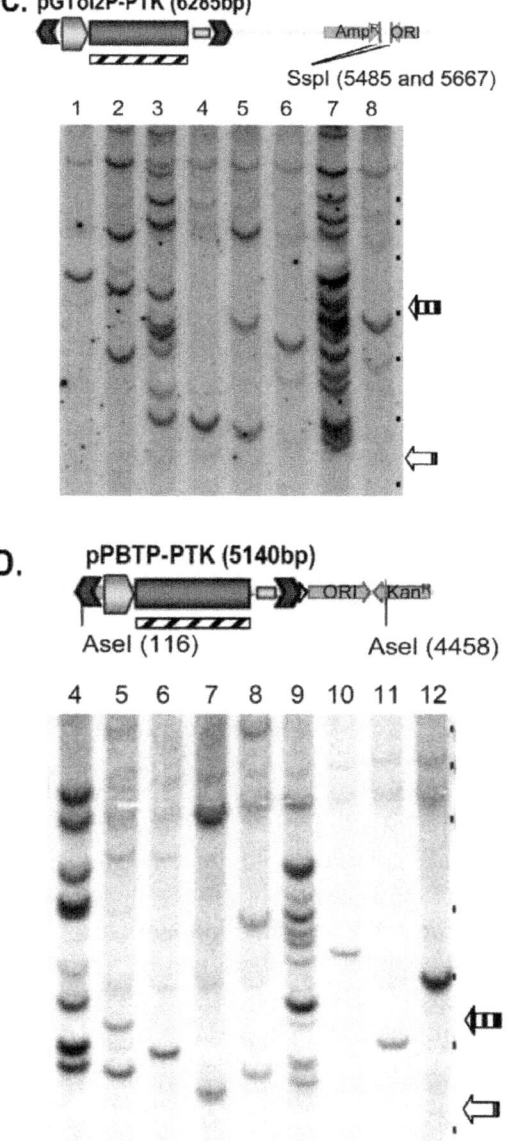

C. pGTol2P-PTK (6285bp)

AmpR ORI

SspI (5485 and 5667)

1 2 3 4 5 6 7 8

D. pPBTP-PTK (5140bp)

AseI (116)   AseI (4458)

4 5 6 7 8 9 10 11 12

Figure 4: Southern Blot of PEGE Clones. Individual puromycin resistant PEGE colonies were isolated and expanded for Southern analysis A) SB B) PP C) *Tol2*, and D) PB. Each transposon donor plasmid transfected into PEGE cells is diagramed with restriction endonuclease sites used for DNA digestion and the probe fragment indicated (diagonal lined rectangle). Expected concatemer sizes (vertical lined arrow)/ smallest possible transposition event (open arrow) for each transposon are 5159/3335 bp, 5083/3275 bp, 6285/3346 bp, and 5140/3320 bp, respectively. The positions of the marker bands are indicated by black dots on the right of each blot with sizes of 12, 10, 8, 6, 5, 4, and 3 kb are shown.

CRE/FLP Activity in Porcine Cells

To test the ability of Cre and/or Flp recombinase to function in porcine cells pT2-FloxP-PTK (Fig. 1A) was transfected into PEGE cells along with SB. These clones were obtained from preliminary transfections that were selected under very stringent drug conditions that favored high-copy integrations, particularly non-transposition events. DNA from puromycin resistant clones was isolated and analyzed by Southern hybridization. Isolated clones contained multiple copies of the PTK transgene due to non-transpositional integration, as indicated by concatemers and concatemer junction bands (Fig 5). PTK transgenic clones were subsequently transfected with pPGK-nlsCre, pKT2P-nlsFlp, or pKT2C-EGFP. Excision of the PTK cassette was detectable in transiently transfected cells by PCR, and the sequence of the excision product confirmed by sequencing (data not shown). Transfected cells were placed under selection with gancyclovir for 10–14 days and colonies counted (Fig. 5C). Only cells that had excised the PTK gene could withstand gancyclovir selection. As expected for concatemers, we observed a low level of transgene instability as evidenced by the appearance of gancyclovir resistant clones upon transfection with pKT2C-EGFP. A much more pronounced recombinase stimulated elimination of the PTK cassette was demonstrated by elevated resistant colony formation for 7 out of 8 of the clones transfected with either pPGK-nlsCre or pKT2P-nlsFlp. While Cre and Flp are both active in PEGE cells, in all cases Cre mediated recombination/excision matched or exceeded that observed for Flp. A single clone (#6) never showed evidence of PTK elimination. The Southern analysis (Fig. 5B), revealed a fragment of pT2-FloxP-PTK likely resulting from the integration of a shortened PTK expression cassette lacking at least one flanking RRS. This clipped PTK transgene is therefore unable to be removed by recombinase-mediated excision.

pT2-FloxP-PTK (6186bp)

A.

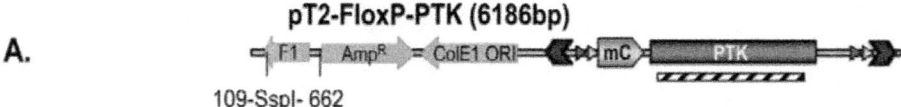

109-SspI- 662

B.

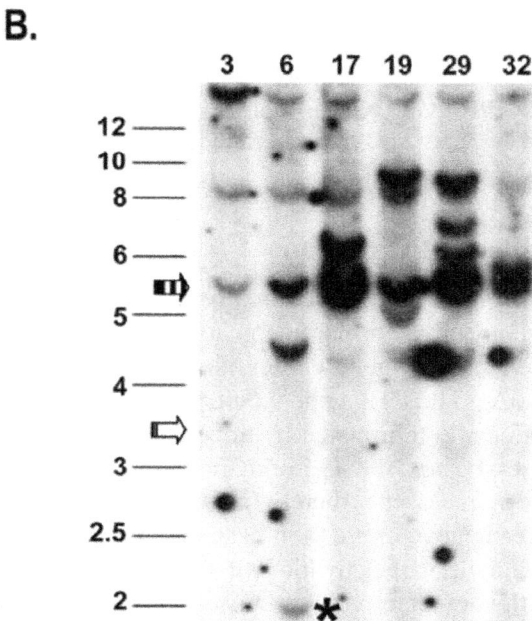

C.

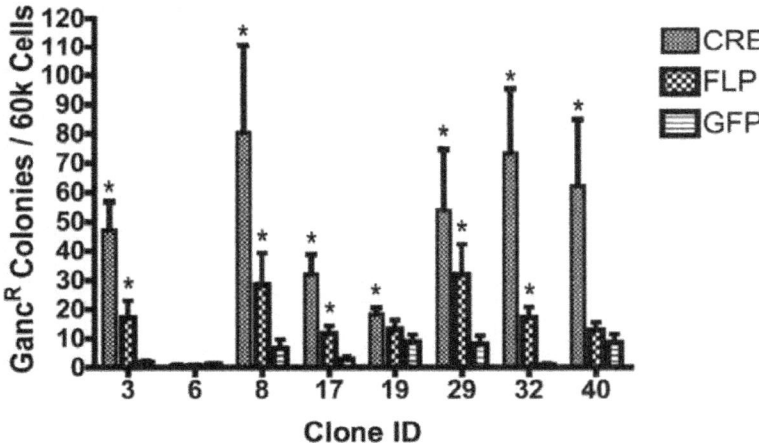

Figure 5: Cre/Flp Activity in Pig Cells. Individual puromycin resistant PEGE colonies were isolated and expanded for analysis. A) A diagram of the pT2-FloxP-PTK vector showing the location of restriction enzyme sites for SspI and the location of the PTK probe (diagonal lined rectangle). B) Southern analysis shows the number and size of vector inserts in several PEGE clones. The expected concatemer size of 5.6 kb (vertical lined arrow) as well as the smallest possible transposition event (open arrow) of

3.3 kb are indicated on the left of the image. An asterisk is placed to the right of a band slightly smaller than 2 kb in lane 2 (Clone #6). C) The rate of gancyclovir resistant colony formation after transfection of PEGE clones with pPGK-nlsCre (CRE), pKT2-nlsFlp (FLP), or pKT2C-EGFP (GFP). Values that are significantly different from the background (GFP) as determined by an unpaired t-test (p = 0.05) are designated with an asterik (*).

CRE-Activated Gene Expression

To further demonstrate the functionality of the transposon based Cre recombinase system for use in porcine genome engineering, a SB transposon containing a Cre-activated gene expression cassette was constructed- pTC-loxPTK-G (Fig. 6A). The PTK gene would be transcribed by the mini-CAGs promoter and efficiently terminated by three complete poly(A) signals (triple stop) in the intact pTC-loxPTK-G [60]. Cre recombination results in deletion of the PTK/triple-stop cassette, thereby juxtaposing the mini-CAGS promoter and the downstream gene expression cassette and enabling transcription of the GFP gene. Conditional activation of GFP expression was assessed by microscopy and flow cytometry after transient transfection of PEGE cells with pTC-loxPTK-G in the presence or absence of pPGK-nlsCRE (Fig. 6B). There was no GFP observed in cells transfected with pTC-loxPTK-G alone, whereas about 10–12% of the cells were GFP+ when transfected with pPGK-nlsCre. This corresponds well with the average transfection efficiency of PEGE cells, indicating that the Cre excision reaction is very efficient in transiently transfected cells.

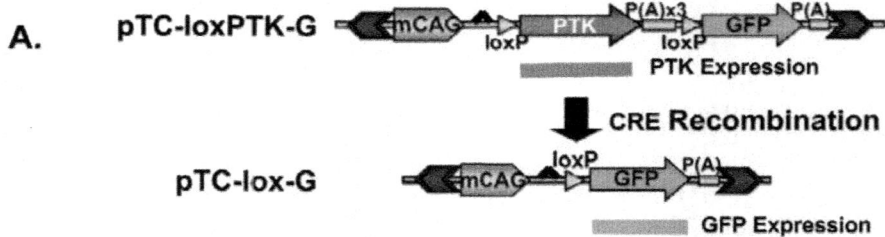

B.

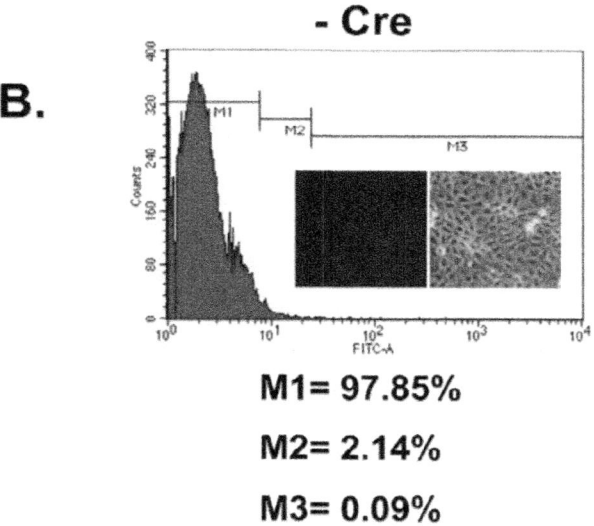

- Cre

M1= 97.85%

M2= 2.14%

M3= 0.09%

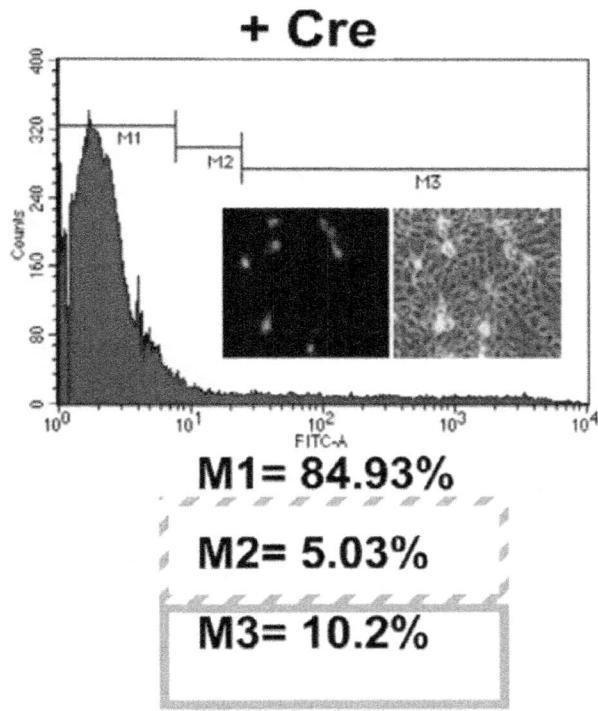

+ Cre

M1= 84.93%

M2= 5.03%

M3= 10.2%

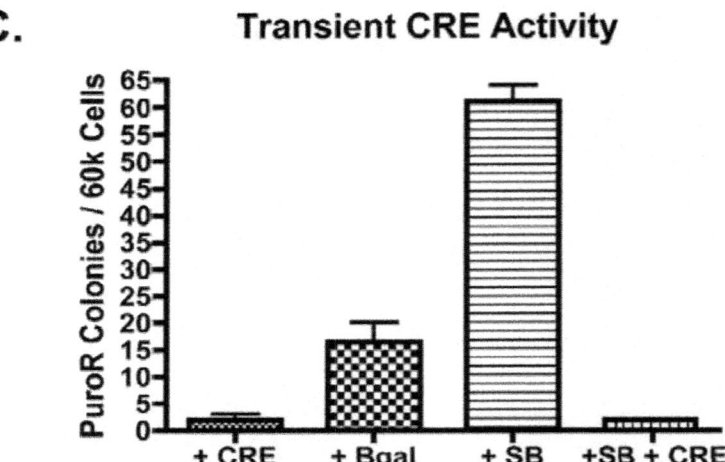

Figure 6: A CRE-Activated Transgene. A) An illustration of the Cre-activated transgene vector. The full vector, pTC-loxPTK-G, produces PTK from the mini-CAGs promoter. Transcriptional leakage into the downstream gene, GFP, is limited due to the incorporation of three full poly-adenylation signals, a so-called triple-stop. Recombination by Cre eliminates PTK and triple-stop, activating GFP expression from pTC-lox-G. B) pTC-loxPTK-G was transfected into PEGE cells with (+Cre) or without (-Cre) pPGK-nlsCre. Cells were monitored for GFP expression by fluorescent microscopy (image inserts) and flow cytometry. The percentage of cells expressing GFP was dependent on co-transfection with pPGK-nlsCre. C) PEGE cells were transfected with pTC-loxPTK-G along with pPGK-nlsCre (+Cre), pCMV-β (+βgal), pKUb-SB11 (+SB), or pKUb-SB11 and pPGK-nlsCre (+SB +Cre). The cells were plated in puromycin selective media and colonies were counted.

To further examine the efficiency of Cre recombinase in transiently transfected cells, conditional removal of the PTK/triple stop expression cassette was assessed by selection in puromycin following co-transfection of PEGE cells with pTC-loxPTK-G and either Cre, β-galactosidase, SB, or Cre + SB. Transfected cells were plated under puromycin selection for 9–12 days, stained with methylene blue, and enumerated to quantify the efficiency of PTK/triple stop elimination prior to or after integration into the genome (Fig. 6C). Addition of pPGK-nlsCRE to the transfection, alone or in combination with pKUb-SB11 reduced puromycin-resistant colony counts to levels significantly lower than that observed for pKUb-SB11 or pCMV-β, which alone result in

TnT and non-transpositional transgenesis with an intact PTK gene expression cassette, respectively. Therefore, Cre recombinase excision activity in transiently transfected PEGE cells approaches 100%, especially with regard to plasmids available for transposition by SB transposase.

Although this particular co-transfection with pTC-loxPTK-G and SB suffered from a low transfection efficiency (~5%) that reduced TnT (compare Fig 6C to 1B), puromycin resistant clones were expanded for characterization by Southern hybridization (Fig 7). Analysis indicated TnT with 1 to 4 transposon integrations per clone. Although, clones 7, 10 and perhaps 11 contained hybridizing species near what would be expected for non-transpositional integration, their molar representation was equal to that of single copy inserts, not multicopy concatemers. Clones 7 and 10 also harbored hybridizing species smaller than was expected for transposition. These fragments likely represent non-transposase mediated DNA recombination events. The proportion of non-transpositional integrations detected by Southern analysis (1 in 4) corresponds well with the observed unfacilitated rate of transgenesis as determined by colony count for this transfection.

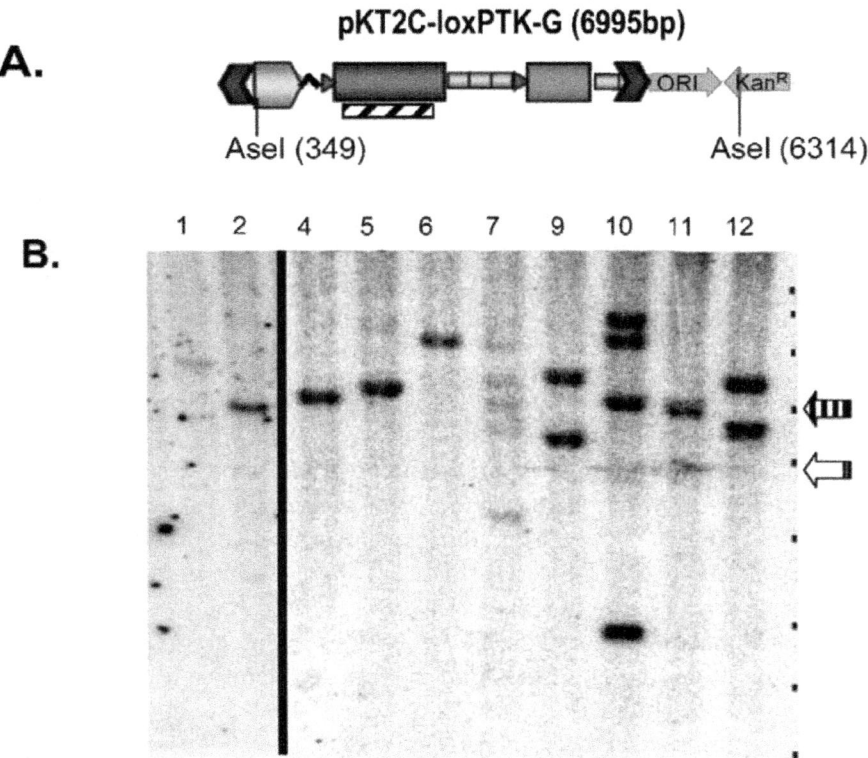

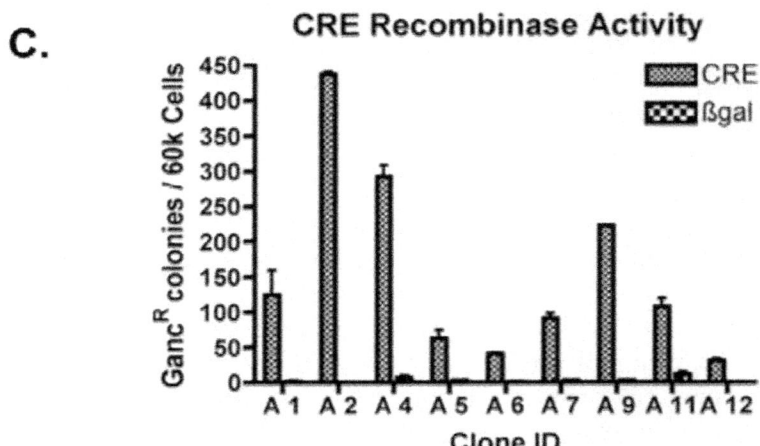

Figure 7: Conditional gene-activation of integrated transposons. Colonies from the transfection of pTC-loxPTK-G with pKUb-SB11 (Fig 5C) were expanded in selective media containing puromycin. DNA from these transgenic colonies was isolated and analyzed by Southern hybridization. A) A schematic of pKT2C-loxPTK-G that shows the AseI restriction sites and the location of the PTK hybridization probe (diagonal lined rectangle) used for Southern analysis. B) A Southern blot of pKT2C-loxPTK-G colonies. The clones were analyzed without Cre excision, so integrants that result from transposition should be equal to or greater than the transposon size of 4.9 kb (open arrow). Whereas, bands associated with concatemer formation are found at 6.0 kb (vertical line arrow). The positions of the DNA marker bands of the 1 kb Quanti-Marker from ISC Bioexpress (Kaysville, Utah), are indicated by black dots on the right of each blot with sizes of 12, 10, 8, 6, 5, 4, 3, 2.5, and 2 kb shown. C) pKT2C-loxPTK-G colonies were transfected with pPGK-nlsCre and plated under gancyclovir selection. Clones with PTK eliminated by recombination became gancyclovir resistant and were counted. Cre-activation of all clones was determined to be significant ($p < 0.5$).

pTC-loxPTK-G clones were generated to analyze the efficiency of recombinase-directed selection-cassette recycling and the conditional activation of gene expression from a variety of porcine genomic loci. Puromycin resistant clones were transfected with pPGK-nlsCRE and scored for gancylovir resistance (Fig 7C). All gancylovir resistant clones expressed GFP, although variation in the intensity of GFP was observed (data not shown) depending on the parental clone source. This expression variance is expected due to the influence of porcine sequence adjacent to the sites of transposon integration, a phenomenon commonly referred to as "position effect". A significant increase in the efficiency of selection cassette recycling was demonstrated in the presence of single copy inserts when compared to multicopy concatemers (Fig 7C vs 5C). In addition, activation of GFP expression upon recombinase-based excision from integrated transposons demonstrates the efficacy of Cre-dependent conditional gene expression in transgenic porcine cells.

DISCUSSION

Multiple Transposons and Recombinases are Active in Porcine Cells

This work demonstrates for the first time the capability of four DNA transposon systems, SB, *Tol2*, PB, and PP, to enzymatically facilitate precise and efficient transpositional transgenesis in porcine cells. We have also established for the first time that Cre and Flp DNA recombinases are active in porcine cells. The combination of these DNA directed enzyme systems provides for the delivery and removal of gene expression cassettes to the porcine genome for the purpose of cellular transgenesis, selection cassette recycling and conditional gene expression based on transposons and recombinases. In these studies, the *Tol2* and PB transposon systems were more efficient than SB, which was more efficient than PP at mediating TnT in PEGE cells, although these relative efficiencies should not be over-interpreted. Although we used favorable conditions for each transposon system by our selection of promoters and transposase/transposon ratios, our focus here was on testing their function, not on determining their relative activities in PEGE cells, an immortalized cell line unsuitable for generating pigs by SCNT. Indeed, it is well established that the rate of TnT in any cell type is likely to depend not only on intrinsic transposase activity, but also on the presence or absence of cellular co-factors and DNA repair enzymes, the method of DNA introduction, and the amount of transposase produced/provided in the specific cell type. Transposon activity varies not only between cells from different species, but also between different cell types from the same species [26, 27, 34]. Future studies will focus on the efficiency of different transposon systems and recombinases in pig fibroblasts (applicable to SCNT), pig stem cells (for functional genomics and SCNT), and porcine embryos (for direct transgenesis by PNI).

In addition to potential differences in efficiency, the integration behavior of each transposon may be an important factor in determining the appropriate transposon system for a specific task. For instance, PB appears to preferentially integrate into transcription units [32, 61]. Consistent with this observation, in our limited examination of integration sites in the pig genome, flanking sequence from two of seven PB insertions matched porcine cDNAs. In addition, PB primarily leaves no footprint when remobilized [33]. Therefore, PB may be most suitable for functional genomics studies in pigs or pig cells, where mutations due to the interruption of genes, and the potential for precise transposition-based rescue is desirable [62–65]. SB does not integrate into transcription units at a rate much higher than what would be expected by random integration [66], so it may represent a better choice for animal transgenesis, transposon-based DNA vaccination, or other somatic therapies. Alternatively transposon systems engineered to target specific genomic locations may be developed and could provide the safest choice for these applications [34, 67, 68]. The integration profiles of *Tol2* and PP are not well characterized in any organism or cell type, and the integration predilections of any transposon system remains to be addressed in specific swine cells being considered for engineering.

Advantages of Transposition for Pig Transgenesis and Genetics

There are several advantages of transpositional versus unguided transgenesis. First, the enzymatic activity of the transposase increases the efficiency of transgene integration (Fig. 1 and 2). Secondly, transposase-mediated transgenesis precisely integrates a single copy of the transposon into one or more locations in the genome. Consequently, transposition avoids the integration of G/C-rich prokaryotic elements of the vector and avoids transgene concatemerization, both of which can lead to shutdown of gene expression [5, 6]. In addition, concatemerization is problematic for selection cassette recycling (Fig. 5) and the implementation of more complex genetic rearrangements with recombinases. We propose the use of transposon systems for transgenesis of porcine cells prior to their use for the creation of pigs by SCNT to enable increased efficiency, better precision, reliable expression, and selection cassette recycling. In addition, SB and PB dramatically improved the transgenesis rate in mice by PNI [32, 69], providing a clear rationale for improving the efficiency of transgenic pig production via this method.

Recombinases in Swine Genetics- Selection-Cassette Recycling and Conditional Alleles

There are several immediate applications for recombinases in swine genetics. First, as shown in Fig 5, 6, 7, recombinases can be combined with a positive/negative selectable marker like PTK for selection cassette recycling [52]. Currently, most, if not all transgenic animals produced by SCNT contain a selectable marker (e.g. neo^R, puro^R, GFP) in addition to an experimental transgene. This selectable marker is useful for the proper identification of nuclear donor cells, but generally is undesirable in the transgenic animal. This could be particularly important for removal of xenogenic elements after gene knockout or manipulation preceding the introgression of engineered germplasm into agricultural production herds. The flanking of selectable markers with RRS provides the opportunity to eliminate them in culture or by breeding to Cre expressing pigs, leaving behind only a single 34-basepair RRS footprint.

Recombinases also permit the creation of conditional alleles for activation or inhibition of gene function in response to Cre or Flp recombinase activity, as illustrated in Figures 6, 7[51]. In addition, the effectiveness of homologous recombination constructs can be improved to allow selection cassette recycling, thereby avoiding 'selection cassette interference', whereby the exogenous regulatory elements in the selection cassette can interfere with the expression of genes in the vicinity of the targeted mutation [52]. As has been elegantly demonstrated in mice, recombinases can also be used to create conditional knock-outs in pigs when tissue specific ablation is desired, or when traditional knockout results in embryonic lethality. The availability of an assortment of transposon and recombinase systems should also permit serial cellular transgenesis and recombination to achieve complex genomic rearrangements in the pig. Serial transgenesis provides a direct method for the production of pigs that express several gene products. Given the dramatic long-range conservation of synteny between pig and human genomes, far more extensive than for mouse and human, engineered chromosomal rearrangements between serially provided RRS in the pig could provide superior large animal models of human congenital and cancer related chromosomal abnormalities [53].

CONCLUSION

Pork represents the single most economically important meat product in the world and pigs are playing an increasingly critical role in biomedicine. An armamentarium of effective genetic tools will be required to capture the value and potential of this species for human nutrition and health. Here we have tested four transposon and two recombinase systems for activity in pig cells. SB, PP, *Tol2*, and PB and transposons are capable of precise transpositional transgenesis of porcine cells, increasing efficiency by 4–28 fold. We have also demonstrated that Cre and Flp recombinases function efficiently in the nucleus of pig cells for selection-cassette recycling and conditional regulation of transgene expression. The combination of these tools will significantly improve the efficiency and sophistication of porcine genetic manipulation for enhancing pig production and human nutrition, as well as modeling and treating human disease.

METHODS

Vector Construction

Sequence information, maps, and material requests for these constructs can be found on our web site [70].

pT2-Floxp-PTK- To generate a multiple cloning sequence flanked by *FRT* and *loxP* recombinase recognition sequences (*FRT-loxP* MCS), two oligonucleotides with overlapping sequence (shown in bold) were designed, *FRT-loxP* Upper [ATACCGGCCGGAAGTTCCTATTCCGAAGT TCCTATTCTCTAGAAAGTATAGGAACTTCATAACTTCGTATAATGTA TGCTATACGAAGTTATCTCGAGAA**TTCCCGGGAGGCCTACTAGT**], and *FRT-loxP* Lower [GTATTCATGAGAAGTTCCTATACTTTCTAGAGA ATAGGAACTTCGGAATAGGAACTTCATAACTTCGTATAGCATACAT TATACGAAGTTATCCATGG**ACTAGTAGGCCTCCCGGGAA**]. These oligonucleotides were annealed and elongated by PCR using Pwo polymerase. The 218 base pair PCR fragment was cloned into pCR4 using the ZERO Blunt TOPO PCR Cloning Kit (Invitrogen, USA) to create pCR4 *FRT-loxP* MCS, and its sequence was verified. *FRT-loxP* MCS was subsequently excised with EagI and BspHI and cloned into pT2/BH [71] cleaved with EagI and NcoI to produce pT2-*FRT-loxP* MCS. Finally, a completely filled XhoI fragment, containing the mouse PGK promoter, the PTK fusion protein, and bovine growth hormone poly(A) signal from YTC37, a kind gift from the laboratory of A. Bradley [54], was cloned into SmaI cleaved pT2-*FRT-loxP* MCS to produce pT2-FloxP-PTK.

pKUb-SB11- A 1.0 kb fragment of the SB11 transposase from pCMV-SB11 [72], which had been amplified with CDS-SB11-F1 [CACCATGGGAAAATCAAAAGAAATCAGCC] and CDS-SB11-R1 [GGATCCCAATTTAAAGGCAATGCTACCAAATACTAG] primers and subcloned into an intermediate vector adding a 5' BglII site and the sequence [AGATCTGAT], was cloned into the BamHI site of pKUb to make pKUb-SB11. pKUb was made by cloning nucleotides 3561–4771 of the human UbC gene (genbank accession D63791), which contains the UbC promoter, non-coding exon 1, and intron 1, into pK-SV40(A) between intact BglII and NheI restriction endonuclease sites. pK-SV40(A) was made by cloning a single copy of the SV40 poly(A) signal amplified by PCR with oligos KJC-SV40(A)-F1 [CATTGATGAGTTTGGACAAACCACA] and KJC-SV40(A)-R1 [ACCACATTTGTAGAGGTTTTACTTGCT] into pK-A10 opened with XmnI. pK-A10 was made by cloning KJC-Adapter 10 [CTGAGATCTTAAGC TAGCAGGATCCAGAATTCATTCAG] into pK digested with PvuII creating a multiple cloning site with PvuII, BglII, AflII, NheI, BamHI, EcoRI, XmnI, and PvuII recognition sites. pK was made by joining an 0.8 kb PCR product of pBluescriptSK- (Stratagene), containing the pUC_ORI amplified with oligos KJC-pUC_ORI-F1 [CTGTTCCGCTTCCTCGCTCACTGACT] and KJC-pUC_ORI-R1 [AAAAGGATCTAGGTGAAGATCCTTTTTGAT], to a 0.9 kb PCR product of pENTR-D-TOPO (Invitrogen), which contains the kanamycin resistance gene amplified by oligos KJC-KanR-F1 [CTGCATCATGAACAATAAAACTGTCTGCT] and KJC-KanR-R1 [TGCCAGTGTTACAACCAATTAACCAAT]. The junction of ORI-F1 to KanR-R1 created a single PvuII site.

pCMV-β is available from Clontech (Mountainview, CA).

pPGK(nls)CRE was a kind gift of Dr. David Largaespada's lab at the University of Minnesota.

pKT2P-(nls)FLP- A Flp open reading frame containing the large T antigen nuclear localization signal (bold) and a Kozak consensuses sequence was generated by amplifying the Flp open reading frame using primers CDS Kozak-NLS Flp 5' [ATATCTCGAGG CCACCATGGC **TCCCAAGAAGAAGAGGAAGGTG**ATGAGTCAAT TTGATATATTATGTAAAAC] and CDS Flp 3' [ATATAGATCTTTATATGCGTCTATTTATGTAGG] using pOG44 (Invitrogen, USA) as template. The resulting PCR product was cloned into pCR4 using the ZERO Blunt TOPO PCR Cloning Kit (Invitrogen, USA) creating pCR4-nlsFlp. The nlsFlp open reading frame was subsequently excised with XhoI and BglII and inserted into XhoI-BglII cleaved pKT2-PGKi to produce pKT2P-nlsFlp. pKT2-PGKi contains the human PGK promoter, a kind gift of Dr. Scott McIvor (University of Minnesota) in front of the mini-intron, MCS, and rabbit beta-globin 3'UTR found in mini-CAGs.

pKT2C-EGFP was made by cloning a 0.7 kb XhoI to BglII fragment of pKT2P-GeN into pKT2-mCAG opened from BglII to XhoI. pKT2-mCAG was made by cloning a 2.2 kb BamHI to KpnI fragment of pSBT-mCAG [73] into pK-A3 opened from BamHI to KpnI. pKT2P-GeN was made by cloning EGFP as a 0.75 kb EcoRI fragment from pCR4-EGFP into the EcoRI site of pKT2P-eNeo. pCR4-EGFP was made by cloning a PCR fragment of EGFP from pEGFP-N1 (Clontech) amplified with primers KJC-EGFP-F3 [CCGA ATTCTACCATGGTGAGCAAGGGCGAG] and KJC-EGFP-R2 [CCAGA TCTTTACTTGTACAGCTCGTCCATGC] into pCR4-TOPO (Invitrogen). pKT2P-eNeo contains the encephalomyocarditis virus internal ribosome entry site and neomycin resistance gene amplified from pGT-Neo [62] with KJC-BactinSA-F1 [CACTGAAGTGTTGACTTCCCTGACAGC] and KJC-

Bgeo-R1   [TTCAATTGTTAGAAGAACTCGTCAAGAAGGCGA].   The
eNeo cassette was subcloned and acquired a modified sequence at the 3' end
[GTTAACTT] to [GTTAAGTCTAGA] including a BglII site. The 1.4 kb
eNeo cassette was isolated with EcoRI and BglII and moved into pKT2-PGKi
opened from BglII to EcoRI.

pKT2P-PTK was made by cloning a 2.7 kb PvuII fragment from pKP-
PTK_TS into pKT2-RV opened with EcoRV. pKT2-RV was made by cloning
a 0.6 kb BamHI to KpnI fragment of pSBT-RV [73] into pK-A3 opened with
BamHI and KpnI. pK-A3 was made by opening pK with PvuII and inserting
KJC-Adapter 3 [CTGGATCCAGATCTGGTACCATTTAAAT] creating a
small multiple cloning site with PvuII, BamHI, BglII, KpnI, and SwaI sites.
pKP-PTK_TS was made by cloning a 2.3 kb BglII to EcoRI fragment of pCR4-
PGK-PTK into the MCS of pK-SV40(×2) opened with EcoRI and BglII. pCR4-
PGK-PTK was made by cloning a 2.3 kb PCR product of pT2-FloxP-PTK
amplified with PuroΔTK-F1 [TTAGATCTGGCCTCGCACACATTCCACAT]
and PuroΔTK-R1 [TGGTTCTTTCCGCCTCAGAAGCCAT] into pCR4-
TOPO (Invitrogen). pK-SV40(×2) was made by cloning two copies of
the SV40 poly(A) signal amplified by PCR with oligos KJC-SV40(A)-F1
[CATTGATGAGTTTGGACAAACCACA]   and   KJC-SV40(A)-R1
[ACCACATTTGTAGAGGTTTTACTTGCT] into pK-A10 opened with
XmnI.

pTol2-PTK- The mini *Tol2* transposon donor plasmid was constructed
by inserting the PvuII fragment of pKP-PTK-TS into pGemT-Tol2 [74, 75]
opened from SwaI to HindIII (filled) to produce pGTol2P_PTK.

pCMV-Tol2 was constructed as indicated [74] from previously described
materials [75].

pPBTP-PTK was made by cloning a 2.7 kb PvuII fragment of pKP-
PTK_TS into pPBT-SE opened from SmaI to EcoRV. pPBT-SE was made by
cloning the 102 bp PCR product containing an outward facing T7 polymerase
site, the SE multiple cloning site, and an outward facing T3 polymerase site
into pPBT cut with MscI. The PCR product was amplified from pKT2-SE
using T7-RevComp [TCTCCCTATAGTGAGTCGTATTA] and T3-RevComp
[TCTCCCTTTAGTGAGGGTTAATT] primers. pPBT was made by cloning
the PB LTR1 and LTR2 into pKT2-SE from KpnI to BamHI. LTR1 and LTR
2 from PB were amplified from pXL-Bac-II, a kind gift of Malcolm Fraser
(Notre Dame University), using PB-LTR1-F1 [TGGATCCCAATCCTTAAC
CCTAGAAAGATAATCATATTG] and PB-LTR1-R1 [GTGGCCATAAAAG
TTTTGTTACTTTATAGAAG] or PB-LTR2-F1 [TTGGCCATAAGTTATCA
CGTAAGTAGAACATG] and PB-LTR2-R1 [TGGTACCTAGATTAACCCT
AGAAAGATAGTCTG], respectively. LTR1 and LTR2 PCR products were

cloned into pCR4 vector (Invitrogen) and subsequently excised by BamHI and MscI or MscI and KpnI digestion, respectively. pKT2-SE was made by cloning the 0.7 kb BamHI to KpnI fragment containing the SB inverted repeats and SE multiple cloning site from pSBT-SE [73] into pK-A3 opened from KpnI to BamHI.

pKC-PB was made by inserting the 2.1 kb NheI to BamHI fragment of p3XP3-DsRed, a kind gift of Dr. Malcolm Fraser (Notre Dame University), containing the PB transposase coding sequence into the 3.2 kb BamHI to NheI fragment of pKC-SB11, which resulted in the exchange of SB11 with PB transposase.

pPTnP-PTK- A 2.7 kb PvuII to PvuII fragment of pKP-PTK_TS was cloned into the EcoRV site of pPTn2-RV to make pPTnP-PTK. pPTn2-RV was made by cloning KJC-Adapter 4 [TCTCCCT TTAGTG AGGGTTAATTGATATCTAATACGACTCACTATAGGGAGA] into the MscI site of prePTn2(-1) creating T7 and T3 polymerase binding sites orientated out towards the inverted repeats of the PTn transposon and separated by an EcoRV site. prePPTn2(-1) was made by cloning a 0.5 kb BamHI to KpnI fragment of pCR4-PPTN2A into pK-A3 opened from KpnI to BamHI. pCR4-PPTN2A was created by topo cloning a 0.5 kb PCR product amplified from prePPTN2(-2) using oligos PPTN-F1 (BamHI) [AAGGATCCGATTACAGTGCCTTGCATAAGTAT] and PPTN-R1 (KpnI) [AAGGTACCGATTACAGTGCCTTGCATAAGTATTC] into pCR4-Topo (Invitrogen). prePPTN2(-2) was created by amplifying the majority of pBluKS-PPTN5 [29], a kind gift of Dr. Michael Leaver (University of Stirling, UK), with oligos PPTN-OL2 [CCATCTTTGTTAGGGGTTTCACAGTA] and PPTN-OR1 [CCAGGTTCTACCAAGTATTGACACA]. The PCR fragment was then self-ligated to produce an empty transposon with a single MscI site in its interior.

pKC-PTs1 was made by cloning a 1.0 kb NheI to EcoRI fragment of pKUb-PTs1 that contained the PPTN transposase (PTs) into pK-mCAG opened from EcoRI to NheI. pK-mCAG was made by cloning the mCAG promoter from pSBT-mCAG [73] as a 0.96 kb SmaI to EcoRI (filled) fragment into pK-SV40(A) × 2 opened with AflII (filled). pKUb-PTs1 was made by replacing the SB11 gene from pKUb-SB11 with PTs by cloning a 1.0 kb BamHI to NheI fragment from pCR4-PPTs1B into pKUb-SB11 from NheI to BamHI. pCR4-PPTs1B was made by cloning a PCR fragment of pBluKS-PPTN4 [29], a kind gift of Dr. Michael Leaver (University of Stirling, UK), amplified with primers CDS-PPTs-F1   [AAAGCTAGCATGAAGACCAAGGAGCTCACC]   and CDS-PPTs-R1 [AAGGATCCTCAATACTTGGTAGAACC] into pCR4-Topo (Invitrogen).

pKT2C-loxPTK-G was made by cloning a 2.3 kb PvuII fragment of pK-PTK_TS into the MscI site of pKT2C-lox-GFP. pK-PTK_TS was made by cloning a 1.9 kb BglII to EcoRI fragment of pCR4-PTK into the MCS of pK-SV40(×2) opened with EcoRI and BglII. pCR4-PTK was made by cloning a 1.9 kb PCR product of pT2-FloxP-PTK using oligos PuroΔTK-F2 [TTAGATCTACCATGACCGAGTACAAGCCCA] and PuroΔTK-R1 [TGGTTCTTTCCGCCTCAGAAGCCAT] into pCR4-TOPO (Invitrogen). pKT2C-lox-GFP was made by cloning 0.1 kb EcoRI fragment of pCR4-loxP, which contains two direct repeat loxP sites separated with a MscI site, into pKT2C-EGFP opened with EcoRI. pCR4-loxP was made by topo cloning the annealed and extended oligos loxP-F1 [ATAACTTCGTATAATGTATG CTATACGAAGTTATCTCGAGTGGCCA] and loxP-R1 [ATAACTTCGTA TAGCATACATTATACGAAGTTATTGGCCACTCGAG] into pCR4-TOPO (Invitrogen).

Cell Culture and Transposition/Recombinase Assays

Pig fibroblasts were isolated from 43 day old embryos. The tissue was dissociated using a collagenase/DNAse I treatment as well as mechanical disruption. The cells from the female piglet #8 were cultured in DMEM enriched with 10%FBS and 2× antibiotic/antimycotic solution (Gibco #15240-022). The cells were passaged in DMEM high glucose media enriched with 10% FBS, 2 mm L-glutamine, 1× P/S until spontaneously establishing line PF8. A subpopulation of porcine endometrial gland epithelium cells [76] were spontaneously immortalized, strain PEGE. The PEGE cells were maintained in DMEM supplemented with 10% FCS, 1× Penn/Strep, 10 μg/ml Insulin (Sigma, USA), and 1× L-Glutamine.

For transposition assays cells were plated in each well of a six well plate to achieve 60–80% confluence within 6–24 hours. Cells were transfected using *Trans*IT-LT1 (Mirus Bio Corporation, WI) transfection reagent according to the manufactures instructions with a ratio of 3:1 lipid: μg DNA. Each transfection contained a total of 1.15 to 1.5 μg of plasmid DNA. Wells 1–3 contained transposon plus transposase, well 4 contained transposon with no transposase, well 5 contained SB plus SB transposase and well 6 contained pKT2C-EGFP only. Molar amounts of each transposon were fixed at 1.5×10^{-13} moles of transposon (0.75×10^{-13} Moles for *Tol2*) while transposase plasmid was added at a molar ratio of 1:1 for SB, *Tol2*, and *PB*, and 1:0.5 for PP. The choice of the promoters and transfection ratios for SB and PP was based on the highest transposition activity observed in human HT1080 cells (data not shown). Strong promoters (CMV & miniCAGs) and transfection conditions for *Tol2* and PB were selected based on previously published data and the observation that these transposon systems seem less susceptible to overexpression inhibition than SB and PP.[34, 61, 74] Total DNA weight was adjusted using pCMV-β plasmid. Forty-eight hours after transfection cells were trypsinized, and two replicates of 60,000 cells were plated onto 100 mm plates in media containing 0.3 μg/ml puromycin and selected for 9–12 days. Colonies were visualized by methylene blue staining and counted. A minimum of two six-well plates were transfected for each experiment. The mean colony number and standard error are shown in figures.

Southern Hybridizations

Several independent puromycin resistant PEGE foci for each transposon were aspirated and grown to confluence on a 100 mm plate. Genomic DNA was extracted using standard methods and approximately 10 μg was digested with SspI (*Tol2*clones) or AseI (SB, PB, and PP) clones. Digested DNA was separated on 0.7% agarose gel and transferred to positively charged nylon membranes (GE Osmotics, USA). Membranes were probed with a random primed 1524 bp XmaI fragment of pKP-PTK-TS that contained the bulk of the PTK gene and visualized by autoradiography or phosphor imaging.

Cloning Transposon Junctions

Genomic DNA was isolated from pooled, fixed, and stained puromycin resistant clones for each transposon. For splinkerette PCR DNA was cut with Sau3AI or NlaIII and junctions were cloned as described [69]. For blocked linker-mediated PCR, DNA was cut with NspI for Tol2 and SB, and a cocktail of enzymes including XbaI, AvrII, NheI and SpeI for PB and PP. The NspI digested DNA was ligated to the blocked linker-SphI that was created by annealing primerette-long [CCTCCACTACGACTCAC TGAAGGGCA AGCAGTCCTAACAACCATG] and blink-SphI [5'P-GTTGTTAGGACTGCTTGC-3'P]. Whereas the DNA digested with the cocktail was ligated to the blocked linker-XbaI that was produced by annealing primerette long to blink-XbaI [5'P-CTAGCATGGTTGTTAGGACTGCTTGC-3'P]. Following ligation the junction sequences were amplified by nested PCR. The primary PCR used the common primer primerette-short [CCTCCACTACGACTCACTGAAGGGC] with transposon-specific primers SB_IRDR(L)-O1   [ATTTTCCAAGCTGTTTAAAGGCACAGTCAAC], Tol2(L)-O1   [AATTAAACTGGGCATCAGCGCAATT],   PB-LTR(R)-O1   [ACAGACCGATAAAACACATGCGTCAA],   and   PTn-IRDR(R)-O1   [GGGTGAATACTTATGCACCCAACAGATG].   The secondary PCR reactions used the common primer primerette-nested [GGGCAAGCAGTCCTAACAACCATG] with transposon-specific primers

SB_IRDR(L)-O2   [GACTTGTGTCATGCACAAAGTAGATGTCCT], To 2(L)-O2 [GCGCAATTCAATTGGTTTGGTAATAGC], PB-LTR(R)-O2 [TCCTAAATGCACAGCGACGGATTC], and   PTn-IRDR(R)-O2 [CAGTACATAATGGGAAAAAGTCCAAGGG]. To   generate   unique sequences serial dilutions (1:50 and 1:500) of the ligation reaction were used as template for the primary PCR. The primary PCR was diluted 1:50 and used as template in the secondary PCR reaction. The PCR fragments were shotgun cloned and sequenced.

ABBREVIATIONS

PEGE: Porcine endometrial glandular epithelium

GFP: Green fluorescent protein

RRS: Recombinase recognition site

SCNT: Somatic cell nuclear transfer

PNI: Pronuclear microinjection

ITR: Inverted terminal repeat

SB: *Sleeping Beauty*

PP: *Passport*

PB: *piggyBac*

TnT: transpositional transgenesis

ACKNOWLEDGEMENTS

We would like to thank Mr. Andrew R. Bents for assistance with tissue culture and blocked linker-mediated PCR. This work was supported by grants from the Juvenile Diabetes Research Foundation (JDRF#7-2005-1167) and a faculty development grant from the Academic Health Center of the University of Minnesota (SCF).

AUTHORS' CONTRIBUTIONS

KJC designed experiments and transposon vectors and together with DFC constructed transposons, performed experiments, and conducted molecular analysis. LKF optimized and conducted tissue culture experiments. BWK performed preliminary transfections in pig cells and was mentored by DNF. SCF conceived the study and mentored KJC, DFC, and LKF in experimental design and data analysis. KJC drafted the manuscript and along with DFC and SCF completed manuscript preparation. All authors read and approved the final manuscript.

REFERENCES

1. Lai L, Prather RS: Creating genetically modified pigs by using nuclear transfer. Reprod Biol Endocrinol. 2003, 1: 82-10.1186/1477-7827-1-82.

2. Zeng L, Rahrmann E, Hu Q, Lund T, Sandquist L, Felten M, O'Brien TD, Zhang J, Verfaillie C: Multipotent adult progenitor cells from swine bone marrow. Stem cells (Dayton, Ohio). 2006, 24 (11): 2355-2366. 10.1634/stemcells.2005-0551.

3. Price EM, Prather RS, Foley CM: Multipotent adult progenitor cell lines originating from the peripheral blood of green fluorescent protein transgenic Swine. Stem Cells Dev. 2006, 15 (4): 507-522. 10.1089/scd.2006.15.507.

4. Pravtcheva DD, Wise TL: Transgene instability in mice injected with an in vitro methylated Igf2 gene. Mutat Res. 2003, 529 (1–2): 35-50.

5. Scrable H, Stambrook PJ: A genetic program for deletion of foreign DNA from the mammalian genome. Mutat Res. 1999, 429 (2): 225-237.

6. Garrick D, Fiering S, Martin DI, Whitelaw E: Repeat-induced gene silencing in mammals. Nat Genet. 1998, 18 (1): 56-59. 10.1038/ng0198-56.

7. Henikoff S: Conspiracy of silence among repeated transgenes. Bioessays. 1998,20(7):532-535.10.1002/(SICI)1521-1878(199807)20:7<532::AID-BIES3>3.0.CO;2-M.

8. Dalle B, Rubin JE, Alkan O, Sukonnik T, Pasceri P, Yao S, Pawliuk R, Leboulch P, Ellis J: eGFP reporter genes silence LCRbeta-globin transgene expression via CpG dinucleotides. Mol Ther. 2005, 11 (4): 591-599. 10.1016/j.ymthe.2004.11.012.

9. Hammer RE, Pursel VG, Rexroad CE, Wall RJ, Bolt DJ, Ebert KM, Palmiter RD, Brinster RL: Production of transgenic rabbits, sheep and pigs by microinjection. Nature. 1985, 315 (6021): 680-683. 10.1038/315680a0.

10. Krumlauf R, Hammer RE, Tilghman SM, Brinster RL: Developmental regulation of alpha-fetoprotein genes in transgenic mice. Mol Cell Biol. 1985, 5 (7): 1639-1648.

11. Ornitz DM, Palmiter RD, Hammer RE, Brinster RL, Swift GH, MacDonald RJ: Specific expression of an elastase-human growth hormone fusion gene in pancreatic acinar cells of transgenic mice. Nature. 1985, 313 (6003): 600-602. 10.1038/313600a0.

12. Palmiter RD, Norstedt G, Gelinas RE, Hammer RE, Brinster RL: Metallothionein-human GH fusion genes stimulate growth of mice.

Science. 1983, 222 (4625): 809-814. 10.1126/science.6356363.

13. Townes TM, Chen HY, Lingrel JB, Palmiter RD, Brinster RL: Expression of human beta-globin genes in transgenic mice: effects of a flanking metallothionein-human growth hormone fusion gene. Mol Cell Biol. 1985, 5 (8): 1977-1983.

14. Mark WH, Signorelli K, Blum M, Kwee L, Lacy E: Genomic structure of the locus associated with an insertional mutation in line 4 transgenic mice. Genomics. 1992, 13 (1): 159-166. 10.1016/0888-7543(92)90216-F.

15. Chen CM, Choo KB, Cheng WT: Frequent deletions and sequence aberrations at the transgene junctions of transgenic mice carrying the papillomavirus regulatory and the SV40 TAg gene sequences. Transgenic research. 1995, 4 (1): 52-59. 10.1007/BF01976502.

16. Pravtcheva DD, Wise TL: A postimplantation lethal mutation induced by transgene insertion on mouse chromosome 8. Genomics. 1995, 30 (3): 529-544. 10.1006/geno.1995.1274.

17. Nakanishi T, Kuroiwa A, Yamada S, Isotani A, Yamashita A, Tairaka A, Hayashi T, Takagi T, Ikawa M, Matsuda Y, et al: FISH analysis of 142 EGFP transgene integration sites into the mouse genome. Genomics. 2002, 80 (6): 564-574. 10.1006/geno.2002.7008.

18. Bishop JO: Chromosomal Insertion of Foreign DNA. Transgenic animals: generation and use. Edited by: Houdebine L-M. 1997, Amsterdam, the Netherlands: Harwood Academic Publishers, 219-223.

19. Covarrubias L, Nishida Y, Mintz B: Early postimplantation embryo lethality due to DNA rearrangements in a transgenic mouse strain. Proc Natl Acad Sci USA. 1986, 83 (16): 6020-6024. 10.1073/pnas.83.16.6020.

20. Gordon JW, Ruddle FH: DNA-mediated genetic transformation of mouse embryos and bone marrow – a review. Gene. 1985, 33 (2): 121-136. 10.1016/0378-1119(85)90087-3.

21. Hamada T, Sasaki H, Seki R, Sakaki Y: Mechanism of chromosomal integration of transgenes in microinjected mouse eggs: sequence analysis of genome-transgene and transgene-transgene junctions at two loci. Gene. 1993, 128 (2): 197-202. 10.1016/0378-1119(93)90563-I.

22. Overbeek PA, Lai SP, Van Quill KR, Westphal H: Tissue-specific expression in transgenic mice of a fused gene containing RSV terminal sequences. Science. 1986, 231 (4745): 1574-1577. 10.1126/science.3006249.

23. Rohan RM, King D, Frels WI: Direct sequencing of PCR-amplified

junction fragments from tandemly repeated transgenes. Nucleic acids research. 1990, 18 (20): 6089-6095. 10.1093/nar/18.20.6089.

24. Takano M, Egawa H, Ikeda JE, Wakasa K: The structures of integration sites in transgenic rice. Plant J. 1997, 11 (3): 353-361. 10.1046/j.1365-313X.1997.11030353.x.

25. Wilkie TM, Palmiter RD: Analysis of the integrant in MyK-103 transgenic mice in which males fail to transmit the integrant. Mol Cell Biol. 1987, 7 (5): 1646-1655.

26. Ivics Z, Hackett PB, Plasterk RH, Izsvak Z: Molecular reconstruction of Sleeping Beauty, a Tc1-like transposon from fish, and its transposition in human cells. Cell. 1997, 91 (4): 501-510. 10.1016/S0092-8674(00)80436-5.

27. Izsvak Z, Ivics Z, Plasterk RH: Sleeping Beauty, a wide host-range transposon vector for genetic transformation in vertebrates. J Mol Biol. 2000, 302 (1): 93-102. 10.1006/jmbi.2000.4047.

28. Clark KJ, Leaver MJ, Foster LK, Carlson DF, Fahrenkrug SC: Passport, a native Tc1/mariner transposon from Pleuronectes platessa, functions in vertebrate cells. in preparation.

29. Leaver MJ: A family of Tc1-like transposons from the genomes of fishes and frogs: evidence for horizontal transmission. Gene. 2001, 271 (2): 203-214. 10.1016/S0378-1119(01)00530-3.

30. Koga A, Iida A, Kamiya M, Hayashi R, Hori H, Ishikawa Y, Tachibana A: The medaka fish Tol2 transposable element can undergo excision in human and mouse cells. J Hum Genet. 2003, 48 (5): 231-235. 10.1007/s10038-003-0016-4.

31. Koga A, Suzuki M, Inagaki H, Bessho Y, Hori H: Transposable element in fish. Nature. 1996, 383 (6595): 30-10.1038/383030a0.

32. Ding S, Wu X, Li G, Han M, Zhuang Y, Xu T: Efficient transposition of the piggyBac (PB) transposon in mammalian cells and mice. Cell. 2005, 122 (3): 473-483. 10.1016/j.cell.2005.07.013.

33. Fraser MJ, Ciszczon T, Elick T, Bauser C: Precise excision of TTAA-specific lepidopteran transposons piggyBac (IFP2) and tagalong (TFP3) from the baculovirus genome in cell lines from two species of Lepidoptera. Insect Mol Biol. 1996, 5 (2): 141-151.

34. Wu SC, Meir YJ, Coates CJ, Handler AM, Pelczar P, Moisyadi S, Kaminski JM: piggyBac is a flexible and highly active transposon as compared to sleeping beauty, Tol2, and Mos1 in mammalian cells. Proc Natl Acad Sci USA. 2006, 103 (41): 15008-15013. 10.1073/pnas.0606979103.

35. Davidson AE, Balciunas D, Mohn D, Shaffer J, Hermanson S, Sivasubbu S, Cliff MP, Hackett PB, Ekker SC: Efficient gene delivery and gene expression in zebrafish using the Sleeping Beauty transposon. Dev Biol. 2003, 263 (2): 191-202. 10.1016/j.ydbio.2003.07.013.

36. Kawakami K, Koga A, Hori H, Shima A: Excision of the tol2 transposable element of the medaka fish, Oryzias latipes, in zebrafish, Danio rerio. Gene. 1998, 225 (1–2): 17-22. 10.1016/S0378-1119(98)00537-X.

37. Kawakami K, Shima A, Kawakami N: Identification of a functional transposase of the Tol2 element, an Ac-like element from the Japanese medaka fish, and its transposition in the zebrafish germ lineage. Proc Natl Acad Sci USA. 2000, 97 (21): 11403-11408. 10.1073/pnas.97.21.11403.

38. Hamlet MR, Yergeau DA, Kuliyev E, Takeda M, Taira M, Kawakami K, Mead PE: Tol2 transposon-mediated transgenesis in Xenopus tropicalis. Genesis. 2006, 44 (9): 438-445. 10.1002/dvg.20234.

39. Kawakami K, Imanaka K, Itoh M, Taira M: Excision of the Tol2 transposable element of the medaka fish Oryzias latipes in Xenopus laevis and Xenopus tropicalis. Gene. 2004, 338 (1): 93-98. 10.1016/j. gene.2004.05.013.

40. Sinzelle L, Vallin J, Coen L, Chesneau A, Du Pasquier D, Pollet N, Demeneix B, Mazabraud A: Generation of trangenic Xenopus laevis using the Sleeping Beauty transposon system. Transgenic research. 2006, 15 (6): 751-760. 10.1007/s11248-006-9014-6.

41. Dupuy AJ, Fritz S, Largaespada DA: Transposition and gene disruption in the male germline of the mouse. Genesis. 2001, 30 (2): 82-88. 10.1002/ gene.1037.

42. Fischer SE, Wienholds E, Plasterk RH: Regulated transposition of a fish transposon in the mouse germ line. Proc Natl Acad Sci USA. 2001, 98 (12): 6759-6764. 10.1073/pnas.121569298.

43. Horie K, Kuroiwa A, Ikawa M, Okabe M, Kondoh G, Matsuda Y, Takeda J: Efficient chromosomal transposition of a Tc1/mariner- like transposon Sleeping Beauty in mice. Proc Natl Acad Sci USA. 2001, 98 (16): 9191-9196. 10.1073/pnas.161071798.

44. Kawakami K, Noda T: Transposition of the Tol2 element, an Ac-like element from the Japanese medaka fish Oryzias latipes, in mouse embryonic stem cells. Genetics. 2004, 166 (2): 895-899. 10.1534/ genetics.166.2.895.

45. Luo G, Ivics Z, Izsvak Z, Bradley A: Chromosomal transposition of a Tc1/mariner-like element in mouse embryonic stem cells. Proc Natl Acad Sci USA. 1998, 95 (18): 10769-10773. 10.1073/pnas.95.18.10769.

46. Yant SR, Meuse L, Chiu W, Ivics Z, Izsvak Z, Kay MA: Somatic integration and long-term transgene expression in normal and haemophilic mice using a DNA transposon system. Nat Genet. 2000, 25 (1): 35-41. 10.1038/75568.

47. Lewandoski M: Conditional control of gene expression in the mouse. Nat Rev Genet. 2001, 2 (10): 743-755. 10.1038/35093537.

48. Gossen M, Bujard H: Tight control of gene expression in mammalian cells by tetracycline-responsive promoters. Proc Natl Acad Sci USA. 1992, 89 (12): 5547-5551. 10.1073/pnas.89.12.5547.

49. Choi BR, Koo BC, Ahn KS, Kwon MS, Kim JH, Cho SK, Kim KM, Kang JH, Shim H, Lee H, et al: Tetracycline-inducible gene expression in nuclear transfer embryos derived from porcine fetal fibroblasts transformed with retrovirus vectors. Mol Reprod Dev. 2006, 73 (10): 1221-1229. 10.1002/mrd.20543.

50. Kues WA, Schwinzer R, Wirth D, Verhoeyen E, Lemme E, Herrmann D, Barg-Kues B, Hauser H, Wonigeit K, Niemann H: Epigenetic silencing and tissue independent expression of a novel tetracycline inducible system in double-transgenic pigs. Faseb J. 2006, 20 (8): 1200-1202. 10.1096/fj.05-5415fje.

51. Branda CS, Dymecki SM: Talking about a revolution: The impact of site-specific recombinases on genetic analyses in mice. Dev Cell. 2004, 6 (1): 7-28. 10.1016/S1534-5807(03)00399-X.

52. Abuin A, Bradley A: Recycling selectable markers in mouse embryonic stem cells. Mol Cell Biol. 1996, 16 (4): 1851-1856.

53. Yu Y, Bradley A: Engineering chromosomal rearrangements in mice. Nat Rev Genet. 2001, 2 (10): 780-790. 10.1038/35093564.

54. Chen YT, Bradley A: A new positive/negative selectable marker, puDeltatk, for use in embryonic stem cells. Genesis. 2000, 28 (1): 31-35. 10.1002/1526-968X(200009)28:1<31::AID-GENE40>3.0.CO;2-K.

55. Plasterk RH, Izsvak Z, Ivics Z: Resident aliens: the Tc1/mariner superfamily of transposable elements. Trends Genet. 1999, 15 (8): 326-332. 10.1016/S0168-9525(99)01777-1.

56. Kempken F, Windhofer F: The hAT family: a versatile transposon group common to plants, fungi, animals, and man. Chromosoma. 2001, 110 (1): 1-9.

57. Sarkar A, Sim C, Hong YS, Hogan JR, Fraser MJ, Robertson HM, Collins FH: Molecular evolutionary analysis of the widespread piggyBac transposon family and related "domesticated" sequences. Mol Genet

Genomics. 2003, 270 (2): 173-180. 10.1007/s00438-003-0909-0.

58. Hackett CS, Geurts AM, Wangensteen KJ, Balciunas D, Ekker SC, Hackett PB: Predicting transposon chromosomal insertion sites: implications for functional genomics and gene therapy. Genome Biology.

59. Altschul SF, Gish W, Miller W, Myers EW, Lipman DJ: Basic local alignment search tool. J Mol Biol. 1990, 215 (3): 403-410.

60. Vallier L, Mancip J, Markossian S, Lukaszewicz A, Dehay C, Metzger D, Chambon P, Samarut J, Savatier P: An efficient system for conditional gene expression in embryonic stem cells and in their in vitro and in vivo differentiated derivatives. Proc Natl Acad Sci USA. 2001, 98 (5): 2467-2472. 10.1073/pnas.041617198.

61. Wilson MH, Coates CJ, George AL: PiggyBac Transposon-mediated Gene Transfer in Human Cells. Mol Ther. 2007, 15 (1): 139-145. 10.1038/sj.mt.6300028.

62. Clark KJ, Geurts AM, Bell JB, Hackett PB: Transposon vectors for gene-trap insertional mutagenesis in vertebrates. Genesis. 2004, 39 (4): 225-233. 10.1002/gene.20049.

63. Collier LS, Carlson CM, Ravimohan S, Dupuy AJ, Largaespada DA: Cancer gene discovery in solid tumours using transposon-based somatic mutagenesis in the mouse. Nature. 2005, 436 (7048): 272-276. 10.1038/nature03681.

64. Dupuy AJ, Akagi K, Largaespada DA, Copeland NG, Jenkins NA: Mammalian mutagenesis using a highly mobile somatic Sleeping Beauty transposon system. Nature. 2005, 436 (7048): 221-226. 10.1038/nature03691.

65. Geurts AM, Wilber A, Carlson CM, Lobitz PD, Clark KJ, Hackett PB, McIvor RS, Largaespada DA: Conditional gene expression in the mouse using a Sleeping Beauty gene-trap transposon. BMC Biotechnol. 2006, 6: 30-10.1186/1472-6750-6-30.

66. Yant SR, Wu X, Huang Y, Garrison B, Burgess SM, Kay MA: High-resolution genome-wide mapping of transposon integration in mammals. Mol Cell Biol. 2005, 25 (6): 2085-2094. 10.1128/MCB.25.6.2085-2094.2005.

67. Ivics Z, Katzer A, Stuwe EE, Fiedler D, Knespel S, Izsvak Z: Targeted sleeping beauty transposition in human cells. Mol Ther. 2007, 15 (6): 1137-1144.

68. Yant SR, Huang Y, Akache B, Kay MA: Site-directed transposon integration in human cells. Nucleic acids research. 2007, 35 (7): e50-

10.1093/nar/gkm089.

69. Dupuy AJ, Clark K, Carlson CM, Fritz S, Davidson AE, Markley KM, Finley K, Fletcher CF, Ekker SC, Hackett PB, et al: Mammalian germline transgenesis by transposition. Proc Natl Acad Sci USA. 2002, 99 (7): 4495-4499. 10.1073/pnas.062630599.

70. Fahrenkrug Lab Home Page. [http://primer.ansci.umn.edu/Fahrenkruglab]

71. Hackett Lab Plasmid Info. [http://www.cbs.umn.edu/labs/perry/plasmids/plasmid.html]

72. Geurts AM, Yang Y, Clark KJ, Liu G, Cui Z, Dupuy AJ, Bell JB, Largaespada DA, Hackett PB: Gene transfer into genomes of human cells by the sleeping beauty transposon system. Mol Ther. 2003, 8 (1): 108-117. 10.1016/S1525-0016(03)00099-6.

73. Ohlfest JR, Frandsen JL, Fritz S, Lobitz PD, Perkinson SG, Clark KJ, Nelsestuen G, Key NS, McIvor RS, Hackett PB, et al: Phenotypic correction and long-term expression of factor VIII in hemophilic mice by immunotolerization and nonviral gene transfer using the Sleeping Beauty transposon system. Blood. 2005, 105 (7): 2691-2698. 10.1182/blood-2004-09-3496.

74. Balciunas D, Wangensteen KJ, Wilber A, Bell J, Geurts A, Sivasubbu S, Wang X, Hackett PB, Largaespada DA, McIvor RS, et al: Harnessing a high cargo-capacity transposon for genetic applications in vertebrates. PLoS Genet. 2006, 2 (11): e169-10.1371/journal.pgen.0020169.

75. Parinov S, Kondrichin I, Korzh V, Emelyanov A: Tol2 transposon-mediated enhancer trap to identify developmentally regulated zebrafish genes in vivo. Dev Dyn. 2004, 231 (2): 449-459. 10.1002/dvdy.20157.

76. Deachapunya C, Palmer-Densmore M, O'Grady SM: Insulin stimulates transepithelial sodium transport by activation of a protein phosphatase that increases Na-K ATPase activity in endometrial epithelial cells. J Gen Physiol. 1999, 114 (4): 561-574. 10.1085/jgp.114.4.561.

Chapter 7

DECONSTRUCTING THE GENETIC BASIS OF SPENT SULPHITE LIQUOR TOLERANCE USING DEEP SEQUENCING OF GENOME-SHUFFLED YEAST

Dominic Pinel[1,3,] David Colatriano[1], Heng Jiang[1,4], Hung Lee[2] and Vincent JJ Martin[1]

[1]Department of Biology, Centre for Structural and Functional Genomics, Concordia University, 7141 Sherbrooke Street West, Montréal, Québec H4B 1R6, Canada

[2] School of Environmental Sciences, University of Guelph, Guelph, Ontario N1G 2 W1, Canada

[3] Energy Biosciences Institute, University of California, Berkeley, Berkeley, CA 94704, USA

[4] Crabtree Nutrition Laboratories, McGill University Health Center, Montreal, Quebec H3A 1A1, Canada.

ABSTRACT

Background

Identifying the genetic basis of complex microbial phenotypes is currently a major barrier to our understanding of multigenic traits and our ability to rationally design biocatalysts with highly specific attributes for the biotechnology industry. Here, we demonstrate that strain evolution by meiotic recombination-based genome shuffling coupled with deep sequencing can be used to deconstruct complex phenotypes and explore the nature of multigenic traits, while providing concrete targets for strain development.

Results

We determined genomic variations found within*Saccharomyces cerevisiae*previously evolved in our laboratory by genome shuffling for tolerance to spent sulphite liquor. The representation of these variations was backtracked through parental mutant pools and cross-referenced with RNA-seq gene expression analysis to elucidate the importance of single mutations and key biological processes that play a role in our trait of interest. Our findings pinpoint novel genes and biological determinants of lignocellulosic hydrolysate inhibitor tolerance in yeast. These include the following: protein homeostasis constituents, including Ubp7p and Art5p, related to ubiquitin-mediated proteolysis; stress response transcriptional repressor, Nrg1p; and NADPH-dependent glutamate dehydrogenase, Gdh1p. Reverse engineering a prominent mutation in ubiquitin-specific protease gene*UBP7*in a laboratory*S. cerevisiae*strain effectively increased spent sulphite liquor tolerance.

Conclusions

This study advances understanding of yeast tolerance mechanisms to inhibitory substrates and biocatalyst design for a biomass-to-biofuel/biochemical industry, while providing insights into the process of mutation accumulation that occurs during genome shuffling.

BACKGROUND

Mapping genotype to phenotype for complex traits and using these data for the rational design of biocatalysts is a natural progression in an increasingly sophisticated biotechnology industry. Unfortunately, current technologies do not allow for the rapid creation of industrially relevant microorganisms or the ability to access and understand multigenic phenotypic traits. Traditionally, strain improvement has been based on a repetitive cycle of random mutagenesis and selection to improve the phenotypic traits of industrial microbes [1]. Advanced DNA sequencing technology now allows for rapid sequencing of the genomes of these industrial strains to identify the mutations that confer improved phenotypes. However, in resequencing the genomes of randomly evolved strains, a small number of potentially productive mutations are often accompanied by a background of non-productive mutations [2-4]. Extensive functional characterizations of individual genotypic variations are therefore needed to unravel which mutations are associated with the phenotype of interest. Furthermore, our ability to deconstruct complex, multigenic traits is still limited. Possible solutions to these problems include sequencing pools of independent mutants [5], backcrossing non-productive mutations prior to genome resequencing, or combining intercrossing with pool sequencing to assign quantitative trait loci [6] in order to hone in on productive mutations. Nonetheless, resolving such data into manageable and testable hypotheses can be insurmountably challenging.

In this study, we aimed to mitigate these shortcomings by sequencing a strain created by genome shuffling from a known background strain and tracking mutations throughout the evolving population. This allows for important mutations to be ranked and novel gene targets to be acquired from background mutations. Moreover, genome shuffling (GS) is an alternative to classical strain improvement that is a means to accelerate the evolution of industrial strains in the laboratory and minimize the accumulation of non-productive mutations. The rationale behind GS is to rapidly combine beneficial mutations and cross out deleterious ones, which can be achieved in *Saccharomyces cerevisiae* by recursive pool-wise mating of mutant populations (Figure1A) [7-10]. This strain engineering technique is particularly powerful to address multigenic, complex phenotypes such as resistance to ethanol, lactic acid, heat and low pH or production of compounds like tylosin or taxol (reviewed in [11]). Theoretically, the background of non-productive or deleterious mutations can be minimized by attenuating mutagen dosage, screening for parental strains that contain productive mutations, followed by trait-enhanced mutant strain recombination to combine mainly productive mutations into a single strain. Furthermore, by its very nature, GS brings interacting mutations together into single strains. Although the utility of GS has been demonstrated repeatedly through phenotypic observation, the nature of the mutations accumulated during the strain evolution has not been tracked through genome resequencing. Sequencing GS isolates, therefore, should yield access to determinants of multigenic traits at single nucleotide resolution, while minimizing non-productive variation discovery. Tracking mutations throughout the population of genome-shuffled strains can then be used to further increase the possibility of finding productive mutations.

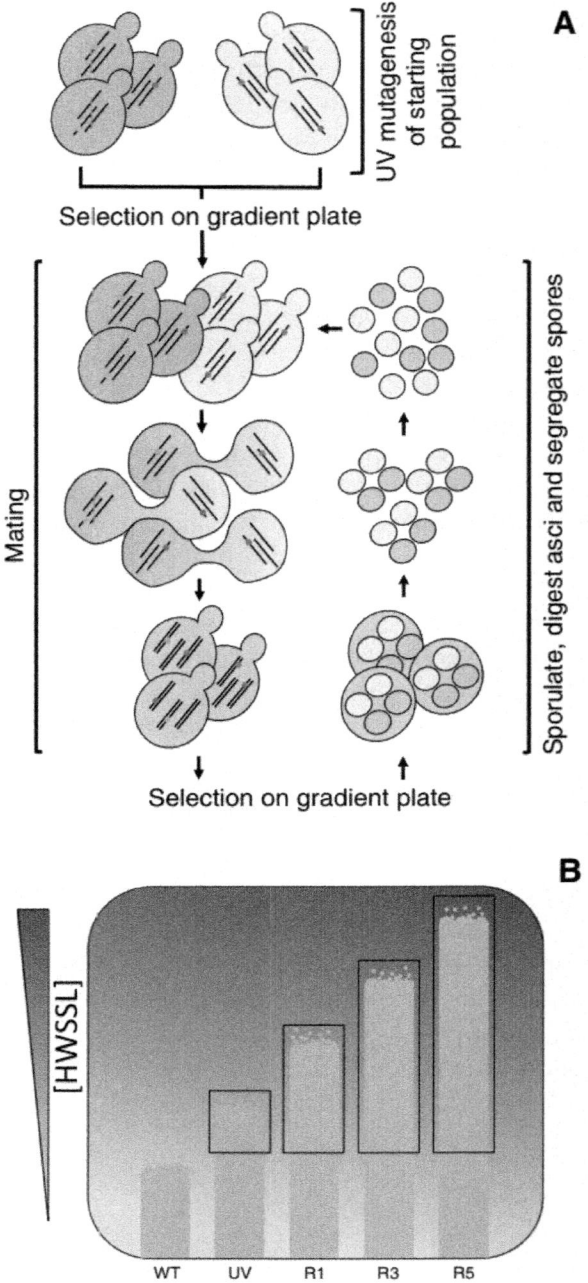

Figure 1: Meiotic recombination-mediated genome shuffling by recursive breeding for HWSSL tolerance. (A)A recursive mating methodology was used to create the

HWSSL strains and populations used in this study. Large pools of UV mutants and genome-shuffled populations were screened on HWSSL gradient agar plates prior to each round of shuffling.(**B**)Portions of each population that showed more tolerance than the reference (WT) (black boxes) were scraped from gradient plates and used for genome shuffling (different rounds of genome shuffling are depicted - round 1 (R1), round 3 (R3) and round 5 (R5)). Initial UV mutant populations (UV) of each haploid mating type showing enhanced HWSSL tolerance were scraped and used to begin the recursive breeding methodology. Selection on HWSSL gradient plates was carried out between each round of GS in order to enrich the mating pool for strains showing the tolerance phenotype. A portion of each mutant pool (UV through five rounds of GS) was frozen for population sequencing (see 'Results and discussion'). Individual colonies showing the highest tolerance to HWSSL were isolated from the frontier of growth. HWSSL, hardwood spent sulphite liquor.

Microbial tolerance to lignocellulosic hydrolysates is a complex, multigenic trait that is of significant importance to a biomass-to-fuel/chemical industry. The pretreatment of lignocellulose to fermentable sugars yields many by-products that are inhibitory to fermenting yeasts. The main sources of inhibition come from osmotic pressure, reactive oxygen species (ROS) damage or compounds that include furan aldehydes, primarily furfural and 5-(hydroxymethyl)-2-furaldehyde (HMF), phenolics and organic acids, especially acetic, formic and levulinic acids [12-16]. The biological factors implicated in the tolerance of yeast to lignocellulose fermentation inhibitors have been reviewed [12,13,17]. Ultimately, engineering productive industrial biocatalysts with tolerance traits will be a pervasive biotechnological problem, and rationally engineering these traits will require an understanding of interacting genes and biological processes that affect tolerance. Currently, a lack of knowledge on the multiple cellular processes and genes involved in microbial tolerance to lignocellulosic hydrolysates makes rational engineering of strains resistant to these substrates implausible [8,18,19].

In a previous study [8], we evolved a strain of *S. cerevisiae*, R57, through genome shuffling (Figure1A,B) that is capable of survival, growth and ethanol productivity in hardwood spent sulphite liquor (HWSSL), a highly inhibitory lignocellulosic substrate generated by the acid bisulphite pulping process [20,21], to levels of tolerance previously unreported. HWSSL contains sugars, lignosulphonates, inhibitory compounds, residual pulping chemicals, ammonia and sulphite [22] and can contain heavy metal ions (iron, chromium, nickel and copper) that originate from the corrosion of pulping and bleaching equipment [23,24]. Some of the major constituents are approximated at the following levels (%w/v): 0.83% to 1.45% hexose sugars, 1.7% to 2.1% pentose sugars, 0.18% to 0.5% furfural, 0.9% to 1.0% acetic acid, 0.5% to 0.7% sulphate, 1% ammonia and 17% lignosulphonate [8,25]. In evolving R57 through GS, it was hypothesized that beneficial, tolerance-conferring mutations were combined through recursive population-wise meiotic recombination, which yielded a progression towards higher tolerance of the hydrolysate displayed with each subsequent round of GS (Figure1) [8]. Strain R57 showed improved cross-tolerance to several known inhibitors of lignocellulosic hydrolysates [8], supporting the theory that R57 harbours multiple tolerance-conferring mutations. In this study, we resequenced the genome of R57 in order to discover the mutations that were accumulated to confer hydrolysate tolerance, combined with profiling the relative abundance of R57 mutations in the heterogeneous parental GS populations to probe for the relative phenotypic effect of each discovered mutation and explore mutation recombination in GS-evolved pools.

This study describes a relatively cost-effective way to explore combinatorial space of productive mutations within a single genome. To our knowledge, this is the first study to resequence the genome of a strain evolved through GS and the first resequencing project for a yeast strain specifically evolved for its tolerance to a complex mixture of inhibitors in lignocellulosic substrates.

RESULTS AND DISCUSSION

Genome Sequencing of GS-evolved Strain R57

The genome of *S. cerevisiae* strain R57 was resequenced in an effort to pinpoint genetic changes associated with its tolerance to HWSSL. Both the parental haploid CEN.PK113-7D and mutant diploid R57 were sequenced and compared at approximately 100-fold and approximately 350-fold coverage per nucleotide (Additional file1: Data S1), respectively, which allows for meaningful mutation prediction [2]. The relative level of sequence read coverage per chromosome between the strains is similar and suggests the absence of aneuploidy (Additional file1: Data S1). Insertion/deletion (indel) and copy number variation (CNV) analysis returned no detectable differences between the wild type (WT) and R57 after visual inspection. All of the mutations discovered from the mutation analysis are single nucleotide polymorphisms (SNPs) and were confirmed by Sanger sequencing.

Twenty-one point mutations were found that could affect at least 17 genes, based on location within open reading frames (ORFs) or untranslated regions (UTRs). These include 16 SNPs affecting 12 ORFs, 14 of which lead to missense mutations, with the remaining 2 leading to silent mutations (Table1). The five mutations not found within ORFs are all located in 5' or 3' UTRs. A heterogeneous SNP lies 43 bps 3' of *BCS1* and is predicted to be part of the 5' UTR of *YDR374C* [26] and in the 3' UTR of *WIP1* [27]. This mutation is not included in subsequent analyses due to the ambiguity of the affected gene. Gene ontology categories and interaction maps for the affected genes were generated (Additional file2: Methods and Additional file3: Figures S1 and S2), and the mutation analysis results are summarized in Table1.

Table 1: Point mutations discovered in GS-evolved strain R57

Gene	Chr	Mutation	Gene function	Genotype	SIFT score
NRG1	IV	137C > A (P46Q)	Transcriptional repressor, stress tolerance	Homo	0
UBP7[a]	IX	2466 T > A (N822K)	Ubiquitin-specific prote-ase	Homo	0.33
ART5[a]	VII	454C > A (L152I)	Regulates endocytosis and turnover of cell-surface proteins by targeted ubiquitination	Hetero	0.17
SSA1[a]	I	91C > A (Q31K)	ATP-ase, protein folding, heat shock, HSP70	Hetero	0
GDH1[b]	XV	47C > T (S16F)	Glutamate synthesis from ammonia	Hetero	0
GDH1[b]		68 T > G (F23C)		Hetero	0
ARO1[b]	IV	1283C > T (S428F)	Catalyzes biosynthesis of chorismate leading to aromatic amino acids	Hetero	0
ARO1[b]		1284C > T (Silent)		Hetero	-
STE5[b]	IV	512C > T (S171F)	Pheromone-response scaffold protein, forms MAPK cascade complex	Hetero	0
STE5[b]		2649 T > C (Silent)		Hetero	-
MAL11[b]	VII	310C > T (P104S)	Alpha-glucoside sym-porter, with high affinity for trehalose	Hetero	0
MAL11[b]		482 T > A (M161K)		Hetero	0.02
GSH1[c]	X	T > A (73 bp 5′ UTR)	Glutamylcysteine synthe-tase, glutathione biosyn-thesis	Hetero	-
PBP1[c]	VII	T > C (191 bp 5′ UTR)	Controls mRNA poly(A), stress granule formation and translation control	Hetero	-

FIT3[c]	XV	C > T (42 bp 3' UTR)	Iron transport	Hetero	-
NOP58[c]	XV	A > T (25 bp 3' UTR)	Pre-rRNA processing and rRNA synthesis	Hetero	-
YNL058C[d]	XIV	7A > G (K3E)	Unknown function	Hetero	0.42
DOP1[d]	IV	40A > T (N14Y)	Endosome to Golgi transport, ER organization, cell polarity and morphogenesis	Hetero	0.05
TOF2[d]	XI	2141 C > T (S714L)	rDNA silencing, stimulates Cdc14p for mitotic rDNA separation	Hetero	0.27
SGO1[d]	XV	575C > A (S192Y)	Chromosomal segregation and stability	Hetero	0.03

Bold font represents alleles that gain in frequency over GS evolution. [a]Gene group containing genes that are related to protein homeostasis.[b]Gene group containing genes bearing more than one mutation in R57.[c]Gene group containing UTR mutations.[d]Gene group containing alleles with limited evidence for a phenotypic linkage to HWSSL tolerance.

Predicting Protein Function and Phenotypic Effect for Missense Mutations in R57

Functional prediction of altered primary protein structure was carried out using the SIFT (Sorting Intolerant From Tolerant) algorithm [28-30] (Table1). Ten of the 14 missense mutations are expected to affect protein function, with 4 predicted as tolerated by the protein. All but two of the mutations are heterozygous and therefore are expected to have a dominant effect if they contribute to the R57 phenotype. The two homozygous mutations are located withinNRG1andUBP7. Homozygous mutations at these loci suggest that they have been enriched at a high enough population density during GS evolution that they were able to mate with the opposite mating type and may therefore be important to the HWSSL tolerance trait or that spores were insufficiently segregated during GS (Figure1A), promoting homozygosity.

Several genes contain multiple mutations in R57. Genes bearing more than one mutation in R57 suggest that the affected gene is important to the phenotype and the mutations have accumulated due to GS evolution and screening. Two missense mutations affect bothGDH1andMAL11. Gdh1p bears two mutations in R57, S16F and F23C. Visual inspection of the mapping alignment shows that the two mutations are never located on the same sequence read, and therefore, both versions of *GDH1* are mutated in R57. The close proximity of these two mutations suggests that a mutation in this region of *GDH1* may yield a phenotypic trait that has been selected for through GS evolution. The R57*MAL11* gene contains two missense mutations (leading to P104S and M161K). Cloning and sequencing of *MAL11* from R57 shows that the two mutations are not located on the same allele. *STE5* and *ARO1* each bear a second silent mutation that may be present due to close genetic linkage to the productive mutation or lead to a non-obvious phenotypic modification, such as altered mRNA stability, and thereby arise in R57 through selection.

Analysis of Mutation Loci in GS-evolved Heterogeneous Populations

We deeply sequenced the R57 mutation loci within the pooled mutant populations generated during the genome shuffling experiments [8]. The mutation loci were PCR amplified from DNA extracted from the heterogeneous populations and sequenced. There were a total of 3.76×10^6 reads with an average read length of 104 bp and an average fold coverage of 1.35×10^5 reads per nucleotide sequenced. Eighteen SNPs were called (Figure2). We assessed the frequency of each sequence read that contains a mutation within the total number of sequence reads spanning each locus (Figure2). These data were used to predict mutations that may have arisen in single parental strains, due to similar prevalence within a population, or may be epistatic, and to probe for changes in SNP frequency between populations that could indicate the relative influence of those alleles on phenotype. Several mutant alleles increase in frequency over GS evolution, and we hypothesize that this increase is due to a beneficial phenotypic effect that is generated through GS evolution and repeated selection. These include the *ART5,UBP7,SSA1,STE5,NRG1,MAL11*(1 eading to Mal11p - P104S) and both *GDH1* mutations (Figure2). This analysis also yields an additional SNP that falls within the area of PCR amplification, affecting *SSA1* 26 bp upstream of the R57 mutation and leading to a D22G missense mutation, which further supports a determinant role for *SSA1* in the observed phenotype. The *NOP58,STE5*(silent),*DOP1,FIT33'* UTR mutations were not located at a high enough density within the sequenced mutant populations to surpass our detection threshold.

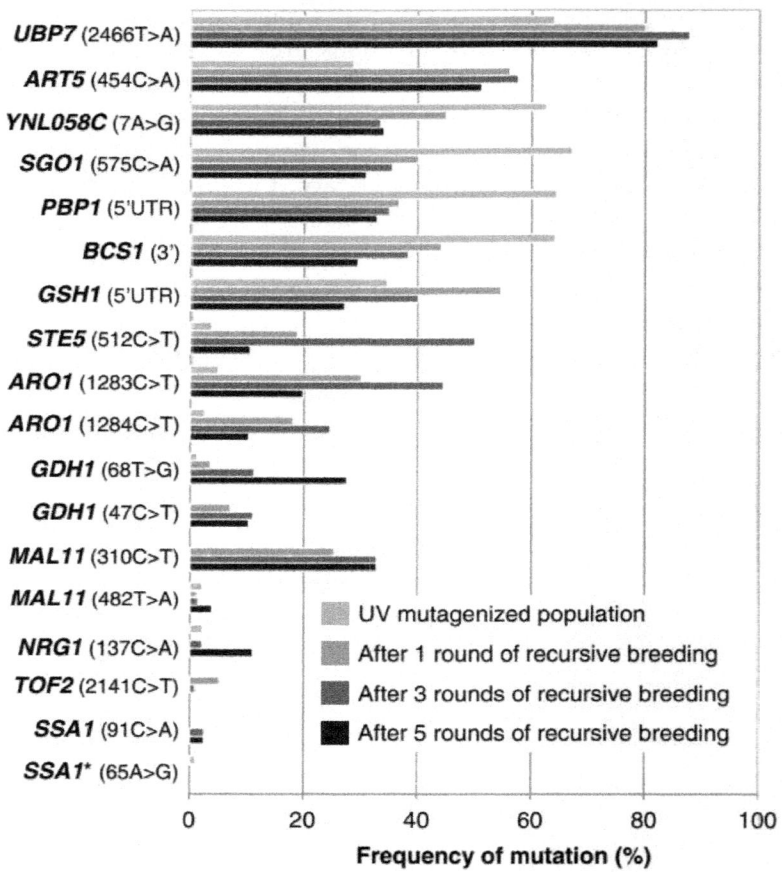

Figure 2: Frequency of R57 mutations throughout GS evolution. The haploid UV mutant population (UV), generated from a single round of UV mutagenesis, and rounds 1, 3 and 5 of GS evolution (R1, R3 and R5, respectively) were PCR amplified at the R57 mutation loci and deep sequenced to determine the relative frequency of sequence reads bearing mutation (above a threshold of 1% of total sequence reads). The asterisk denotes mutations discovered that are not found in R57. UV, ultraviolet.

Five mutations (located upstream of*BCS1*and*PBP1*and in the ORFs of YNL058C,*SGO1*and*UBP7*) are represented at approximately 60% frequency in the UV mutant population. Due to the virtually identical representation of these five mutations within the UV mutant population, we hypothesized that they may arise from a single, highly tolerant haploid strain. The mutated loci identified in R57 were sequenced from seven random, discernible colonies selected from the frontier of growth for the UV mutant populations on HW3SL gradient agar plates (Figure1B). Each of the single colonies contained the*BCS1*,*PBP1*,   YNL058C,*SGO1*and*UBP7*mutations,   corroborating   our hypothesis that a single mutant strain present after UV mutagenesis was likely enriched to represent a large portion of that population. Assuming this mutant strain harboured at least one particularly productive mutation, propagation of these alleles to strain R57 may be a likely outcome. Heterogeneous population sequencing shows that of these five mutations, only the*UBP7*mutation increased in frequency through GS evolution while the other four decreased in frequency (Figure2). We hypothesize that this finding is indicative of a set of four non-productive or less important mutations found with the productive*UBP7*mutation within a single genome, and when meiotic recombination occurs, linkage to the productive mutation diminishes until it reaches a steady state within the evolving population.

The mutation in the 5′ UTR of*GSH1*was identified in approximately 30% of sequence reads generated for this locus (Figure2) in the UV population and was also identified in each of the seven UV mutant isolates that contain*BCS1*,*PBP1*, YNL058C,*SGO1*and*UBP7*mutations. It is therefore likely that all six mutations arose in a single mutant strain. The reason for the discrepancy in allele frequency between*GSH1*and the other five mutations from UV mutant population sequencing is unknown but may be a population sequencing artefact. Unlike*BCS1*,*PBP1*, YNL058C and*SGO1*, the 5′ UTR*GSH1*mutation increases in frequency in the first three rounds of evolution (Figure2) and therefore, as with*UBP7*, more likely contributes to the tolerance phenotype than the other four mutations.

Several mutations that comprise a large part of the first three rounds of GS, and are therefore likely playing a determinant role in HWSSL tolerance, decrease in population frequency in the fifth round of GS (*UBP7*,*ART5*,*ARO1*). We hypothesize this occurs due to competition from strains bearing mutations that were rare in the initial mutant pools (that is,*NRG1*and*GDH1*) or strains harbouring rare recombination events of multiple mutations that have resulted in augmented fitness.

Analysis of R57 SNPs in Isolates of GS round 5 Heterogeneous Population

To identify possible combinations or permutations of mutations enriched through evolution, 20 strains isolated from the growth frontier of the fifth round of recursive GS (Figure1B lane R5) were sequenced via the Sanger method at each of the mutation loci. All of the strains show heterozygosity in at least one of the mutation loci. The results show a heterogeneous population of mutations that are found together in one strain (Figure3). Of these isolates, ≥70% contain at least one mutation in*UBP7*,*ART5*and either of the*GDH1*mutation loci, which further supports determinant roles for these genes on the phenotype. Several of the strains contain very few of the R57 mutations, which may indicate that not all R57 mutations are needed for HWSSL tolerance and likely that other unidentified mutations present within the round 5 population contribute to the tolerance trait. The percentages of mutated alleles within mutant populations as enumerated by population sequencing and single colony sequencing are similar (Figure4); discrepancies in these frequencies may be due to the relatively large difference in sample size. General trends as to allele frequency are easily apparent and support our ability to generalize allele frequency within GS mutant populations by sequencing PCR-amplified mutation loci. The GS population sequencing data support determinant roles in HWSSL tolerance for a large portion of the mutated R57 genes based on their pervasiveness throughout the evolving populations. However, the most highly enriched mutations that increase over the strain evolution are of particular interest for reverse engineering studies. These include*UBP7,GDH1,ART5,ARO1,STE5*and*MAL11*.

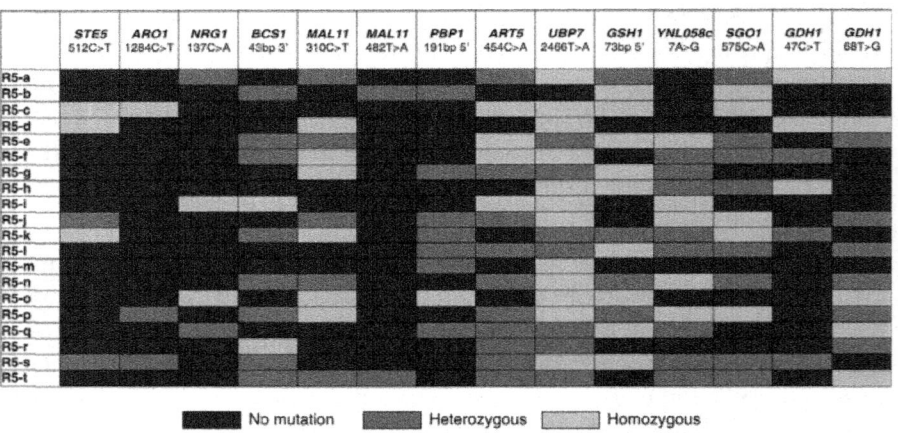

Figure 3: Prevalent mutation loci found in 20 individual isolates from the fifth round of GS.Twenty (R5-a-t) strains from the fifth round (R5) of GS were isolated, and the most prevalent mutation loci were PCR amplified and sequenced to determine their presence within each strain. The presence of homozygous mutation (green), heterozygous mutation (red), as determined by a mixed Sanger sequence at the nucleotide of interest, or no mutation (brown) are depicted for each strain.

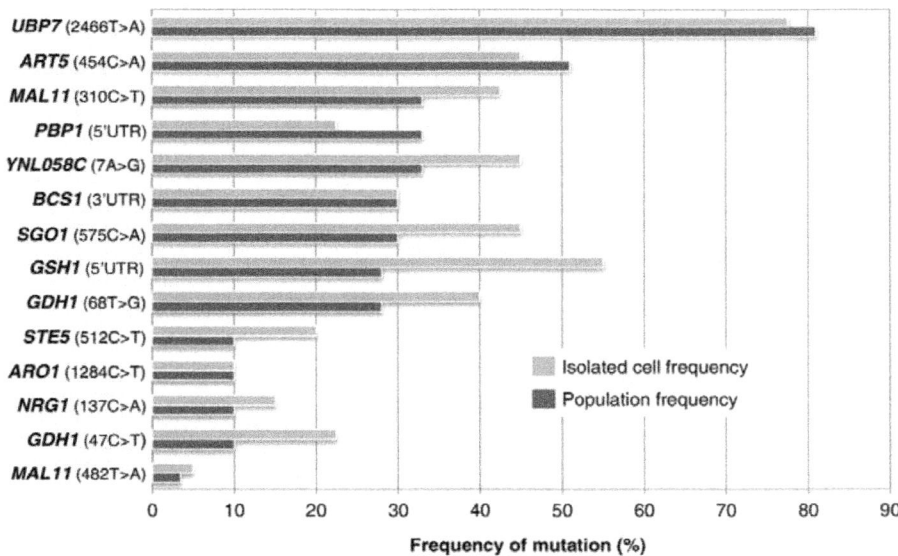

Figure 4: Comparison of mutation frequencies between population and single colony at mutation loci. Twenty single colonies were sequenced using the Sanger method at the most prevalent R57 mutation loci for comparison to deep sequencing results of mutation loci within the entire GS-evolved population.

RNA-Seq Gene Expression Analysis

To measure the impact of the mutations on gene expression in strain R57 and to probe for biological processes related to HWSSL tolerance, the gene transcription profile of strain R57 was compared to the WT diploid under control conditions (growth in defined medium, see 'Materials and methods') (Additional file4: Data S2). Functional clustering was performed on the differentially expressed genes to discover enriched functional roles of gene products and biological pathways of interest (Figure5).

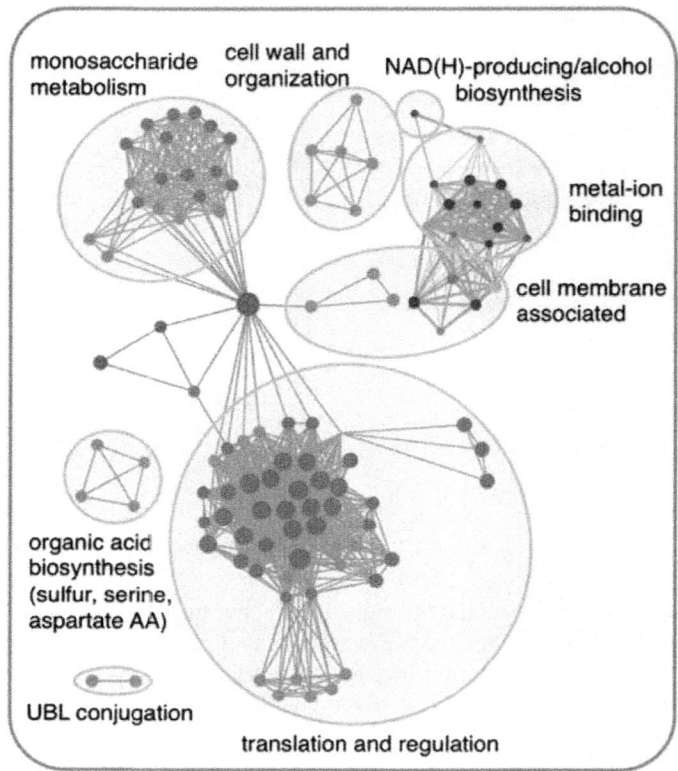

monosaccharide metabolism

cell wall and organization

NAD(H)-producing/alcohol biosynthesis

metal-ion binding

cell membrane associated

organic acid biosynthesis (sulfur, serine, aspartate AA)

UBL conjugation

translation and regulation

Figure 5: Enrichment clustering of differentially expressed genes from RNA-seq analysis. WT and R57 were compared for constitutive differential expression of genes when grown in SD medium. Gene lists were compiled for significantly ($P < 0.05$) upregulated >2-fold or significantly downregulated <2-fold differentially expressed genes. Red colours represent upregulation, while blue colours represent downregulation. Darker shades represent a relatively higher confidence of enrichment score. Larger node sizes represent relatively larger numbers of differentially regulated genes associated with the given ontology category as compared to the full gene. Smaller distance between nodes denotes a higher degree of relationship between ontology categories, while thicker edge lines (green) denote a relatively higher degree of similarity between category nodes in terms of the degree of overlap between the specific gene sets they are associated with. AA, amino acid; UBL, ubiquitin-like.

These analyses identified 149 differentially expressed genes (>2-fold) (Additional file3: Figure S3 and Additional file4: Data S2). None of the 16 genes harbouring a mutation (Table1) are found in this group with the exception of *NRG1*, which is upregulated 3-fold. Clustering of the 131 upregulated R57 genes as compared to the WT includes the major cluster of translation-related genes, mainly associated with ribosome biogenesis and translation regulation and 15 genes related to monosaccharide metabolism. These findings suggest a more active metabolism of R57 in early stationary phase, which may be related to growth differences between the WT and R57 [8]. Indeed, R57 displays a similar growth rate to the WT but reaches a lower optical density (OD) at stationary phase under non-inhibitory conditions with residual glucose remaining in the R57 medium (Additional file3: Figure S4). The remaining upregulated enrichment clusters include genes related to cell wall organization, the cell membrane, ubiquitin-like (UBL) conjugation and organic acid synthesis pathways. Only 18 genes were downregulated under non-inhibitory conditions, resulting in 3 clusters of genes that are highly enriched (Figure5). These include genes related to NADH/alcohol metabolism and metal-ion metabolism or are associated with cellular membrane transport or lipid metabolism.

*UBP7*and the Ubiquitin-Mediated Protein Homeostasis Machinery are Determinants of HWSSL Tolerance

*UBP7*bears a mutation that gains in frequency to represent a large portion of the GS-evolved population (Figure2). Functional analysis suggests highly probable effects on protein structure and function due to this mutation, and*UBP7*is known to have a high degree of interaction with mutated R57 genes (Table1, Additional file2: Methods, Additional file3: Figures S1 and S2). The RNA-seq data also shows enrichment clustering for increased expression of genes that encode proteins related to the ubiquitin-mediated proteolytic machinery (UBL conjugation) (Figure5). However, the genes within these UBL-conjugated gene clusters are associated with diverse biological processes, many of which do not play a direct role in the ubiquitination of proteins or ubiquitin-mediated proteolysis. Two notable exceptions,*UBI2*and*UBI3*, which show increased expression in R57 relative to the WT (2.7- and 2.8-fold, respectively; Additional file4: Data S2), encode ubiquitin fused to ribosomal proteins [31] and are responsible for generating ubiquitin as a fusion protein that is then cleaved to yield free ubiquitin by deubiquitinases. Most of the UBL conjugation cluster genes encode proteins that are regulated by this mechanism. Many of these genes are stress-tolerance related (*RHR2,PUN1,ENA5,PDR5,PDR12*). This suggests that stress tolerance genes showing increased expression due to HWSSL exposure may be also differentially controlled at the protein level by a modified ubiquitination machinery. Altogether, a significant portion of the HWSSL tolerance trait shown by R57 seems to be a direct result of changes in ubiquitin-mediated proteolytic pathways.

Protein damage and aggregation are likely a source of toxicity in cells exposed to lignocellulosic hydrolysates and have at least been partially shown to arise due to ROS damage from furan aldehyde exposure [32]. Cells regulate protein quality through destruction of misfolded or damaged polypeptides largely through selective, energy-dependent labelling with ubiquitin leading to digestion by the 26S proteasome complex [33].*UBP7*encodes a ubiquitin-specific protease that cleaves ubiquitin-protein fusions [34], and as such, it is part of this ubiquitin-induced signalling machinery of the cell [35-37]. The cell's requirement for available ubiquitin increases during stress exposure [38]. Deubiquitinating enzymes act to recover ubiquitin from ubiquitin-protein conjugates and may therefore have a direct bearing on cellular protein and ubiquitin homeostasis [37]. It has already been shown that mutations within a deubiquitinase enzyme,*UBP6*, can dramatically change steady-state ubiquitin levels within a cell [39], which is known to affect tolerance to a variety of stressors [40-42] and yeast prion toxicity [42]. Furthermore, upregulation of*UBP13*, another yeast deubiquitinating enzyme, is beneficial to cells under cold stress and suggests that altering ubiquitin-induced signalling may be a viable path towards other forms of stress tolerance [43].

In order to test the role of the*UBP7*2466 T > A mutation in hydrolysate tolerance, we replaced both WT copies of*UBP7*with this gene variant in a diploid WT CEN.PK background. The homozygous*UBP7*2466 T > A strain was able to colonize a higher concentration of HWSSL on gradient agar plate screening (Figure6) compared to the WT, but its tolerance to HWSSL is still below that of R57. The phenotype conferred by the*UBP7*mutant does not reconstitute the full HWSSL tolerance displayed by R57 and supports the hypothesis that the high level of tolerance shown by R57 is a result of several mutations incorporated through GS. One role of ubiquitination is the internalization of cell surface proteins [44-46]. This function relates to Art5p, which belongs to the ART (arrestin-related trafficking) family of proteins that are believed to function as adaptors for Rsp5p, a ubiquitin ligase that promotes endocytosis of plasma membrane proteins, including transporters, targeting damaged or unneeded plasma membrane proteins for vacuolar degradation [47]. Mutation of*ART5*may represent a way for R57 to regulate destruction of proteins damaged by HWSSL stress or direct changes to the plasma membrane in order to respond more efficiently to the toxic HWSSL environment. As might be expected of a leucine to isoleucine mutation, like that found in the R57 Art5p L152I protein, this change is expected to be tolerated by Art5p (Table1). Nevertheless, the high and increasing frequency of the*ART5*454C > A mutation shown in the mutant pool sequencing experiment (Figure2), the differential regulation of cell surface remodelling genes between the WT and R57 (Additional file4: Data S2) and the proven role of ubiquitin-mediated degradation-machinery gene*UBP7*in HWSSL tolerance suggest that this mutation might also play a determinant role in HWSSL tolerance

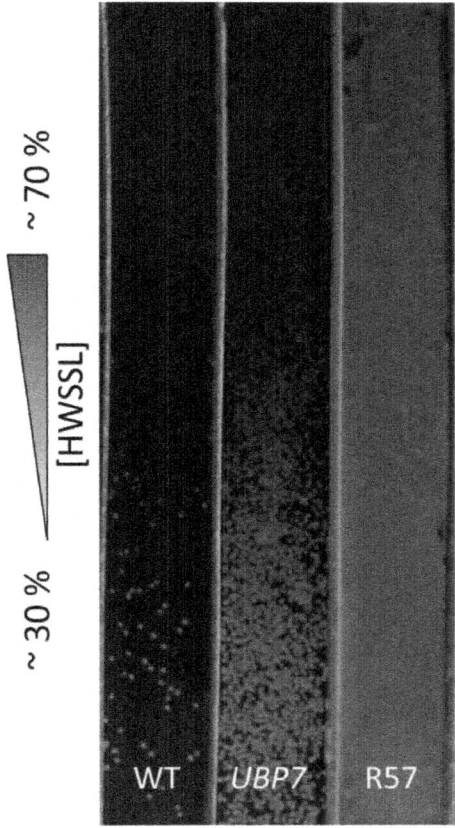

Figure 6: Testing of*UBP7*mutation using HWSSL gradient agar plate screening.A HWSSL gradient agar plate (approximately 30% to 70% HWSSL from bottom to top of plate) was spread with cells from cultures of (from left to right lanes) CEN.PK 113-7D diploid (WT), CEN.PK 113-7D diploid bearing the homozygous*UBP7*mutation from R57 (*UBP7*), and strain R57 (R57). HWSSL, hardwood spent sulphite liquor; WT, wild-type.

.Nrg1p as a Determinant in the Inhibitor Tolerance Trait

As the sole transcription factor-encoding gene located amongst R57 mutations, the*NRG1*137C > A may result in the most pervasive phenotypic consequences. The homozygous P46Q mutation of Nrg1p was predicted to be non-tolerated with a SIFT score of 0.00 (Table1), as proline is strictly conserved at this position and its substitution would result in altered protein function. SNP analysis of the GS-evolved populations showed a diminished prevalence of this mutation after mating of the UV-treated haploid population (Figure2), leading to the hypothesis that the*NRG1*mutation results in a recessive, loss-of-function mutation. The*NRG1*137C > A allele gains prevalence in the GS populations after that point, present at a frequency of approximately 2% after three rounds and approximately 11% after five rounds of pool-wise recursive breeding, when two copies of the allele are more likely present in single strains.

Nrg1p recruits the Tup1p-Cyc8p complex to repress gene expression. Therefore, the Nrg1p P46Q mutation may decrease repression of Nrg1p-controlled genes.*NRG1*, which self-regulates its transcription [48-51], was upregulated 3-fold in R57 over WT (Additional file4: Data S2). The closely related transcription factor*NRG2*was similarly upregulated approximately 5-fold in R57 over WT. In addition, four of the five most highly upregulated genes in R57 are known to be regulated by Nrg1p. These genes include*CWP1*(approximately 13-fold upregulated), YLR015C (approximately 44-fold upregulated),*GAT3*(approximately 23-fold upregulated) and*TDA6*(approximately 14-fold upregulated) (Additional file4: Data S2). In*S. cerevisiae*, the most significant transcriptional responses governed by Nrg1p are all related to a multitude of stress conditions [52]. One of the main functional gene categories controlled by Nrg1p is related to peroxide tolerance, which is a specific trait of R57 [8]. Additionally, Vyas*et al.* reported that Nrg1-2p regulates a set of stress response genes and *nrg1 nrg2*deletion mutants exhibit tolerance to oxidative stress and salt exposure [53], a trait that is also shared by R57 [8].

R57 also displays acetic acid tolerance after pre-exposure to HWSSL [8]. Nrg1p can directly repress genes that are activated by the downstream action of protein kinase Snf1p [54], which stimulates upregulation of stress responsive genes [55] and the transcription activator Haa1p, which imparts acetic acid tolerance [56]. Many of the genes highlighted as members of enriched clusters that show increased expression in R57 are known constituents of the Haa1p regulon including*AQR1,HSP26,MSN4,PDR12,PDR16,SPI1,SUR2,SSE2,TDA6,TPO1,TPO2*and*TPO3*(Additional file4: Data S2). Genes like Haa1p-dependent*TPO2*and*TPO3*(Nrg1p-regulated [49]) show a prominent effect on acetic acid tolerance [57]. Overall, our data support a strong role for Nrg1p in the control of hydrolysate inhibitor tolerance.

A Determinant Role for*GDH1*in HWSSL Tolerance

The presence of mutations predicted as non-tolerated in both copies of*GDH1*and their close proximity (Table1) strongly suggest a determinant role for*GDH1*in HWSSL tolerance. Furthermore, population sequencing shows a steady increase in mutant allele frequency at both loci (Figure2).*GDH1*encodes an NAD(P)H-dependent glutamate dehydrogenase that catalyses the reductive amination of α-ketoglutarate to yield glutamate, responsible for the majority of cellular nitrogen in *S. cerevisiae*via ammonium assimilation [58].*GDH1*is recognized as a determinant of resistance to acetic acid [59,60]. Likewise, recent proteomics studies using a strain that is tolerant of furfural, phenol and acetic acid show a downregulated nitrogen assimilation machinery, including Gdh1p [59]. It is believed this occurs in order to slow growth and allow stress tolerance mechanisms to protect the cell more effectively. A decrease in biomass yield exhibited by R57 relative to the WT under permissive conditions (Additional file3: Figure S4) suggests that metabolism has shifted to a state of decreased resource utilization efficiency. As*GDH1*is a central hub of nitrogen metabolism, microbial substrate tolerance engineering studies that focus on this gene are warranted, especially on the HWSSL substrate that was used to evolve strain R57, which is generally high in ammonia content (approximately 1%*w/v*[22]) and could lead to ammonia toxicity [61]. Bayer*et al.* recently showed that by increasing expression variability of*GDH1*alone, one can tune the metabolism of a cell so that it responds more efficiently to limiting or toxic levels of ammonia [62].

The NADPH cofactor requirement of*GDH1*is also a major consideration in attempting to explain the consequences of the*GDH1*mutations. Yeast detoxifies the furan aldehyde inhibitors found in lignocellulosic hydrolysates by way of NADH/NADPH requiring enzymes [13]. Differential gene expression analysis between the WT and R57 show increased expression of*GRE3*(Additional file4: Data S2) in R57, encoding for a methylglyoxal reductase that can reduce furan aldehyde inhibitors via NADH [14]. The NAD(P)H cofactor usage of R57 may be modulated by a modified GDH1p, providing the reducing equivalents needed to detoxify the HWSSL furan aldehyde inhibitors.

Sequencing Supports Determinant HWSSL Tolerance Roles for Mutated Genes*SSA1,ARO1,MAL11*and*GSH1*

Given the GS methodology, it is likely that several of the accumulated mutations influence hydrolysate tolerance. This study was able to generate more restricted evidence that the*SSA1,ARO1,MAL11*and*GSH1*mutations found in R57 may be affecting HWSSL tolerance.

SSA1, also related to protein homeostasis, bears a mutation in R57. Ssa1p is a member of the heat shock 70 (Hsp70) family of proteins, which consists of highly conserved, broad specificity, essential protein chaperones (for reviews, see [63,64]). The tendency of harsh conditions to damage proteins and lead to aggregation [33] likely makes the role of Hsp-encoding genes important for HWSSL tolerance. The R57 Q31K mutation of Ssa1p is located at a highly conserved residue, as predicted by SIFT (Table1). Although not highly represented in population sequencing, the discovery of an adjacent*SSA1* mutation (D22G) that is also present within the tolerant population (Figure2) suggests that this region of Ssa1p may have bearing on the R57 phenotype. The structure of Ssa1p contains two distinct domains, the nucleotide-binding domain (NBD) that is responsible for binding and hydrolyzing ATP and the substrate-binding domain which can bind short hydrophobic segments of incompletely folded or unfolded polypeptides, in order to prevent adverse aggregation [63]. The Q31K mutation of Ssa1p is located in the NBD, and when the *Escherichia coli* Hsp70 homolog DnaK is used as a structural reference, the mutation is shown to lie adjacent to the ATP-binding pocket [65]. Although the Q31K and D22G mutations in Ssa1p have not been studied thus far, residues in this area of the protein have been shown to influence its NBD function and folding activity [65,66].

Furthermore, Aro1p, which catalyses steps 2 to 6 of the chorismate pathway leading to synthesis of aromatic amino acids [67], harbours a predicted phenotype-conferring mutation that gains in frequency within sequenced mutant pools. After an acetic acid challenge, aromatic amino acid synthesis and tryptophan synthesis in particular are pathways that are found to be upregulated [60], while mutants auxotrophic for aromatic amino acid synthesis show acetic acid sensitivity [68] and deletion of *ARO1* leads to sensitivity to osmotic and ethanol stress [69].

Mal11p bears two mutations that pervade and increase in the GS-evolved HWSSL-tolerant population. Mal11p is a trehalose-H$^+$symporter [70,71] and could be related to osmotic stress protection via trehalose transport or pH stasis due to its proton requirement. Although the effect of *MAL11* on tolerance to industrial processes has not been demonstrated, it is a common trend for the genome of industrial yeast strains to show a loss or reduction of *MAL11* genes [72-74].

Finally, the mutation in the 5′ UTR of *GSH1* may potentially be affecting redox homeostasis by influencing glutathione levels; *GSH1* can influence tolerance to lignocellulosic hydrolysate inhibitors in this way [75]. Glutathione is comprised of glycine, cysteine and glutamate and is a major redox buffer of the cell, cycling between its reduced and oxidized form, relying on NADP(H) for recycling. Hydrogen peroxide induces *GSH1* transcription but relies on the presence of intercellular amino acid pools, namely glutamate, glutamine and lysine to induce glutathione production [76]. Therefore, modified glutamate assimilation via *GDH1* mutation and the 5′ UTR mutation in *GSH1* both potentially affect intercellular glutamate pools and concomitant expression of*GSH1*. Between the WT and R57, genes that lead to cysteine biosynthesis (*HOM2,MET16,MET17*and*MET3*), along with genes leading to glutamate and lysine biosynthesis (*HOM3,LYS9*and*LYS12*), are upregulated (Additional file4: Data S2). This finding suggests that R57 has upregulated pathways towards glutathione precursor generation as part of its physiology and constitutes a possible link to the*GSH1*5′ UTR mutation.

GS Population Sequencing Provides Evidence against Prominent Phenotypic Roles of Isolated Mutations

Although some of the remaining R57 mutations may be of interest based on the known functions of the affected genes, sequencing of GS populations did not support a determinant effect on HWSSL tolerance for every mutation. Namely, the mutations located in the 5′ UTR of*PBP1*, 3′ of*BCS1*and in*SGO1*and YNL058C seem to be linked in a single mutant that comprised a large proportion of the UV mutant population, but decreased in frequency throughout GS evolution. The verified tolerance-conferring effect of the*UBP7*mutation to which they were initially linked, and which gained in frequency over population evolution, suggests that these linked mutations were merely carried through to strain R57 from UV mutagenesis. The*TOF2*mutation gained in frequency within the GS-evolved strains at round 1 but diminished in later rounds, while the*DOP1*and*FIT3*3′ UTR mutations were not highly represented in the GS populations and may be relatively specific to R57, which devalues these genes as potential tolerance-associated determinants. One exception is the mutation in Ste5p, which increased in frequency throughout GS, suggesting that it may play a role in the evolution of R57. However, since Ste5p is involved in mating, which is essential to the GS method used to generate R57, a strain with a modified mating behaviour may play a role in GS evolution but is likely not linked to stress tolerance.

CONCLUSIONS

GS theory asserts that beneficial mutations accumulate within strains to rapidly evolve a trait of interest while simultaneously eliminating detrimental mutations. Our data suggest that during GS evolution, there is accumulation of complementary mutations in key cellular processes through recursive genetic recombination or by the accumulation of single mutations in crucial genes that confer a fitness advantage. Mutations that lead to a large fitness advantage, such as*UBP7*, may become highly represented in the initial mutant pool and lead to over-representation of non-productive mutations found in the same strain through genetic linkage. The decrease in the frequency of these non-productive mutations during GS evolution suggests that mutations of lesser or no impact on the trait of interest can be crossed out of final strains.

As a workflow, meiotic recombination-mediated GS of *S. cerevisiae*, combined with genome resequencing, population sequencing of mutation loci and RNA-seq transcriptional profiling, generated complementary results that provided novel insights into tolerance to a specific lignocellulosic hydrolysate, along with gene targets that can be used for strain engineering. The assortment of processes and genes involved in inhibitor tolerance could not have been rationally determined prior to this study. This study provides insights into the multiple biological processes that act in concert to establish tolerance to a multi-inhibitory substrate. Our strongest evidence supports determinant roles for Ubp7p and Art5p, related to ubiquitin-mediated proteolysis; stress response transcriptional repressor, Nrg1p; and NADPH-dependent glutamate dehydrogenase, Gdh1p. However, important roles in hydrolysate tolerance are supported for several of the mutations discovered in the GS-evolved strains and populations (*SSA1,ARO1,MAL11*and*GSH1*), and the potential phenotypic impact of each mutation has not been ruled out. Therefore, a subsequent study is ongoing in our lab to examine the effect of each mutation discovered in R57 and explore potential epistatic effects between the mutations reported here. As whole genome sequencing becomes ever more accessible, and as the biotechnology industry requires biocatalysts with increasingly complex traits, GS followed by analyses like those carried out in this study stands to have a rapid and profound effect on our understanding of complex multigenic traits.

MATERIALS AND METHODS

Strains and Materials

The*S. cerevisiae*CEN.PK strains, supplied by EUROSCARF (Institute for Molecular Biosciences, Frankfurt, Germany), were used as the WT reference and progenitor strains for genome-shuffled mutant populations, including prototrophic diploid strain CEN.PK 122 and haploid strains CEN.PK 113-1A (*MATα*) and 113-7D (*MATa*). The haploid CEN.PK strains were used in a previous study to generate HWSSL-tolerant strain R57.

The HWSSL used for all experiments was provided by Tembec Inc. (Temiscaming, Quebec, Canada). HWSSL was adjusted to pH 5.5 with 10 M NaOH and contained, on average (*w/v*), 0.076% arabinose, 2% xylose, 0.16% galactose, 0.24% glucose, 0.43% mannose, 1% acetic acid, 0.18% furfural and 0.11% HMF.

Sequencing of WT and R57 Strains

Genomic DNA of WT haploid CEN.PK113-7D and the CEN.PK-derived R57 diploid yeast strains were isolated from overnight cultures grown in YPD using the DNeasy Blood & Tissue Kit (Qiagen, Toronto, Ontario, Canada). Library construction and sequencing were done at the Michael Smith Genome Sciences Centre using an Illumina 1G Genome Analyzer (Illumina, Inc., San Diego, CA, USA). Two lanes were sequenced for the WT strain and four for strain R57. All genome sequence data and RNA-seq read data from this publication have been submitted to the National Center for Biotechnology Information (NCBI) sequence read archive under BioProject # PRJNA231093.

Sequencing Alignment and Mutation Calling

Sequencing reads were aligned to the WT CEN.PK113-7D reference genome obtained from the NCBI (PRJNA52955) [77] and cross-referenced with read alignments obtained in our laboratory using our WT consensus sequence that was created using the S288c genome sequence as an alignment backbone. Alignments were performed both with Bowtie using standard parameters [78] and CLC Genomics Workbench version 5.1 with default parameters and ignoring non-specific matches. SNP and indel calling were both performed with Maq version 0.7.1 [79] and CLC Genomics Workbench for verification and visualization. To eliminate false positives in mutation calling, the DNA sequencing reads obtained from the WT were subjected to the same variation calling protocol as strain R57. Variations that were called when CEN.PK113-7D Illumina reads were aligned onto the CEN.PK113-7D consensus genome sequence and those that corresponded to variations called for R57 reads aligned onto the CEN.PK113-7D consensus sequence were discarded as false positives. False positive SNP calls were likely derived from sequence-specific miscall errors [80] or due to alignments in non-specific regions of the genome or to areas of low complexity [81]. The coverage requirement for SNP calling for WT reads was also lowered to ≤5-fold in order to ensure that SNPs that may result due to misalignment could be easily identified. These miscalls were edited out of the final mutation list by manual inspection and visualization with CLC Genomics Workbench. Maq SNP analysis was performed using the cns2snp command, and SNPs were called if the region upon which they were mapped returned a genome copy score = 1 and carried a Phred-like quality score ≥40. SNP analysis was corroborated with CLC Genomics Workbench SNP Detection function with a quality score of ≥40 for the central base and ≥30 for the surrounding bases. The threshold for variation at a specific base in the genome needed for SNP calling was lowered to ≥10% of reads, yielding an aberrant base call for the CEN.PK113-7D read alignment from $\geq$35% for R57 SNP calling in order to maintain stringency on positive variation calling. Therefore, if a SNP was called for more than 35% of the reads in R57 but was also called for 10% or more of the reads for the WT control, it was discarded as a false SNP call. Indel analysis was also performed with Maq using the Indelpe command and verified with CLC Genomics Workbench, both under standard parameters and compared to WT reads for control, as described above. Copy number variation was assessed using CNV-seq [82] to compare WT to R57 reads with a log2 threshold of 0.75, below the level used to detect reliable CNVs in CEN.PK [77]. The mutations identified were verified by Sanger sequencing and compared to the other Mat parental WT, CEN.PK113-1A, to ensure the mutations were not present in either of the WT parental haploid strains. Locations of mutations were assessed using the CEN.PK consensus sequence as a guide.

Protein Impact Assessment of ORF-Located Mutations

Mutations in ORFs were examined by translating DNA sequences using ExPASy translate [83] and performing a BLAST comparison using the BLAST2Seq software program hosted by NCBI. Mutational impact assessment was done with the SIFT program [28,29] using recommended best practices [28]. Homologous proteins with <90% identity were chosen for comparison of the degree of conservation of the amino acid position in question for up to 100 homologous proteins. Amino acid substitutions were predicted as leading to a phenotype if the SIFT score was $\leq$0.05 and tolerated if the score is >0.05.

Sequencing of R57 SNPs in GS Heterogeneous Populations

The following heterogeneous populations of cells were used to track R57 SNP frequency through GS evolution: WT diploid CEN.PK122, a pooled UV mutagenized population (three pools of CEN.PK113-1A and two pools CEN.PK113-7D UV mutants), along with cells from the population of the first, third and fifth rounds of recursive GS obtained from our previous study that generated R57 [8]. Cells from each population were selected from above the frontier of WT growth as observed by screening on HWSSL gradient agar plates (Figure1B), yielding populations that were enriched for tolerance to HWSSL, as described [8]. A sample of 10^9 cells, suspended in phosphate-buffered saline (PBS), from each population was spread onto a single lane of the gradient plate and incubated for 6 days at 30°C. Each plate contained two lanes spread with CEN.PK 122 cells for comparison to each individual GS population, in duplicate. Each plate was screened in biological duplicates.

Cells that grew to higher HWSSL concentrations than the WT on the gradient plates were scraped, suspended in PBS and adjusted to approximately 4×10^8 cells/mL using a haemocytometer. For each population, DNA from approximately 4×10^7 cells was extracted using a DNeasy Blood & Tissue Kit (Qiagen, Toronto, Ontario, Canada). Five microlitres of each genomic DNA preparation was used as template for PCR with primers specific for each of the 20 SNP regions, located at approximately 50 bp from either end of the SNP. The primers were designed with sequencing adapter attachment for use with the Ion Torrent Personal Genome Machine (Life Technologies, Carlsbad, CA, USA) according to the manufacturer's instructions (Additional file2: Table S1). All PCR products were gel purified and quantified in triplicate using the Promega Quantifluor dsDNA system (Promega Corporation, Madison, WI, USA). The PCR products were diluted to 16 pmol, and 20-µL samples from each reaction were pooled to make four pools (one UV mutagenized and three GS pools). Pools of PCR products were sequenced using a 316 chip with the 200-bp kit (following the Ion Torrent protocols).

Sequencing SNPs from Isolates of the Round 5 GS Heterogeneous Population

Twenty isolated colonies picked from the growth frontier of a HWSSL gradient plate of the round 5 GS population were streak purified on 50% (v/v) HWSSL and 2% agar (w/v) Petri plates. Colonies isolated from the plates were grown overnight in 5 mL YPD broth at 30°C, and DNA was extracted with a DNeasy Blood & Tissue Kit for use as a template in the PCR amplification of the mutated gene region. Primers used for amplification are described in Additional file2: Table S2, and PCR products were sequenced using the Sanger method.

Transcriptome Analysis

WT diploid strain CEN.PK122 and HWSSL-tolerant R57 were used in RNA-seq experiments. The WT and R57 strains were grown in 50 mL synthetic defined (SD) medium (yeast nitrogen base without amino acids 0.17%w/v, ammonium sulphate 0.5%w/v, glucose 2%w/v) overnight at 30°C under semi-fermentative conditions (sealed 125-mL flasks shaken at 100 rpm) to early stationary phase (Additional file3: Figure S4). Cultures were normalized to an $OD_{600\,nm}$ of 3, and two independently grown cell samples of these cultures were used for RNA extraction. For each sample, cells from 5 mL of culture were harvested by centrifugation at 1,800 $\times g$ and 4°C and frozen in liquid N_2 until RNA isolation. RNA extracts were prepared using the RNeasy Plant Mini Kit (Qiagen) according to manufacturer's specifications for use with yeast, in which frozen cells were suspended in lysis buffer and disrupted with a mini bead beater (Precellys 24, Bertin Technologies, Montigny-le-Bretonneux, France) at 4°C. Prior to sequencing, RNA quality was confirmed using an Agilent 2100 Bioanalyzer (Agilent Technologies, Santa Clara, CA, USA). RNA sequencing was performed at the McGill/Genome Quebec Innovation Centre in duplicate on the Illumina Genome Analyzer*IIx*and Illumina HiSeq 2000 for the WT*vs.* R57. DNA libraries were subjected to 36 or 50 cycles of sequencing on the Illumina Genome Analyzer*IIx*and Illumina HiSeq 2000, respectively. RNA-seq differential transcription analysis and statistical comparisons were performed with CLC Genomics Workbench version 5.1. The cDNA sequence reads from RNA-seq were trimmed to remove Illumina sequencing adaptors as well as unreliable read ends, and alignments were performed using the CEN.PK113-7D genome sequence and associated GTF file [77] as the backbone for alignment mapping and quantitation. Significance values for differential expression were computed using Baggerly's test [84]. The samples were then FDR-corrected in order to eliminate non-productive leads from the expression results. Gene transcripts showing differential expression with a corrected*P*value of <0.05 and a >2-fold increase were used for functional clustering and enrichment mapping of differentially expressed genes.

Functional annotation clustering was executed with DAVID Bioinformatics Resources 6.7 [85]. Clusters of up or down expressed genes with gene ontology (GO) term enrichment scores of≥1.3 (equivalent to a non-log scale value of 0.05) are reported, unless stated otherwise. Enrichment maps of ontology categories from clustering were generated with the Enrichment Map 1.2 software plug-in for Cytoscape 2.8 [86,87]. All functional annotations presented were derived from SGD [88] or the DAVID server unless otherwise referenced. Transcription factor binding analysis was done through the YEASTRACT database [89-91].

Reconstitution of the*UBP7*Mutation in WT

To determine if the*UBP7*mutation was contributing to the HWSSL tolerance phenotype of R57 as predicted, the mutation was introduced into WT and the resulting strain was tested for growth on HWSSL. The WT allele was replaced in CEN.PK113-7D via homologous recombination of a DNA cassette containing the mutated*UBP7*sequence flanked by a kanamycin resistance marker. Sanger sequencing of the PCR-amplified region was used to confirm that the transformants harboured the mutation. Homozygous diploid strains of the WT and*UBP7*mutant were created by mating type switching using the YCp50::HO plasmid [92] and mating haploid strains of opposite mating type. The*UBP7*homozygous diploid mutant and WT diploid strains along with R57 were tested for their tolerance to HWSSL in parallel, as previously described [8].

ABBREVIATIONS

CNV:copy number variation

GS:genome shuffling

HMF:5-(hydroxymethyl)-2-furaldehyde

HWSSL:hardwood spent sulphite liquor

indel:insertion/deletion

NBD:nucleotide-binding domain

UBL:ubiquitin-like

ACKNOWLEDGEMENTS

An NSERC Strategic Project (GHGPJ322381), the NSERC Bioconversion Network (NETGP350246-07), the AAFC Agricultural Bioproducts Innovation Program (ABTP_000159), BioFuelNet and a Canada Research Chair to V.J.J.M supported this research. Dominic Pinel was supported by a graduate scholarship from Le Fonds Québécois de la Récherche sur la Nature et les Technologies.

ADDITIONAL FILES

Additional File 1

Data S1.A summary mapping report for CEN.PK 113-7D.

Summary statistics

	Count	Average length	Total bases
Reads	33307144	50	1.7E+09
Matched	26438855	50	1.3E+09
Not matcl	6868289	50	3.4E+08
Reference	17	616363	1E+07

General algorithm parameters

Paramet er	Value
Conflict re	Vote (A, C, G, T)

Distribution of read length

Length	Count
50	33307144

Sequence read coverage based on CEN.PK 113-7D consensus backbone

Chromoson	Consensus	Total read	Average c	Reference	Coverage per chromosome/average coverage (excluding chromosome 0)
0	55178	6077471	4,584.45	66026	46.6006
1	177803	411354	91.83	223180	0.93345
2	781415	1580620	97.5	807753	0.99108
3	290216	612115	99.31	307146	1.00948
4	1424349	2694117	89.95	1492465	0.91434
5	415125	889284	104.76	422853	1.06488
6	237770	536807	107.95	247663	1.0973
7	1049983	2129560	97.86	1084284	0.99474
8	269032	537237	96.42	277614	0.9801
9	278627	562859	91.71	305804	0.93223
10	690216	1378147	94.3	728208	0.95855
11	253171	488892	95.7	254544	0.97278
12	478072	1224227	116.56	522948	1.18482
13	898265	1764151	94.66	928646	0.96221
14	733668	1489813	95.43	777945	0.97004
15	1053344	2046778	94.04	1084527	0.95591
16	893915	2015423	106.06	946580	1.07809
average coverage			98.3775		

Summary mapping report for strain R57

Summary statistics

	Count	Average length	Total bases
Reads	106607552	50	5.33E+09
Matched	86829570	50	4.34E+09
Not matche	19777982	50	9.89E+08

General algorithm parameters

Parameter	Value
Conflict res	Vote (A, C, G, T)

Distribution of read length

Length	Count
50	106607552

Distribution of non-matched read length

Length	Count
50	19777982

Sequence read coverage based on CEN.PK 113-7D consensus backbone

Chromoson	Consensus l	Total read c	Average co	Reference l	Coverage per chromosome/average coverage (excluding chromosome 0)
0	55075	13246017	9,991.79	66026	28.71261
1	177791	1230165	274.88	223180	0.789901
2	781485	5763868	356.01	807753	1.023038
3	290333	2057057	334.09	307146	0.960048
4	1424502	10276738	343.58	1492465	0.987318
5	415107	3037046	358.26	422853	1.029503
6	237793	1792261	360.84	247663	1.036917
7	1050015	7948486	365.74	1084284	1.050998
8	269024	1909347	343.1	277614	0.985939
9	278681	1961656	320.02	305804	0.919616
10	690291	5075244	347.73	728208	0.999244
11	253125	1801511	353.13	254544	1.014761
12	478044	4083343	389.19	522948	1.118384
13	898330	6525008	350.58	928646	1.007434
14	733632	5377044	344.85	777945	0.990968
15	1053499	7643184	351.65	1084527	1.010509
16	894147	7101595	374.24	946580	1.075424
average coverage			347.9931		

Additional File 2

Supplemental Methods

A Document on Assessing Functional Interaction of Mutations

Analysis of potential functional interaction was carried out on the 16 genes non-ambiguously affected by mutations through analysis with the BiNGO version 2.44 [1] software plugin for the Cytoscape 2.8 network analysis software program [2-4]. Mutant genes were annotated with the complete BiNGO Gene Ontology (GO) file to visualize the full range of known functional relationships between the genes of interest (Supplemental Fig. 6) or the GOSlim yeast annotation file for broader visualization of affected functional categories. Protein-protein, genetic, co-expression, co-localization, predicted functional relationship, and shared protein domain interactions between products of the genes affected by mutation were explored using the GeneMANIA version 3.2 [5, 6] software plugin for Cytoscape 2.8. The 20 most related genes as determined by GeneMANIA were included within the interaction network to explore non-direct interrelationships. Weighting for the GeneMANIA generated interaction map (Supplemental Fig. 7) was based on GO molecular functional relationships, while evidence scoring was based on assigned relevance of the sources used to generate interactivity by the software. All annotations of genes are based on the *Saccharomyces* Genome Database (SGD) unless otherwise referenced.

Supplemental Table S1. Primers used for PGM sequencing. Primers used to amplify mutated regions and add the adapter regions and key sequence for Ion Torrent sequencing. Adapter region P1 is depicted in bold in the forward primers while adapter region A is depicted in bold in the reverse primers. The key sequence is italicized in the reverse primers

Primer Name	Primer
SSA1_F_PGM	**CCTCTCTATGGGCAGTCGGTGAT**GCATCACCAATCAATCTTTCAGTGTC
SSA1_R_PGM	**CCATCTCATCCCTGCGTGTCTCCGAC***TCAG*CGTGTGTTGCTCACTTTGCTAATG
TOF2_F_PGM	**CCTCTCTATGGGCAGTCGGTGAT**CAGCTGGTCCATGACTTGTTTTG
TOF2_R_PGM	**CCATCTCATCCCTGCGTGTCTCCGAC***TCAG*CTTCCATGTCACTTGAATCTTCATTG
UBP7_F_PGM	**CCTCTCTATGGGCAGTCGGTGAT**CCATGAAGTTGAGTAAACTTGGTAGG
UBP7_R_PGM	**CCATCTCATCCCTGCGTGTCTCCGAC***TCAG*GGGCAATCCCATGCATTTTCACC
YNL058c_F_PGM	**CCTCTCTATGGGCAGTCGGTGAT**GGAAGGTCTCTTCTGACCAGC
YNL058c_R_PGM	**CCATCTCATCCCTGCGTGTCTCCGAC***TCAG*CTAGACTCATACCTTGTCAAAAGTTC
MAL11a_F_PGM	**CCTCTCTATGGGCAGTCGGTGAT**CCTTCCATAACCAGGGTAGTAG
MAL11a_R_PGM	**CCATCTCATCCCTGCGTGTCTCCGAC***TCAG*GCTAACAGCGAGGAAAAAAGCATG
MAL11b_F_PGM	**CCTCTCTATGGGCAGTCGGTGAT**CCATATAAGTCGTGATTTGCAAACC
MAL11b_R_PGM	**CCATCTCATCCCTGCGTGTCTCCGAC***TCAG*GGGAGGGTTCTTACGAAATTACTTCC
GDE1a_F_PGM	**CCTCTCTATGGGCAGTCGGTGAT**CCTCTTTGGAAGACTCTACTCTTTTC
GDE1a_R_PGM	**CCATCTCATCCCTGCGTGTCTCCGAC***TCAG*CCCAGGTGACTCTGAATTGTATG
GDE1b_F_PGM	**CCTCTCTATGGGCAGTCGGTGAT**GTCAGAGCCAGAATTTCAACAAGC
GDE1b_R_PGM	**CCATCTCATCCCTGCGTGTCTCCGAC***TCAG*GGCAAAACCTTTCTGTATTCTGGGTG
GSH1_F_PGM	**CCTCTCTATGGGCAGTCGGTGAT**GTATATTTTCGATACTCTAAACCACCC
GSH1_R_PGM	**CCATCTCATCCCTGCGTGTCTCCGAC***TCAG*GCTGGAGTAGTTGGATCTTTCC
DOP1_F_PGM	**CCTCTCTATGGGCAGTCGGTGAT**GCTCGACATTAGCACGAAACTTC
DOP1_R_PGM	**CCATCTCATCCCTGCGTGTCTCCGAC***TCAG*GGGATCGACAAAAAATGTCCTTACCAC
SGO1_F_PGM	**CCTCTCTATGGGCAGTCGGTGAT**GCATGAATCAAGTTTTAACAAGGACG
SGO1_R_PGM	**CCATCTCATCCCTGCGTGTCTCCGAC***TCAG*CGGTTTCGTCTTCAGGTTCTAAAC
BCS1_F_PGM	**CCTCTCTATGGGCAGTCGGTGAT**CGACATGGTAGATTGAGGGC
BCS1_R_PGM	**CCATCTCATCCCTGCGTGTCTCCGAC***TCAG*GTAAAGATTTCCACTCTTCTATATTTGC
RPB1_F_PGM	**CCTCTCTATGGGCAGTCGGTGAT**CGGATAGCAGCATTATCCATGAC
RPB1_R_PGM	**CCATCTCATCCCTGCGTGTCTCCGAC***TCAG*GAATTTACAGACTACTGGAGAGGG
FIT3_F_PGM	**CCTCTCTATGGGCAGTCGGTGAT**GTGTGTTATTAAAATTTTTTATTCTAACATAACTTCG
FIT3_R_PGM	**CCATCTCATCCCTGCGTGTCTCCGAC***TCAG*CCGTCATGTTATTGTAAATGATATGTG
ARO1_F_PGM	**CCTCTCTATGGGCAGTCGGTGAT**CCCTGCTGATCAACAGAAAGTTG
ARO1_R_PGM	**CCATCTCATCCCTGCGTGTCTCCGAC***TCAG*GATTTTACATTGACCTTCACCGAGG
ART5_F_PGM	**CCTCTCTATGGGCAGTCGGTGAT**GCAGTACAAGCAACCAAGATATGG
ART5_R_PGM	**CCATCTCATCCCTGCGTGTCTCCGAC***TCAG*CTTCGAATTATTTTGTTGAAAACAGGG
NRG1_F_PGM	**CCTCTCTATGGGCAGTCGGTGAT**GCAGTCTTATTTAATTTGTGTTTTAGTTCATTC
NRG1_R_PGM	**CCATCTCATCCCTGCGTGTCTCCGAC***TCAG*GGGAAACGTTGAAATAAGCCCGG
PBP1_F_PGM	**CCTCTCTATGGGCAGTCGGTGAT**GGAAAGAACAAAAAGAAAGATAGAAGAAAAC
PBP1_R_PGM	**CCATCTCATCCCTGCGTGTCTCCGAC***TCAG*CGTTTTTGTAAAGCAGTTCTTAAATCG
STE5_F_PGM	**CCTCTCTATGGGCAGTCGGTGAT**GGTCGCTCCATTTGGCTATC
STE5_R_PGM	**CCATCTCATCCCTGCGTGTCTCCGAC***TCAG*CCCTTTCTTCTGTTAGAAATAGGC

Supplemental Table S2. List of primers used to integrate the mutation found in *UBP7* into the chromosome

Primer names	Primer sequence
UBP7mut_1	CCATGAAGTTGAGTAAACTTGGTAGGTCTACTGAGAAAAGAGTTAAGTTAGAGG
UBP7mut_2	ACGAAGTTATATTAAGGGTTGTCGACCTGCAGCGTACGAAGCTTCAGCTGGGTGACAAA GATAACATTCACAAGAG
UBP7mut_3	CTCTTGTGAATGTATCTTTGTACCCCAGCTGAAGCTTCGTACGCTG
UBP7mut_4	GGAACACTGCCAGCGCATC
UBP7mut_5	GATGCGCTGGCAGTGTTCC
UBP7mut_6	CATCAAGACGTTTGGTGTCTAAATCGGCCGCATAGGCCACTAG

Additional File 3

Supplemental Figures

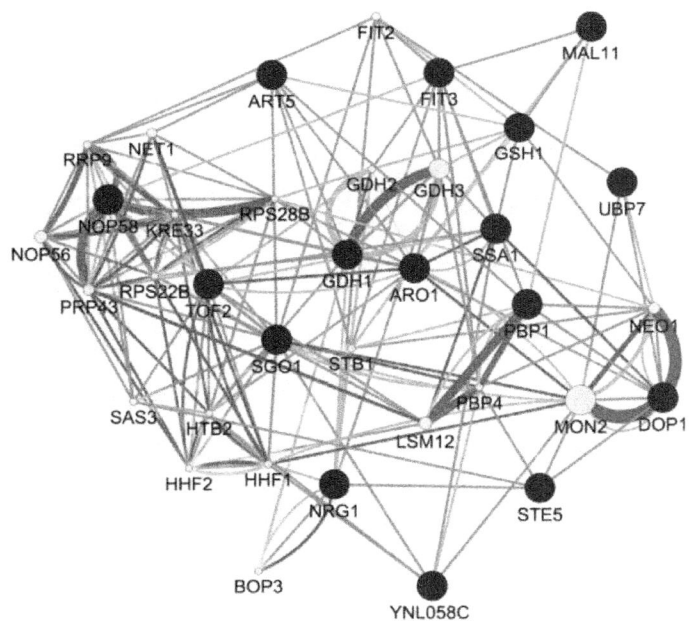

Figure S1. Interaction map of genes affected by mutation in R57. Interaction maps were generated through GeneMANIA, drawing from extensive source annotations built into the software. Blue nodes are the genes affected by mutation in R57 (uniform size), while the yellow nodes are the top 20 interacting genes as determined by GeneMANIA (larger size equals a higher degree of interaction within the network). The edges connecting the nodes represent protein-protein (red lines), genetic (green lines)

and co-expression (light-blue lines) interactions. Thicker lines represent a higher confidence of interaction.

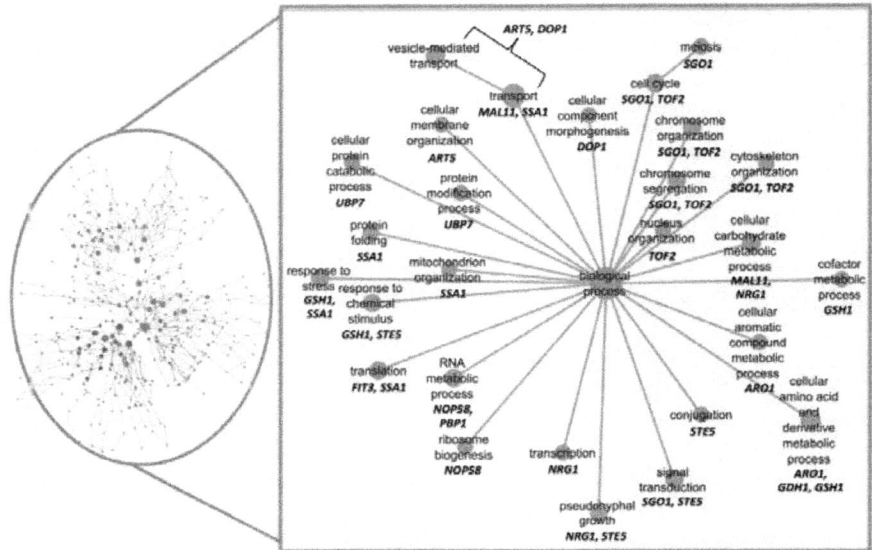

Figure S2. Ontology categories associated with R57 genes affected by mutation. Using the full gene ontology set for yeast, a vast interaction network can be generated of biological processes potentially affected by mutation (left) and summarized using the GOSlim Yeast ontology categories (right). Genes affected by mutation (bold, Table 1) are listed under their associated ontology category (orange node).

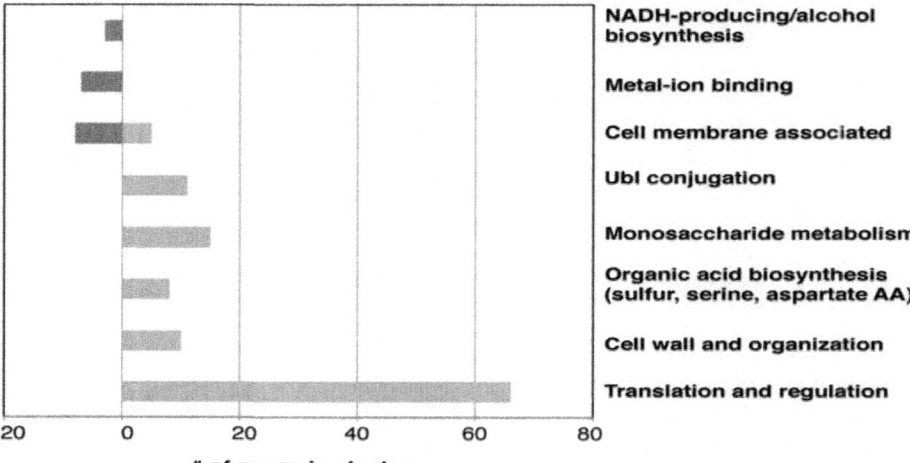

Figure S3. Differentially expressed genes comprising enrichment clusters based on biological function between the WT and R57. The number of upregulated (> 2 fold, green) or downregulated genes (> 2-fold, red) are depicted based on associated GO term enrichment clustering with scores > 1.3.

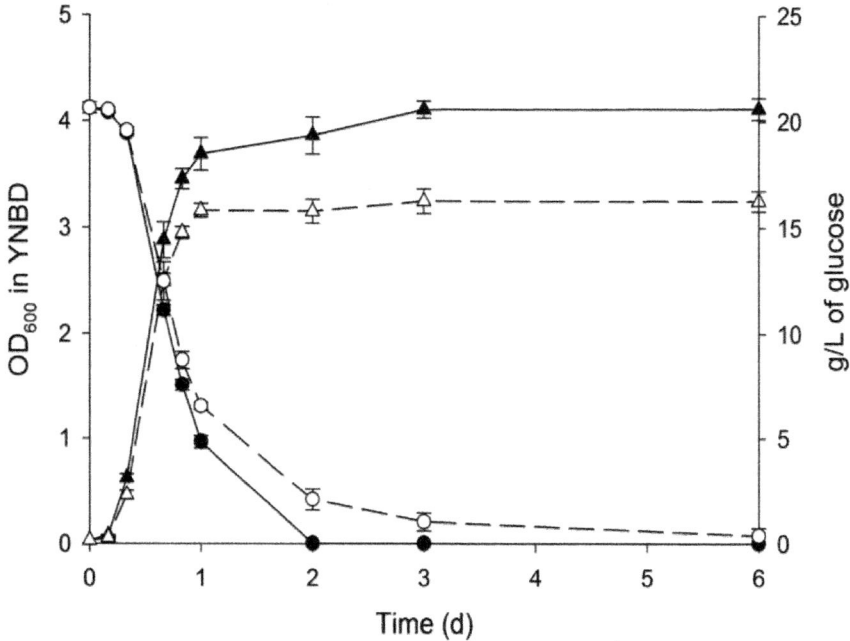

Figure S4. Cell growth and glucose consumption of WT and R57 mutant strains. Overnight SD cultures of WT (solid symbols) and R57 (open symbols) were inoculated into 50 mL of SD medium with an initial OD600 of 0.03 (0 h) and then cells were cultured at 200 rpm at 30 °C in 250-mL flasks. Cell growth (left y-axis, triangles) was monitored by measuring OD600 at 4, 8, 16, 20, 24, 48, 72 and 144 hours. Glucose (right y-axis, circles), concentration in the medium was determined by HPLC.

Additional File 4

Data S2.RNA-seq results for differentially expressed genes between WT*vs*R57 without HWSSL exposure

RNA-seq results for differentially expressed genes between WT vs R57 without HWSSL exposure

Gene name	Gene ID	Expression Fold Change	FDR p-value correction	122T0 (a) Expression values (RPKM)	122T0 (b) Expression values (RPKM)	122T0 - Means	R57T0 (a) Expression values	R57T0 (b) Expression values	R57T0 - Means
LEE1	YPL054W	-4.0	0.00	122.31	81.78	102.04	26.52	24.34	25.43
SMF2	YHR050W	-2.5	0.00	318.20	354.87	336.54	120.51	147.96	134.24
PHO90	YJL198W	-2.5	0.01	35.24	36.71	35.97	20.53	8.38	14.46
SDS23	YGL056C	-2.5	0.00	317.75	364.11	340.93	153.56	123.89	138.73
INO1	YJL153C	-2.5	0.00	2033.25	1944.92	1989.08	663.02	956.89	809.96
MEP3	YPR138C	-2.4	0.00	71.49	77.68	74.58	30.07	32.56	31.32
CAN1	YEL063C	-2.3	0.00	99.96	71.11	85.54	36.57	36.93	36.75
SUR2	YDR297W	-2.3	0.00	276.12	311.05	293.59	140.93	113.80	127.36
ADH2	YMR303C	-2.3	0.05	14.83	19.22	17.03	9.49	5.40	7.44
CYB5	YNL111C	-2.3	0.04	186.44	108.46	147.45	80.73	49.06	64.90
PTK1	YKL198C	-2.2	0.00	282.82	218.48	250.65	144.97	78.53	111.75
PLB3	YOL011W	-2.2	0.00	139.11	109.53	124.32	77.62	36.15	56.89
RCF2	YNR018W	-2.1	0.00	204.94	201.05	203.00	125.59	65.20	95.39
MDH2	YOL126C	-2.1	0.00	191.91	203.35	197.63	99.71	87.95	93.83
PPH3	YDR075W	-2.1	0.04	34.85	37.87	36.36	23.54	11.06	17.30
GEP5	YLR091W	-2.0	0.00	36.19	39.06	37.62	17.61	19.44	18.52
GZF3	YJL110C	-2.0	0.05	27.75	20.06	23.91	14.05	9.71	11.88
ERG11	YHR007C	-2.0	0.00	234.23	260.18	247.20	139.43	109.31	124.37
DIC1	YLR348C	2.0	0.01	25.97	28.46	27.21	47.18	59.29	53.24
RPS8a	YBL072C	2.0	0.00	177.91	113.24	145.58	319.75	251.80	285.78
PDR17	YNL264C	2.0	0.01	23.68	31.62	27.65	60.68	49.45	55.06
RPP2b	YDR382W	2.0	0.00	115.63	75.81	95.72	218.50	163.95	191.22
MET17	YLR303W	2.0	0.00	153.97	94.09	124.03	284.95	211.82	248.38
GRE3	YHR104W	2.0	0.00	411.48	329.88	370.68	711.31	781.83	746.57
RPL10	YLR075W	2.0	0.01	310.04	243.97	277.00	669.00	451.92	560.46
APA1	YCL050C	2.0	0.00	39.05	39.17	39.11	80.66	78.30	79.48
HOM2	YDR158W	2.0	0.00	47.42	42.06	44.74	97.71	84.29	91.00
CYC7	YEL039C	2.0	0.00	3032.78	3405.22	3219.00	7395.29	5704.92	6550.11
IMA2	YOL157C	2.0	0.00	52.37	34.08	43.22	97.39	78.65	88.02
RPL24b	YGR148C	2.1	0.00	92.50	64.58	78.54	186.11	136.11	161.11
RPS4b	YHR203C	2.1	0.01	94.98	66.77	80.88	194.34	137.69	166.02
RPL35a	YDL191W	2.1	0.04	181.04	119.30	150.17	372.12	246.85	309.49
GAS1	YMR307W	2.1	0.00	50.69	48.97	49.83	111.29	95.79	103.54
BFR1	YOR198C	2.1	0.02	19.76	17.25	18.50	36.23	40.92	38.57
MET16	YPR167C	2.1	0.04	18.66	13.22	15.94	38.08	28.64	33.36
GPI18	YBR004C	2.1	0.02	32.34	37.12	34.73	85.54	59.87	72.71
RPS6a	YPL090C	2.1	0.00	126.40	102.53	114.47	271.80	207.62	239.71
ECM33	YBR078W	2.1	0.00	95.57	65.89	80.73	159.95	180.84	170.39
HRI1	YLR301W	2.1	0.02	31.05	17.76	24.41	57.33	46.96	52.14
DIA1	YMR316W	2.1	0.01	58.18	57.02	57.60	145.55	100.59	123.07
RPL32	YBL092W	2.1	0.00	139.28	110.73	125.01	312.03	222.46	267.25
CCW12	YLR110C	2.2	0.00	665.61	545.65	605.63	1311.71	1304.11	1307.91
CDC19	YAL038W	2.2	0.00	366.34	190.20	278.27	665.52	542.74	604.13
YLR179C	YLR179C	2.2	0.05	61.99	36.60	49.29	128.89	87.06	107.97
RPS1a	YLR441C	2.2	0.00	109.26	105.26	107.26	274.48	196.34	235.41
PGK1	YCR012W	2.2	0.00	392.36	293.29	342.82	719.91	792.43	756.17
RPL42a	YNL162W	2.2	0.02	48.86	69.46	59.16	135.37	127.64	131.50
HOM3	YER052C	2.2	0.01	16.44	18.44	17.44	40.57	37.01	38.79
RPS4a	YJR145C	2.2	0.00	72.14	66.09	69.11	156.98	152.97	154.97
RPS13	YDR064W	2.2	0.00	89.08	68.24	78.66	184.39	168.77	176.58
GPM1	YKL152C	2.3	0.00	371.35	280.89	326.12	709.59	758.55	734.07
STM1	YLR150W	2.3	0.00	80.14	65.39	72.76	194.17	139.67	166.92
LYS12	YIL094C	2.3	0.00	29.23	27.87	28.55	66.04	65.23	65.63
ODC2	YOR222W	2.3	0.01	21.47	15.42	18.44	49.00	36.10	42.55

YJL144W	YJL144W	2.3	0.01	848.95	336.69	592.82	1475.15	1261.81	1368.48
RPS2	YGL123W	2.3	0.00	99.74	95.09	97.42	238.97	214.04	226.50
FMP27	YLR454W	2.3	0.02	18.32	26.82	22.57	46.27	58.76	52.52
TYS1	YGR185C	2.3	0.03	13.63	10.42	12.03	28.66	27.41	28.03
RPS24a	YER074W	2.3	0.00	81.50	60.67	71.08	179.41	152.09	165.75
RPS3	YNL178W	2.3	0.00	84.71	84.37	84.54	225.02	169.46	197.24
CMK2	YOL016C	2.3	0.00	180.42	142.45	161.44	378.49	379.72	379.11
RPL29	YFR032C-A	2.4	0.01	64.92	84.15	74.53	210.67	142.06	176.37
RPL11b	YGR085C	2.4	0.00	43.32	59.09	51.20	128.41	114.01	121.21
RPS0a	YGR214W	2.4	0.00	51.25	41.39	46.32	114.42	105.89	110.16
TPI1	YDR050C	2.4	0.00	277.36	175.88	226.62	507.89	571.89	539.89
RPL20b	YOR312C	2.4	0.00	85.07	55.34	70.20	190.24	144.30	167.27
RPL6b	YLR448W	2.4	0.05	52.58	27.32	39.95	114.52	76.31	95.42
RPL2b	YIL018W	2.4	0.00	104.87	75.29	90.08	236.92	193.48	215.20
YDR209C	YDR209C	2.4	0.00	61.37	49.45	55.41	142.33	122.48	132.41
EGD2	YHR193C	2.4	0.01	54.67	38.53	46.60	134.04	90.03	112.04
RPS16a	YMR143W	2.4	0.00	87.14	53.05	70.09	195.07	142.77	168.92
RPL15a	YLR029C	2.4	0.05	74.47	126.35	100.41	232.84	251.18	242.01
FEN1	YCR034W	2.4	0.04	9.99	9.18	9.58	20.00	26.23	23.11
RPS19a	YOL121C	2.4	0.00	38.22	45.67	41.94	116.52	86.89	101.70
RPS28b	YLR264W	2.4	0.00	38.83	39.48	39.15	110.98	79.40	95.19
RPL42b	YHR141C	2.4	0.00	67.92	47.16	57.54	156.53	124.02	140.27
RPS5	YJR123W	2.4	0.00	170.72	121.60	146.16	399.77	313.33	356.55
RPP0	YLR340W	2.5	0.00	93.96	62.00	77.98	221.41	161.25	191.33
YIL055C	YIL055C	2.5	0.00	19.61	16.07	17.84	47.43	40.12	43.78
PUN1	YLR414C	2.5	0.00	60.20	61.86	61.03	165.32	134.34	149.83
RPS17b	YDR447C	2.5	0.00	43.70	56.76	50.23	121.80	125.14	123.47
RCR1	YBR005W	2.5	0.00	59.85	59.84	59.85	171.42	123.47	147.45
MET3	YJR010W	2.5	0.00	24.10	15.87	19.98	55.19	45.54	50.36
SUR4	YLR372W	2.5	0.00	13.60	13.23	13.41	30.35	37.40	33.88
RPL13b	YMR142C	2.5	0.04	89.14	54.38	71.76	225.55	140.09	182.82
TDH3	YGR192C	2.6	0.00	889.26	452.80	671.03	1704.56	1728.25	1716.40
RPS6b	YBR181C	2.6	0.00	52.06	64.51	58.28	165.22	135.88	150.55
SOL1	YNR034W	2.6	0.00	108.05	66.84	87.45	220.23	232.22	226.22
FBA1	YKL060C	2.6	0.00	728.46	470.79	599.62	1615.96	1505.23	1560.60
ARO8	YGL202W	2.6	0.00	23.01	20.65	21.83	53.68	60.29	56.98
YSC84	YHR016C	2.6	0.00	190.40	152.62	171.51	417.50	482.46	449.98
XKS1	YGR194C	2.6	0.00	55.98	24.23	40.10	119.53	91.29	105.41
TEF4	YKL081W	2.6	0.00	17.66	25.16	21.41	52.99	59.57	56.28
RPS19b	YNL302C	2.6	0.00	60.57	50.18	55.37	144.42	147.30	145.86
RPL7a	YGL076C	2.6	0.00	72.13	52.72	62.43	165.30	165.28	165.29
RPS12	YOR369C	2.7	0.00	119.99	107.26	113.63	354.76	251.61	303.18
YOR385W	YOR385W	2.7	0.00	13.74	12.84	13.29	37.67	33.37	35.52
RPS1b	YML063W	2.7	0.00	66.58	64.71	65.65	181.70	169.20	175.45
RPS27b	YHR021C	2.7	0.04	79.68	49.05	64.36	214.93	131.64	173.29
RPL31b	YLR406C	2.7	0.00	47.24	39.54	43.39	114.74	121.09	117.92
RPL40b	YKR094C	2.7	0.01	64.24	55.55	59.89	199.55	127.58	163.56
RPL5	YPL131W	2.7	0.00	76.76	92.53	84.64	252.11	211.93	232.02
RPS7a	YOR096W	2.8	0.00	47.90	45.26	46.58	132.58	124.69	128.63
RPS8b	YER102W	2.8	0.00	81.41	47.08	64.25	212.96	145.83	179.40
RPL21a	YBR191W	2.8	0.00	55.04	43.97	49.51	144.28	132.64	138.45
RPS9b	YBR189W	2.8	0.00	59.62	51.88	55.75	164.49	147.47	155.98
SGA1	YIL099W	2.8	0.04	11.45	6.44	8.95	30.30	20.29	25.30
RPS31	YLR167W	2.8	0.01	186.91	130.08	158.50	551.87	346.14	449.01
RPS18b	YML026C	2.8	0.01	71.31	44.33	57.82	199.21	128.60	163.91
BAT1	YHR208W	2.9	0.04	39.74	31.30	35.52	88.53	114.03	101.28
RPL24a	YGL031C	2.9	0.05	67.53	48.28	57.90	209.22	122.14	165.68
RPL18a	YOL120C	2.9	0.00	66.12	68.21	67.17	180.48	210.08	195.28
YGR137W	YGR137W	2.9	0.04	31.16	25.21	28.18	102.97	61.50	82.23

HTA2	YBL003C	2.9	0.00	49.33	19.70	34.52	115.06	86.45	100.76
RPL8b	YLL045C	2.9	0.00	60.72	50.94	55.83	185.40	142.35	163.87
NRG1	YDR043C	2.9	0.00	137.35	153.76	145.56	504.45	350.13	427.29
ASC1	YMR116C	3.0	0.00	61.76	44.57	53.16	162.33	157.32	159.82
RPS20	YHL015W	3.0	0.00	145.49	86.06	115.77	419.25	280.35	349.80
RPL2a	YFR031C-A	3.0	0.00	48.80	33.72	41.26	130.62	119.05	124.84
RPL13a	YDL082W	3.1	0.00	25.23	24.65	24.94	90.35	62.19	76.27
OAC1	YKL120W	3.1	0.00	8.45	8.27	8.36	27.54	24.35	25.95
RPL26b	YGR034W	3.1	0.01	55.96	55.26	55.61	214.86	132.34	173.60
RPL18b	YNL301C	3.2	0.00	6.71	7.85	7.28	20.96	24.99	22.98
ENA5	YDR038C	3.3	0.00	7.44	9.45	8.45	26.74	28.43	27.58
RPL17b	YJL177W	3.3	0.00	28.97	24.08	26.53	101.19	72.62	86.90
YPS5	YGL259W	3.6	0.05	6.46	16.69	11.57	48.49	34.02	41.26
RPL22b	YFL034C-A	3.6	0.02	7.44	8.98	8.21	36.13	22.75	29.44
HXK2	YGL253W	3.6	0.00	22.97	18.73	20.85	70.16	80.99	75.58
LYS9	YNR050C	3.8	0.00	11.64	6.73	9.18	33.44	36.14	34.79
RPL22a	YLR061W	3.9	0.00	27.43	28.40	27.91	115.98	102.04	109.01
RPS26b	YER131W	4.0	0.01	45.09	25.49	35.29	176.78	104.66	140.72
ACO2	YJL200C	4.1	0.00	7.64	6.36	7.00	28.98	28.25	28.61
ENO2	YHR174W	4.3	0.01	198.32	156.01	177.17	648.49	860.60	754.55
TDH2	YJR009C	4.3	0.00	155.12	107.55	131.33	582.29	548.37	565.33
RPS22a	YJL190C	4.6	0.00	35.52	24.66	30.09	142.26	135.16	138.71
UTR2	YEL040W	4.8	0.00	7.32	3.23	5.27	23.52	27.13	25.32
CIS3	YJL158C	4.9	0.00	19.26	6.53	12.90	73.18	53.11	63.15
FDC1	YLR044C	5.1	0.02	163.01	95.02	129.01	848.31	457.02	652.66
NRG2	YBR066C	5.2	0.00	8.04	12.03	10.04	57.29	46.34	51.82
FHR2	YIL053W	6.5	0.00	21.97	24.88	23.43	172.63	130.68	151.65
SRL1	YOR247W	6.5	0.00	27.63	12.42	20.02	130.84	128.91	129.87
YIL029C	YIL029C	7.8	0.04	0.73	1.31	1.02	8.06	7.94	8.00
YPS6	YIR039C	10.3	0.00	22.79	16.32	19.55	234.37	168.89	201.63
CWP1	YKL096W	13.6	0.00	4.68	8.16	6.42	99.05	76.07	87.56
TDA6	YPR157W	14.7	0.00	1.01	3.59	2.30	27.59	39.79	33.69
YOL014W	YOL014W	16.8	0.00	2.72	1.99	2.36	47.46	31.95	39.70
GAT3	YLR013W	23.3	0.00	1.47	0.58	1.02	23.77	23.94	23.85
YLR012C	YLR012C	44.4	0.05	1.31	2.02	1.66	55.55	92.26	73.90

AUTHORS' CONTRIBUTIONS

DP and VM conceived and designed the study. DP, DC and HJ acquired the data. DP, DC and VM analysed and interpreted the data. Drafting of the manuscript was done by DP and DC. Critical revision of the manuscript for intellectual content was performed by DP, HL and VM. VM and HL obtained funding. All authors have read and approved of the final version of this manuscript.

REFERENCES

1.   Oud B, van Maris AJ, Daran JM, Pronk JT. Genome-wide analytical approaches for reverse metabolic engineering of industrially relevant phenotypes in yeast. FEMS Yeast Res. 2012;12:183–96.

2. Smith DR, Quinlan AR, Peckham HE, Makowsky K, Tao W, Woolf B, et al. Rapid whole-genome mutational profiling using next-generation sequencing technologies. Genome Res. 2008;18:1638–42.

3. Le Crom S, Schackwitz W, Pennacchio L, Magnuson JK, Culley DE, Collett JR, et al. Tracking the roots of cellulase hyperproduction by the fungus Trichoderma reesei using massively parallel DNA sequencing. Proc Natl Acad Sci U S A. 2009;106:16151–6.

4. Sarin S, Bertrand V, Bigelow H, Boyanov A, Doitsidou M, Poole RJ, et al. Analysis of multiple ethyl methanesulfonate-mutagenized*Caenorhabditis elegans*strains by whole-genome sequencing. Genetics. 2010;185:417–30.

5. Harper MA, Chen Z, Toy T, Machado IM, Nelson SF, Liao JC, et al. Phenotype sequencing: identifying the genes that cause a phenotype directly from pooled sequencing of independent mutants. PLoS One. 2011;6:e16517.

6. Parts L, Cubillos FA, Warringer J, Jain K, Salinas F, Bumpstead SJ, et al. Revealing the genetic structure of a trait by sequencing a population under selection. Genome Res. 2011;21:1131–8.

7. Zhang YX, Perry K, Vinci VA, Powell K, Stemmer WPC, del Cardayre SB. Genome shuffling leads to rapid phenotypic improvement in bacteria. Nature. 2002;415:644–6.

8. Pinel D, D'Aoust F, del Cardayre SB, Bajwa PK, Lee H, Martin VJJ. *Saccharomyces cerevisiae*genome shuffling through recursive population mating leads to improved tolerance to spent sulfite liquor. Appl Environ Microbiol. 2011;77:4736–43.

9. Patnaik R, Louie S, Gavrilovic V, Perry K, Stemmer WPC, Ryan CM, et al. Genome shuffling of*Lactobacillus*for improved acid tolerance. Nat Biotechnol. 2002;20:707–12.

10. Dai MH, Copley SD. Genome shuffling improves degradation of the anthropogenic pesticide pentachlorophenol by*Sphingobium chlorophenolicum*ATCC 39723. Appl Environ Microbiol. 2004;70:2391–7.

11. Biot-Pelletier D, Martin VJJ. Evolutionary engineering by genome shuffling. Appl Microbiol Biotechnol. 2014;98(9):3877–87.

12. Almeida JRM, Modig T, Petersson A, Hahn-Hägerdal B, Lidén G, Gorwa-Grauslund MF. Increased tolerance and conversion of inhibitors in lignocellulosic hydrolysates by*Saccharomyces cerevisiae*. J Chem Technol Biotechnol. 2007;82:340–9.

13. Liu ZL. Molecular mechanisms of yeast tolerance and in situ detoxification of lignocellulose hydrolysates. Appl Microbiol Biotechnol. 2011;90:809–25.

14. Liu ZL, Blaschek HP. Lignocellulosic biomass conversion to ethanol by*Saccharomyces*. In: Vertes A, Qureshi N, Yukawa H, Blaschek H, editors. Biomass to biofuels: strategies for global industries. West Sussex, U. K: John Wiley & Sons, Ltd; 2010. p. 17–36.

15. Palmqvist E, Hahn-Hägerdal B. Fermentation of lignocellulosic hydrolysates. I: Inhibition and detoxification. Biores Technol. 2000;74:17–24.

16. Richardson TL, Harner NK, Bajwa PK, Trevors JT, Lee H. Approaches to deal with toxic inhibitors during fermentation of lignocellulosic substrates. Acs Sym Ser. 2011;1067:171–202.

17. Pinel D, Gawand P, Mahadevan R, Martin VJJ. 'Omics' technologies and systems biology for engineering*Saccharomyces cerevisiae*strains for lignocellulosic bioethanol production. Biofuels. 2011;2:659–75.

18. Gorsich SW, Slininger PJ, Liu ZL. Physiological responses to furfural and HMF and the link to other stress pathways. J Biotechnol. 2005;118:S91–1.

19. Petersson A, Almeida JRM, Modig T, Karhumaa K, Hahn-Hägerdal B, Gorwa-Grauslund MF, et al. A 5-hydroxymethyl furfural reducing enzyme encoded by the*Saccharomyces cerevisiae*ADH6 gene conveys HMF tolerance. Yeast. 2006;23:455–64.

20. Keating JD, Panganiban C, Mansfield SD. Tolerance and adaptation of ethanologenic yeasts to lignocellulosic inhibitory compounds. Biotechnol Bioeng. 2006;93:1196–206.

21. Pinel D, Martin VJJ. Meiotic recombination-based genome shuffling of*Saccharomyces cerevisiae*and*Schefferomyces stiptis*for increased inhibitor tolerance to lignocellulosic substrate toxicity. In: Patnaik R, editor. Engineering complex phenotypes in industrial strains. 1st ed. Hoboken, New Jersey: John Wiley & Sons, Inc; 2012. p. 233–50.

22. Helle SS, Murray A, Lam J, Cameron DR, Duff SJ. Xylose fermentation by genetically modified*Saccharomyces cerevisiae*259ST in spent sulfite liquor. Bioresour Technol. 2004;92:163–71.

23. Olsson L, HahnHagerdal B. Fermentation of lignocellulosic hydrolysates for ethanol production. Enzyme Microb Tech. 1996;18:312–31.

24. Parajó JC, Domíngues H, Domínguez JM. Biotechnological production of xylitol. Part 3: operation in culture media made from lignocellulose

hydrolysates. Bioresour Technol. 1998;66:25–40.

25. Helle S, Duff S. Supplementing spent sulfite liquor with a lignocellulosic hydrolysate to increase pentose/hexose co-fermentation efficiency and ethanol yield. Final report-Natural Resources Canada-Tembec Industries; 2004.http://www.lifesciencesbc.ca/files/dufffinal_report.pdf.

26. Yassour M, Kaplan T, Fraser HB, Levin JZ, Pfiffner J, Adiconis X, et al. Ab initio construction of a eukaryotic transcriptome by massively parallel mRNA sequencing. Proc Natl Acad Sci U S A. 2009;106:3264–9.

27. Nagalakshmi U, Wang Z, Waern K, Shou C, Raha D, Gerstein M, et al. The transcriptional landscape of the yeast genome defined by RNA sequencing. Science. 2008;320:1344–9.

28. Kumar P, Henikoff S, Ng PC. Predicting the effects of coding non-synonymous variants on protein function using the SIFT algorithm. Nat Protoc. 2009;4:1073–81.

29. Ng PC, Henikoff S. SIFT: predicting amino acid changes that affect protein function. Nucleic Acids Res. 2003;31:3812–4.

30. Sim NL, Kumar P, Hu J, Henikoff S, Schneider G, Ng PC. SIFT web server: predicting effects of amino acid substitutions on proteins. Nucleic Acids Res. 2012;40:W452–7.

31. Ozkaynak E, Finley D, Solomon MJ, Varshavsky A. The yeast ubiquitin genes: a family of natural gene fusions. EMBO J. 1987;6:1429–39.

32. Modig T, Lidén G, Taherzadeh MJ. Inhibition effects of furfural on alcohol dehydrogenase, aldehyde dehydrogenase and pyruvate dehydrogenase. Biochem J. 2002;363:769–76.

33. Goldberg AL. Protein degradation and protection against misfolded or damaged proteins. Nature. 2003;426:895–9.

34. Hochstrasser M. Ubiquitin-dependent protein degradation. Annu Rev Genet. 1996;30:405–39.

35. Kimura Y, Tanaka K. Regulatory mechanisms involved in the control of ubiquitin homeostasis. J Biochem. 2010;147:793–8.

36. Hershko A, Ciechanover A. The ubiquitin system. Annu Rev Biochem. 1998;67:425–79.

37. Mukhopadhyay D, Riezman H. Proteasome-independent functions of ubiquitin in endocytosis and signaling. Science. 2007;315:201–5.

38. Finley D, Ozkaynak E, Varshavsky A. The yeast polyubiquitin gene is essential for resistance to high temperatures, starvation, and other stresses. Cell. 1987;48:1035–46.

39. Hanna J, Meides A, Zhang DP, Finley D. A ubiquitin stress response

induces altered proteasome composition. Cell. 2007;129:747–59.

40. Leggett DS, Hanna J, Borodovsky A, Crosas B, Schmidt M, Baker RT, et al. Multiple associated proteins regulate proteasome structure and function. Mol Cell. 2002;10:495–507.

41. Hanna J, Leggett DS, Finley D. Ubiquitin depletion as a key mediator of toxicity by translational inhibitors. Mol Cell Biol. 2003;23:9251–61.

42. Chernova TA, Allen KD, Wesoloski LM, Shanks JR, Chernoff YO, Wilkinson KD. Pleiotropic effects of Ubp6 loss on drug sensitivities and yeast prion are due to depletion of the free ubiquitin pool. J Biol Chem. 2003;278:52102–15.

43. Hernández-López MJ, Garcia-Marqués S, Randez-Gil F, Prieto JA. Multicopy suppression screening of Saccharomyces cerevisiae identifies the ubiquitination machinery as a main target for improving growth at low temperatures. Appl Environ Microbiol. 2011;77:7517–25.

44. Kolling R, Hollenberg CP. The ABC-transporter Ste6 accumulates in the plasma membrane in a ubiquitinated form in endocytosis mutants. EMBO J. 1994;13:3261–71.

45. Hein C, Springael JY, Volland C, Haguenauer-Tsapis R, Andre B. NPl1, an essential yeast gene involved in induced degradation of Gap1 and Fur4 permeases, encodes the Rsp5 ubiquitin-protein ligase. Mol Microbiol. 1995;18:77–87.

46. Hicke L, Riezman H. Ubiquitination of a yeast plasma membrane receptor signals its ligand-stimulated endocytosis. Cell. 1996;84:277–87.

47. Lin CH, MacGurn JA, Chu T, Stefan CJ, Emr SD. Arrestin-related ubiquitin-ligase adaptors regulate endocytosis and protein turnover at the cell surface. Cell. 2008;135:714–25.

48. Goh WS, Orlov Y, Li J, Clarke ND. Blurring of high-resolution data shows that the effect of intrinsic nucleosome occupancy on transcription factor binding is mostly regional, not local. PLoS Comput Biol. 2010;6:e1000649.

49. Harbison CT, Gordon DB, Lee TI, Rinaldi NJ, Macisaac KD, Danford TW, et al. Transcriptional regulatory code of a eukaryotic genome. Nature. 2004;431:99–104.

50. Lee TI, Rinaldi NJ, Robert F, Odom DT, Bar-Joseph Z, Gerber GK, et al. Transcriptional regulatory networks in Saccharomyces cerevisiae. Science. 2002;298:799–804.

51. Workman CT, Mak HC, McCuine S, Tagne JB, Agarwal M, Ozier O, et al. A systems approach to mapping DNA damage response pathways.

Science. 2006;312:1054–9.

52. Zhu C, Byers KJ, McCord RP, Shi Z, Berger MF, Newburger DE, et al. High-resolution DNA-binding specificity analysis of yeast transcription factors. Genome Res. 2009;19:556–66.

53. Vyas VK, Berkey CD, Miyao T, Carlson M. Repressors Nrg1 and Nrg2 regulate a set of stress-responsive genes in*Saccharomyces cerevisiae*. Eukaryot Cell. 2005;4:1882–91.

54. Kuchin S, Vyas VK, Carlson M. Snf1 protein kinase and the repressors Nrg1 and Nrg2 regulate FLO11, haploid invasive growth, and diploid pseudohyphal differentiation. Mol Cell Biol. 2002;22:3994–4000.

55. Mayordomo I, Estruch F, Sanz P. Convergence of the target of rapamycin and the Snf1 protein kinase pathways in the regulation of the subcellular localization of Msn2, a transcriptional activator of STRE (Stress Response Element)-regulated genes. J Biol Chem. 2002;277:35650–6.

56. Mira NP, Becker JD, Sá-Correia I. Genomic expression program involving the Haa1p-regulon in*Saccharomyces cerevisiae*response to acetic acid. OMICS. 2010;14:587–601.

57. Fernandes AR, Mira NP, Vargas RC, Canelhas I, Sá-Correia I.*Saccharomyces cerevisiae*adaptation to weak acids involves the transcription factor Haa1p and Haa1p-regulated genes. Biochem Biophys Res Commun. 2005;337:95–103.

58. Magasanik B. Ammonia assimilation by Saccharomyces cerevisiae. Eukaryot Cell. 2003;2:827–9.

59. Ding MZ, Wang X, Liu W, Cheng JS, Yang Y, Yuan YJ. Proteomic research reveals the stress response and detoxification of yeast to combined inhibitors. PLoS One. 2012;7:e43474.

60. Mira NP, Palma M, Guerreiro JF, Sá-Correia I. Genome-wide identification of*Saccharomyces cerevisiae*genes required for tolerance to acetic acid. Microb Cell Fact. 2010;9:79.

61. Hess DC, Lu W, Rabinowitz JD, Botstein D. Ammonium toxicity and potassium limitation in yeast. PLoS Biol. 2006;4:e351.

62. Bayer TS, Hoff KG, Beisel CL, Lee JJ, Smolke CD. Synthetic control of a fitness tradeoff in yeast nitrogen metabolism. J Biol Eng. 2009;3:1.

63. Wegele H, Müller L, Buchner J. Hsp70 and Hsp90–a relay team for protein folding. Rev Physiol Biochem Pharmacol. 2004;151:1–44.

64. Verghese J, Abrams J, Wang Y, Morano KA. Biology of the heat shock response and protein chaperones: budding yeast (*Saccharomyces cerevisiae*) as a model system. Microbiol Mol Biol Rev. 2012;76:115–58.

65. Jones GW, Masison DC.*Saccharomyces cerevisiae*Hsp70 mutations affect [PSI+] prion propagation and cell growth differently and implicate Hsp40 and tetratricopeptide repeat cochaperones in impairment of [PSI+]. Genetics. 2003;163:495–506.

66. Loovers HM, Guinan E, Jones GW. Importance of the Hsp70 ATPase domain in yeast prion propagation. Genetics. 2007;175:621–30.

67. Duncan K, Edwards RM, Coggins JR. The pentafunctional arom enzyme of*Saccharomyces cerevisiae*is a mosaic of monofunctional domains. Biochem J. 1987;246:375–86.

68. Bauer BE, Rossington D, Mollapour M, Mamnun Y, Kuchler K, Piper PW. Weak organic acid stress inhibits aromatic amino acid uptake by yeast, causing a strong influence of amino acid auxotrophies on the phenotypes of membrane transporter mutants. Eur J Biochem. 2003;270:3189–95.

69. Yoshikawa K, Tanaka T, Furusawa C, Nagahisa K, Hirasawa T, Shimizu H. Comprehensive phenotypic analysis for identification of genes affecting growth under ethanol stress in*Saccharomyces cerevisiae*. Fems Yeast Research. 2009;9:32–44.

70. Stambuk BU, Panek AD, Crowe JH, Crowe LM, de Araujo PS. Expression of high-affinity trehalose-H+ symport in*Saccharomyces cerevisiae*. Biochim Biophys Acta. 1998;1379:118–28.

71. Jules M, Guillou V, Francois J, Parrou JL. Two distinct pathways for trehalose assimilation in the yeast*Saccharomyces cerevisiae*. Appl Environ Microbiol. 2004;70:2771–8.

72. Babrzadeh F, Jalili R, Wang C, Shokralla S, Pierce S, Robinson-Mosher A, et al. Whole-genome sequencing of the efficient industrial fuel-ethanol fermentative*Saccharomyces cerevisiae*strain CAT-1. Mol Genet Genomics. 2012;287:485–94.

73. Carreto L, Eiriz MF, Gomes AC, Pereira PM, Schuller D, Santos MA. Comparative genomics of wild type yeast strains unveils important genome diversity. BMC Genomics. 2008;9:524.

74. Dunn B, Levine RP, Sherlock G. Microarray karyotyping of commercial wine yeast strains reveals shared, as well as unique, genomic signatures. BMC Genomics. 2005;6:53.

75. Ask M, Mapelli V, Hock H, Olsson L, Bettiga M. Engineering glutathione biosynthesis of Saccharomyces cerevisiae increases robustness to inhibitors in pretreated lignocellulosic materials. Microb Cell Fact. 2013;12:87.

76. Stephen DW, Jamieson DJ. Amino acid-dependent regulation of

the*Saccharomyces cerevisiae*GSH1 gene by hydrogen peroxide. Mol Microbiol. 1997;23:203–10.

77. Nijkamp JF, van den Broek M, Datema E, de Kok S, Bosman L, Luttik MA, et al. De novo sequencing, assembly and analysis of the genome of the laboratory strain Saccharomyces cerevisiae CEN.PK113-7D, a model for modern industrial biotechnology. Microb Cell Fact. 2012;11:36.

78. Langmead B, Trapnell C, Pop M, Salzberg SL. Ultrafast and memory-efficient alignment of short DNA sequences to the human genome. Genome Biol. 2009;10:R25.

79. Li H, Ruan J, Durbin R. Mapping short DNA sequencing reads and calling variants using mapping quality scores. Genome Res. 2008;18:1851–8.

80. Nakamura K, Oshima T, Morimoto T, Ikeda S, Yoshikawa H, Shiwa Y, et al. Sequence-specific error profile of Illumina sequencers. Nucleic Acids Res. 2011;39:e90.

81. Oyola SO, Otto TD, Gu Y, Maslen G, Manske M, Campino S, et al. Optimizing Illumina next-generation sequencing library preparation for extremely AT-biased genomes. BMC Genomics. 2012;13:1.

82. Xie C, Tammi MT. CNV-seq, a new method to detect copy number variation using high-throughput sequencing. Bmc Bioinformatics. 2009;10:80.

83. Artimo P, Jonnalagedda M, Arnold K, Baratin D, Csardi G, de Castro E, et al. ExPASy: SIB bioinformatics resource portal. Nucleic Acids Res. 2012;40:W597–603.

84. Baggerly KA, Deng L, Morris JS, Aldaz CM. Differential expression in SAGE: accounting for normal between-library variation. Bioinformatics. 2003;19:1477–83.

85. da Huang W, Sherman BT, Lempicki RA. Systematic and integrative analysis of large gene lists using DAVID bioinformatics resources. Nat Protoc. 2009;4:44–57.

86. Merico D, Isserlin R, Bader GD. Visualizing gene-set enrichment results using the Cytoscape plug-in enrichment map. Methods Mol Biol. 2011;781:257–77.

87. Merico D, Isserlin R, Stueker O, Emili A, Bader GD. Enrichment map: a network-based method for gene-set enrichment visualization and interpretation. PLoS One. 2010;5:e13984.

88. Cherry JM, Hong EL, Amundsen C, Balakrishnan R, Binkley G, Chan ET, et al.*Saccharomyces*Genome Database: the genomics resource of budding yeast. Nucleic Acids Res. 2012;40:D700–5.

89. Abdulrehman D, Monteiro PT, Teixeira MC, Mira NP, Lourenco AB, dos Santos SC, et al. YEASTRACT: providing a programmatic access to curated transcriptional regulatory associations in*Saccharomyces cerevisiae*through a web services interface. Nucleic Acids Res. 2011;39:D136–40.

90. Monteiro PT, Mendes ND, Teixeira MC, d'Orey S, Tenreiro S, Mira NP, et al. YEASTRACT-DISCOVERER: new tools to improve the analysis of transcriptional regulatory associations in*Saccharomyces cerevisiae*. Nucleic Acids Res. 2008;36:D132–6.

91. Teixeira MC, Monteiro P, Jain P, Tenreiro S, Fernandes AR, Mira NP, et al. The YEASTRACT database: a tool for the analysis of transcription regulatory associations in*Saccharomyces cerevisiae*. Nucleic Acids Res. 2006;34:D446–51.

92. Russell DW, Jensen R, Zoller MJ, Burke J, Errede B, Smith M, et al. Structure of the*Saccharomyces cerevisiae*HO gene and analysis of its upstream regulatory region. Mol Cell Biol. 1986;6:4281–94.

93. Maere S, Heymans K, Kuiper M: BiNGO: a Cytoscape plugin to assess overrepresentation of gene ontology categories in biological networks. Bioinformatics 2005, 21:3448-3449.

94. Smoot ME, Ono K, Ruscheinski J, Wang PL, Ideker T: Cytoscape 2.8: new features for data integration and network visualization. Bioinformatics 2011, 27:431-432.

95. Shannon P, Markiel A, Ozier O, Baliga NS, Wang JT, Ramage D, Amin N, Schwikowski B, Ideker T: Cytoscape: a software environment for integrated models of biomolecular interaction networks. Genome Res 2003, 13:2498-2504.

96. Cline MS, Smoot M, Cerami E, Kuchinsky A, Landys N, Workman C, Christmas R, Avila-Campilo I, Creech M, Gross B, et al: Integration of biological networks and gene expression data using Cytoscape. Nat Protoc 2007, 2:2366-2382.

97. Montojo J, Zuberi K, Rodriguez H, Kazi F, Wright G, Donaldson SL, Morris Q, Bader GD: GeneMANIA Cytoscape plugin: fast gene function predictions on the desktop. Bioinformatics 2010, 26:2927-2928.

98. Warde-Farley D, Donaldson SL, Comes O, Zuberi K, Badrawi R, Chao P, Franz M, Grouios C, Kazi F, Lopes CT, et al: The GeneMANIA prediction server: biological network integration for gene prioritization and predicting gene function. Nucleic Acids Res 2010, 38:W214-220.

Chapter 8

CONSTRUCTION OF A RESTRICTION-LESS, MARKER-LESS MUTANT USEFUL FOR FUNCTIONAL GENOMIC AND METABOLIC ENGINEERING OF THE BIOFUEL PRODUCER CLOSTRIDIUM ACETOBUTYLICUM

Christian Croux[1], NgocPhuongThao Nguyen[1], Jieun Lee[2], Céline Raynaud[3], Florence SaintPrix[1], Maria GonzalezPajuelo[1], Isabelle Meynial-Salles[1] and Philippe Soucaille[1,3]

[1] LISBP, INSA, University of Toulouse, 135 Avenue de Rangueil, 31077 Toulouse Cedex, France

[2] College of Life Sciences and Biotechnology, Korea University, Seoul, South Korea

[3] Metabolic Explorer, SaintBeauzire, France.

ABSTRACT

Background

Clostridium acetobutylicum is a gram-positive, spore-forming, anaerobic bacterium capable of converting various sugars and polysaccharides into solvents (acetone, butanol, and ethanol). The sequencing of its genome has prompted new approaches to genetic analysis, functional genomics, and metabolic engineering to develop industrial strains for the production of biofuels and bulk chemicals.

Results

The method used in this paper to knock-out or knock-in genes in *C. acetobutylicum* combines the use of an antibiotic-resistance gene for the deletion or replacement of the target gene, the subsequent elimination of the antibiotic-resistance gene with the flippase recombinase system from *Saccharomyces cerevisiae*, and a *C. acetobutylicum* strain that lacks *upp*, which encodes uracil phosphoribosyl-transferase, for subsequent use as a counter-selectable marker. A replicative vector containing (1) a pIMP13 origin of replication from *Bacillus subtilis* that is functional in *Clostridia*, (2) a replacement cassette consisting of an antibiotic resistance gene (*MLS*^{R}) flanked by two FRT sequences, and (3) two sequences homologous to selected regions around target DNA sequence was first constructed. This vector was successfully used to consecutively delete the *Cac*824I restriction endonuclease encoding gene (*CA_C1502*) and the *upp* gene (*CA_C2879*) in the *C. acetobutylicum* ATCC824 chromosome. The resulting *C. acetobutylicum* Δ*cac1502*Δ*upp* strain is marker-less, readily transformable without any previous plasmid methylation and can serve as the host for the "marker-less" genetic exchange system. The third gene, *CA_C3535*, shown in this study to encode for a type II restriction enzyme (*Cac*824II) that recognizes the CTGAAG sequence, was deleted using an *upp*/5-FU counter-selection strategy to improve the efficiency of the method. The restriction-less marker-less strain and the method was successfully used to delete two genes (*ctfAB*) on the pSOL1 megaplasmid and one gene (*ldhA*) on the chromosome to get strains no longer producing acetone or l-lactate.

Conclusions

The restriction-less, marker-less strain described in this study, as well as the maker-less genetic exchange coupled with positive selection, will be useful for functional genomic studies and for the development of industrial strains for the production of biofuels and bulk chemicals.

BACKGROUND

In recent years, *Clostridium acetobutylicum* ATCC824 has been of interest in the postgenomic era due to the complete sequencing and annotation of its genome [1], supplying a wealth of information regarding its protein machinery. This global knowledge has prompted new approaches to genetic analysis, functional genomics, and metabolic engineering in order to develop industrial strains for the production of biofuels and bulk chemicals.

To this end, several reverse genetic tools have been developed for *C. acetobutylicum* ATCC 824, including a gene inactivation system based on non-replicative [2, 3] and replicative plasmids [4–7] and the group II intron gene inactivation system [8, 9]. Among these methods, only the method developed by Al-Hinai et al. [5] allows for *in frame* deletions and/or the introduction of genes at their normal chromosomal context without an antibiotic marker remaining. This system is made of two parts. The first part is a replicative vector containing (1) a pIMP13 origin of replication from *Bacillus subtilis* functional in *Clostridia,* (2) a replacement cassette consisting of an antibiotic resistance gene (*Th*^{R}) flanked by two FRT sequences, (3) two sequences homologous to the selected regions around the target DNA sequence, and (4) a codon-optimized *mazF* toxin gene from *Escherichia coli* under the control of a lactose-inducible promoter from *Clostridium perfringens* to allow for the positive selection of double-crossover allelic exchange mutants. The second part is a plasmid system with inducible segregational instability, enabling efficient deployment of the FLP-FRT system to generate marker-less deletion or integration mutants.

In 2006, our group patented a marker-less, in-frame deletion method [10] similar to the two-part method published by Al-Hinai et al. [5] in 2012. The first part of our method is based on the same replicative plasmid and the same replacement cassette, but it uses the uracil PRTase *upp*/5-fluorouracil (5-FU) system as a counter-selection strategy. The second part is based on a plasmid carrying (1) the FLP-FRT system to generate marker-less deletion and (2) the uracil PRTase *upp*/5-FU system to select for the plasmid loss after marker excision. This method was successfully used by the Metabolic Explorer Company to develop and patent an industrial recombinant strain of *C. acetobutylicum* for *n*-butanol production. As this method has not been described in detail and to make it available to and usable by the scientific community, we report how this method was developed and its use to create a restriction-less, marker-less strain of *C. acetobutylicum*. We show that this strain lacking *upp* (*CA_C2879*, encoding the uracil–phosphoribosyl-transferase), *CA_C1502* encoding *Cac*824I and*CA_C3535* encoding *Cac*824II (the second type II restriction enzyme) can be transformed by non-methylated DNA at very high efficiency and can be used for rapid gene knock-in and knock-out using the *upp*/5-FU counter-selectable system for both functional genomic and metabolic engineering of *C. acetobutylicum*. This strain and the method were further used to delete three genes *ctfAB* and *ldhA* to create strains no longer producing acetone and lactate, respectively.

RESULTS

MGC□*cac1502* Strain, a *C. acetobutylicum* Strain that is Transformable without Previous in Vivo Plasmid Methylation

*Cac*824I, the type II restriction endonuclease encoded by *CA_C1502*, is a major barrier to the electrotransformation of *C. acetobutylicum* with *E. coli*–*C. acetobutylicum* shuttle vectors [11]. The *Cac*824I restriction endonuclease recognition sequence 5′-GCNGC3′, where N can be any nucleotide, occurs infrequently in *C. acetobutylicum* DNA because of the high A + T DNA content (72 % A + T), but the sequence occurs frequently in *E. coli* plasmids. No methyltransferase that can be used in vitro to protect DNA from restriction by *Cac*824I is commercially available. Prior to the transformation of *C. acetobutylicum*, shuttle plasmids have to be methylated in vivo by transformation into *E.coli* ER2275 (pAN1) expressing the *Bacillus subtilis* phage φ3TI methyltransferase, which protects the shuttle plasmids from digestion by the clostridial endonuclease *Cac*824I [11]. This step is time consuming and may be a drawback if the genes to be transferred to *C. acetobutylicum* are toxic when expressed in *E. coli*. Therefore, a *C. acetobutylicum* strain deficient for this particular restriction system would be valuable for efficient electrotransformation without previous treatment of the plasmid to be transformed.

To delete the *Cac*824I encoding gene, the first step is the construction of a shuttle vector carrying the replacement cassette. The *CA_C1502* replacement cassette was cloned into the *Bam*HI site of the pCons2-1 and pCIP2-1 to generate the pREPcac15 and pCIPcac15 plasmids, respectively. The difference between these two plasmids is the origin of replication. The pREPcac15 contains a pIMP13 origin of replication from *B. subtilis* (rolling circle mechanism) functional in *Clostridia*, whereas pCIPcac15 contains the origin of replication of the pSOL1 megaplasmid (θ replication mechanism). The pREPcac15 and pCIPcac15 plasmids were methylated in vivo in *E.coli* ER2275 (pAN1) and were used to transform *C. acetobutylicum* ATCC824 by electroporation. After selection on plates for clones resistant to erythromycin at 40 µg/ml, one colony of each transformant was cultured for 24 h in liquid SM with erythromycin and was then subcultured four times in liquid 2YTG medium without antibiotic (Fig. 1a). To select integrants that lost the pREPcac15 or pCIPcac15 plasmids, 10^3 erythromycin resistant clones were replica plated on both RCA with erythromycin and RCA with thiamphenicol at 50 µg/ml. Whereas several colonies resistant to erythromycin and sensitive to thiamphenicol were obtained with pREPcac15 transformants, no such colonies were obtained with the pCIPcac15 transformants, which indicates that the θ replication mechanism of pCIPcac15 is less favorable for promoting double-crossover in *C. acetobutylicum* than a rolling circle mechanism. The genotype of clones with the desired phenotype was checked by PCR (polymerase chain reaction) analysis (Fig. 2a). The Δ*cac1502::mls* ^{R} strain, which had lost the pREPcac15, was isolated. This strain was transformed with the pCLF1 plasmid expressing the *FLP1* gene of *S. cerevisiae* encoding for the FLP recombinase. The expression of *FLP1* was under the control of the promoter and RBS (ribosome binding site) from the thiolase gene from *C. acetobutylicum*. After transformation and selection for resistance to thiamphenicol at 50 μg/ml, one colony was cultured in liquid SM with thiamphenicol. One hundred thiamphenicol resistant clones were replica plated on both RCA with erythromycin and RCA with thiamphenicol. The genotype of the clones with erythromycin sensitivity and thiamphenicol resistance was checked by PCR analysis with primers CAC 0 and CAC 5 (Fig. 2a). Two successive 24-h liquid cultures of the Δ*cac1502* strain were conducted in the absence of antibiotics to remove pCLF1. The Δ*cac1502* strain that lost pCLF1 was isolated according to its sensitivity to both erythromycin and thiamphenicol. This strain was called MGCΔ*cac1502*.

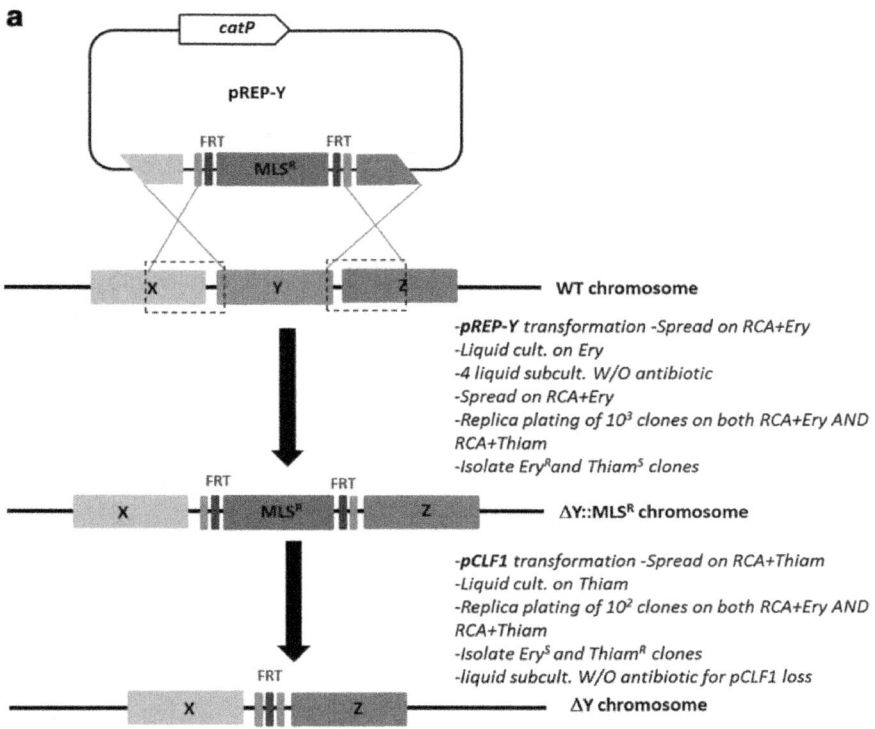

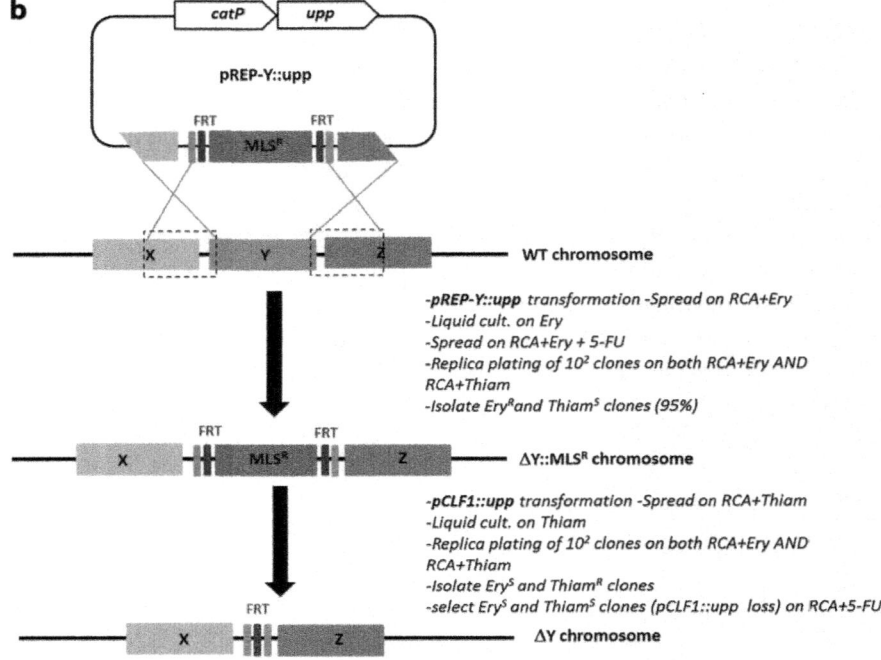

Figure. 1: General diagram representing gene replacement via allelic exchange at the *Y* locus, and excision of the MLS^R marker by the FLP recombinase to create an unmarked *Y* deletion mutant. The *boxed regions of X* and *Z genes* represent approximatively the regions of homology incorporated into the replicative plasmid used for the double-crossover event (~ 1 kbp each). **a** Initial strategy used for the construction of the MGC*Δcac1502* and MGC*Δcac1502Δupp* strains, **b** counter-selection strategy with the 5-FU/*upp* system used for the construction of the MGC*Δcac1502ΔuppΔcac3535* strain.

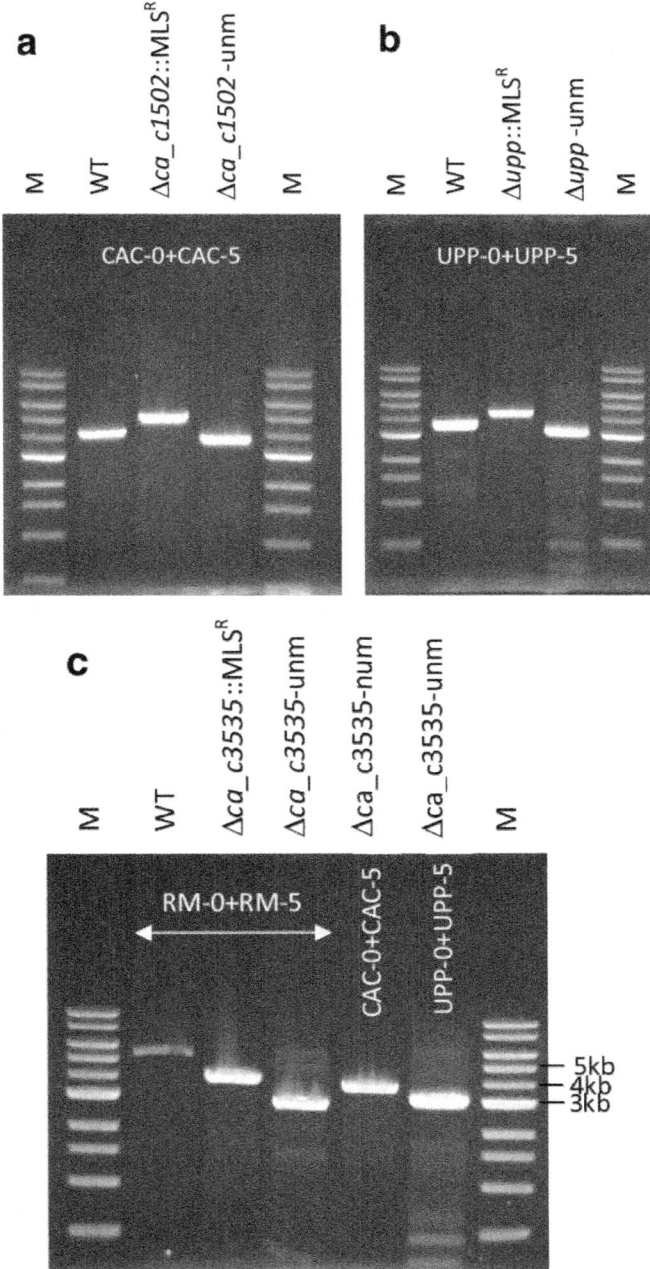

Figure. 2: Gene replacement via allelic exchange at the *ca_c1502*, *upp* and *ca_c3535* loci. PCR confirmation of the different double-crossover deletion mutants using ex-

ternal primers annealing to the chromosome upstream and downstream of each deletion cassette used in the different pREP plasmids: **a** Δ*ca_c1502* deletion mutants with CAC-0 + CAC-5 primers, **b** Δ*upp* deletion mutants with UPP-0 + UPP-5 primers, **c** Δ*ca_c3535* deletion mutants with RM-0 + RM-5 primers (*lanes 1, 2* and *3*). For each experiment, *lanes 2* and *3*refer to before and after excision of the MLS^R marker by the FLP recombinase, respectively, giving finally an unmarked deletion mutant (Δ-unm). The previous unmarked deletions in the final Δ*ca_c1502* Δ*upp* Δ*ca_c3535* were confirmed with CAC-0 + CAC-5 (**c**,*lane 4*) and UPP-0 + UPP-5 (**c**, *lane 5*) primers. *Lane M*, 1 kb DNA ladder (0.5–10 kb) (NEB).

The efficiency of transformation of this strain with methylated and unmethylated pCons2.1 plasmid was evaluated and compared to the wild type strain. Both strains can be transformed with methylated pCons2.1 with similar efficiency, but only MGCΔ*cac1502* can be transformed efficiently with unmethylated DNA (Table 1).

Table 1: Transformation efficiencies of *C. acetobutylicum* ATCC824 and MGCΔ*cac1502* for unmethylated and methylated pCons2.1

	C. acetobutylicum ATCC824	MGCΔ*cac1502*
Unmethylated pCons2.1	0	$0.79 (\pm0.24) \times 10^4$
Methylated pCons2.1	$0.46 (\pm0.11) \times 10^4$	$0.58 (\pm0.18) \times 10^4$

Values are expressed in number of transformants per μg DNA

Mean values and standard deviations from three independent experiments are given

25 ng pCons2.1 was used in each experiment

The following deletions described in this manuscript were conducted in this strain without previous in vivo plasmid methylation.

Construction of the MGCΔcac1502Δupp Strain: the First Marker-Free *C. acetobutylicum* Strain with Two Deleted Genes

To develop a positive screening of integrants, we used the "*upp*/5-FU as counter selection marker" system. The *C. acetobutylicum upp* gene (*CA_C2879*) encodes uracil phosphoribosyl-transferase (UPRTase), which catalyzes the conversion of uracil into UMP, thus allowing the cell to use exogenous uracil [12]. The pyrimidine analog 5-fluoro uracil (5-FU) can be converted by UPRTase into 5-fluoro-UMP, which is metabolized into 5-fluoro-dUMP, an inhibitor of thymidylate synthetase, toxic for the cell. The use of the *upp* expression cassette as a counter-selection marker is linked to the construction of a *C. acetobutylicum* strain deleted for the *upp* gene, thus resistant to 5-FU.

To delete *upp*, the *upp* replacement cassette was cloned into the *Bam*HI site of pCons2-1 to generate the plasmid pREPupp. The plasmid pREPupp was used to transform the MGCΔ*cac1502* strain by electroporation without previous in vivo methylation. After selection on plates for clones resistant to erythromycin at 40 µg/ml, one colony was cultured for 24 h in liquid SM with erythromycin and was then subcultured in liquid 2YTG medium without antibiotic (Fig. 1a). To select integrants having lost the pREPupp plasmid, 10^3 erythromycin resistant clones were replica plated on both RCA with erythromycin and RCA with thiamphenicol. The genotype of the clones resistant to erythromycin and sensitive to thiamphenicol was determined by PCR analysis (Fig. 2b). The MGCΔ*cac1502*Δ*upp::mls* ^{R} strain that lost pREPupp was isolated. When the resistance to 5-FU was analyzed, we showed that this strain was resistant to up to 1 mM 5-FU compared to 50 µM for the MGCΔ*cac1502* strain. This strain was then transformed with the pCLF1 plasmid, and selection of MGC Δ*cac1502*Δ*upp* strain with sensitivity to both erythromycin and thiamphenicol was performed, as previously described for the MGCΔ*cac1502* strain (Fig. 2b).

Deletion of the *CA_C3535* Gene in the MGCΔ*cac1502*Δ*upp* Strain using the *upp*/5-FU System as a Counter-Selectable Marker for the Loss of Plasmid

The *CA_C3535* gene encodes *Cac*824II, a potentially bi-functional enzyme carrying both a type II restriction endonuclease and methylase activities. To delete *CA_C3535*, the *CA_C3535* replacement cassette was cloned into the *Bam*HI site of the pCons::upp to generate the plasmid pREPcac3535::upp. The plasmid pREPcac3535::upp was used to transform the *C. acetobutylicum* MGCΔ*cac1502*Δ*upp* strain by electroporation without previous in vivo methylation.

After plate selection for clones resistant to erythromycin at 40 µg/ml, 100 transformants were replica plated on RCA with erythromycin, RCA with thiamphenicol and RCA with 5-FU at 400 µM (Fig. 1b). All transformants were resistant to erythromycin and thiamphenicol and were sensitive to 5-FU compared to the parental strain, which was resistant to 5-FU. This result demonstrates that the expression of the *upp* gene carried by pREPcac::upp confers sensitivity to 5-FU.

To select for Δ*cac3535::Em* [R] integrants that lost the pREPcac3535::upp plasmid, erythromycin- and 5-FU-resistant clones were selected on RCA plates containing erythromycin and 5-FU from 100 µl of a liquid culture of the MGC Δ*cac1502*Δ*upp*(pREPcac3535::upp) strain. Approximately 500 colonies were obtained, and 100 of them were replica plated on both RCA with erythromycin and RCA with thiamphenicol. Most of the clones (95 %) were resistant to erythromycin and sensitive to thiamphenicol. Four clones were checked by PCR analysis (Fig. 2c) All four clones had the correct phenotype, and one of the clones was selected as the MGC Δ*cac1502*Δ*upp*Δ*cac3535::mls* [R] strain. This strain was then transformed with pCLF::upp, a derivative of the pCLF1 plasmid that also carries the *upp* gene, in order for the positive selection of plasmid loss after the excision of the *mls* [R] marker. After the first selection of clones resistant to thiamphenicol and sensitive to erythromycin, a second selection of clones resistant to 5-FU and sensitive to thiamphenicol was performed to obtain the MGC Δ*cac1502*Δ*upp*Δ*cac3535* strain that was control by PCR (Fig. 2c) for the presence of all the marker-less deletions. Finally, when compared to *C. acetobutylicum* ATCC824 wild type, the growth of the restriction-less marker-less strain in MS medium at pH 4.5 (Fig. 3) was shown to be unaffected by the different deletions.

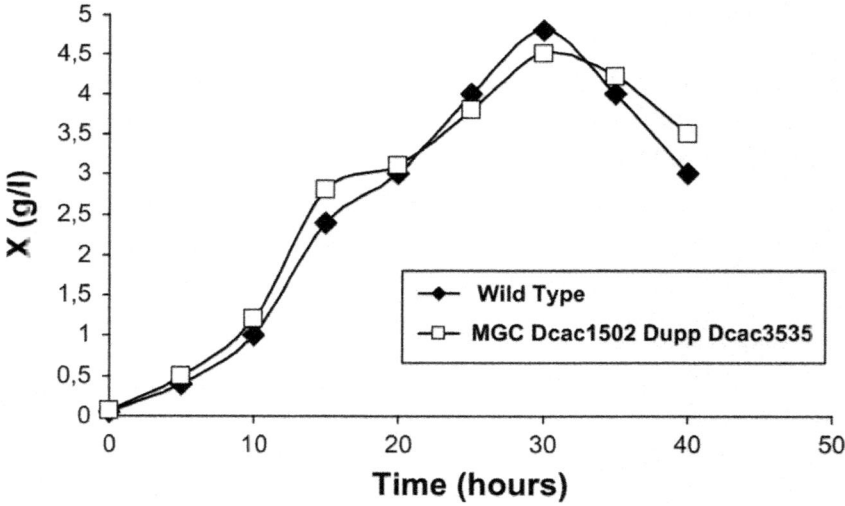

Figure. 3: Growth curves of *C. acetobutylicum* ATCC824 and MGCΔ*cac1502* Δ*upp* Δ*cac3535* at pH 4.5 in SM medium.

The unmethylated pCons2.1 plasmid was used to evaluate the transformation efficiency of the MGCΔcac1502 and the MGCΔcac1502Δupp Δcac3535 strains. The transformation efficiency of MGCΔcac1502ΔuppΔcac3535 for unmethylated pCons2.1 was ~eightfold higher than that of MGCΔcac1502 (Table 2).

Table 2: Transformation efficiencies of MGCΔcac1502 and MGCΔcac1502ΔuppΔcac3535 for unmethylated pCons2.1

	MGCΔcac1502	**MGC Δcac1502Δupp Δcac3535**
Unmethylated pCons2.1	0.79 (±0.24) × 10⁴	6.1 (±3.2) × 10⁴

Values are expressed in number of transformants per μg DNA

Mean values and standard deviations from three independent experiments are given

25 ng pCons2.1 was used in each experiment

Determination of the Recognition Sequence of *Cac824*II Encoded by *CA_C3535*

CA_C3535 encoded a 993 amino acid protein with a calculated molecular mass of 116,842 Da. The amino acid sequence analysis revealed high similarities with two restriction endonucleases: *Acu*I from *Acinetobacter calcoaceticus* SRW4 [13] and*Eco*57I from *E. coli* RFL57 [14] with 44 and 46 % identity, respectively. Both enzymes belong to the IIg family of restriction enzymes and possess both a restriction and methylase activity. To heterologously express the *Acu*I-encoding gene in *E. coli*[13], it was necessary to first express the *Acu*IM methylase-encoding gene because the methylase activity of *Acu*I was not sufficient to protect DNA against its restriction activity. We applied the same strategy for the expression of *CA_C3535*-encoding *Cac824*II: we cloned into the pSOS2K2 gene and expressed in *E. coli* the *CA_C3534* gene that encodes a putative methylase and that is located immediately downstream of *CA_C3535* gene in the *C. acetobutylicum* chromosome. The pSC-CAC3534 plasmid expressing *CA_C3534* has three *Acu*I recognition sites, but when we tried to digest it with *Acu*I, it was completely protected from the activity of this enzyme. To express, purify and determine the recognition sequences of*Cac824*II, we cloned *CA_C3535*in the pPAL vector using the *E. coli* BL21-AI cells containing the pSC-CAC3534 plasmid as host. The *Cac824*II endonuclease was purified, and its activity towards unmethylated pCons2.1

in the presence of SAM was determined. *Acu*I recognizes the 5'-CTGAAG-3' sequence and cuts the pCons2.1 plasmid two times, resulting in two fragments of 2411- and 882-bp. Figure 4 shows that *Cac824*II gives the same restriction pattern as *Acu*I. To confirm that the*Acu*I and *Cac824*II recognition sequences were identical, pCons2.1 was digested by 50 μg of *Cac824*II in the presence of 1 U of *Acu*I. Figure 4 shows that the restriction pattern was unchanged, which definitively confirms that *Acu*I and *Cac824*II are isoschizomers.

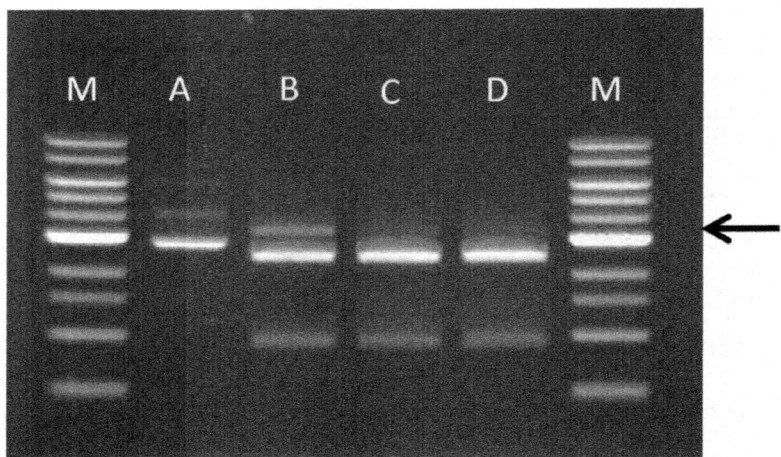

Figure. 4: Digestion properties of recombinant *Cac*824II as compared to commercial *Acu*I (New England Biolabs). 250 ng of unmethylated pCONS2.1 plasmid were incubated for 1 h at 37 °C in a reaction volume of 20 μL containing 50 mM potassium acetate, 20 mM Tris–acetate (pH 7.9), 10 mM Magnesium acetate, 100 μg/mL BSA and 0.04 mM S-adenosyl-methionine with (*A*) No enzyme, (*B*) purified*Cac*824II (50 μg), (*C*) *Acu*I (5U), and (*D*) purified *Cac*824II (50 μg) + *Acu*I (5U). *Lanes M*, 1 kb DNA ladder (0.5–10 kbp, NEB). Reactions products were electrophoresed on a 0.8 % agarose gel. An *arrow* indicates the incomplete digestion product remaining after incubation with *Cac*824II.

Deletion of the *ctfAB* Genes in the MGCΔ*cac1502*Δ*upp*Δ*cac3535* to create a Strain no Longer Producing Acetone

The *ctfAB* genes (*CA_P0163-CA_P0164*) located on the pSOL1 megaplasmid encodes an acetoacetyl-CoA:acyl CoA-transferase involved in the first specific step of acetone formation [15]. To delete *ctfAB*, the *ctfAB* replacement cassette was cloned into the *Bam*HI site of the pCons::upp to generate the plasmid pREPctfAB::upp. The plasmid pREPctfAB::upp was used to transform the *C. acetobutylicum* MGCΔ*cac1502*Δ*upp*Δ*cac3535* strain by electroporation without previous in vivo methylation and cell containing the plasmid were selected on RCA plate with erythromycin at 40 µg/ml. To select for Δ*ctfAB::Em* [R] integrants that lost the pREPctfAB::upp plasmid, erythromycin- and 5-FU-resistant clones were selected on RCA plates containing erythromycin and 5-FU from 100 µl of a liquid culture of the MGC*cac1502ΔuppΔcac3535*(p REPctfAB::upp) strain. Approximately 500 colonies were obtained, and 50 of them were replica plated on both RCA with erythromycin and RCA with thiamphenicol. Most of the clones (90 %) were resistant to erythromycin and sensitive to thiamphenicol. Four clones were checked by PCR analysis (with primers CTF-0 and CTF-5 located outside of the *ctfAB*replacement cassette and primers CTF-D and CTF-R located inside of *ctfAB*). All four clones had the correct phenotype, and one of the clones was selected as the MGC Δ*cac1502ΔuppΔcac3535ΔctfAB::mls* [R] strain. The fermentation profile of this strain was compared to the MGC Δ*cac1502ΔuppΔcac3535* control strain during batch fermentation at pH 4.5 (Fig. 5). The production of acetone was totally abolished but the production of acetic acid was increased more than sixfold while butyric acid was only slightly increased, proving that the acetoacetyl-CoA:acyl CoA-transferase is mainly involved in the consumption of acetic acid.

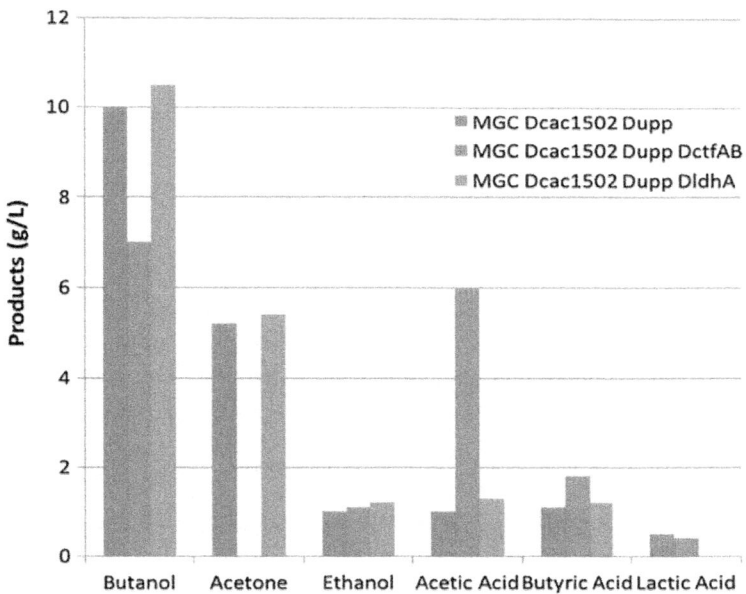

Figure. 5: Solvent and acid production of MGCΔ*cac1502 Δupp Δcac3535,* MGCΔ*cac1502 Δupp Δcac3535Δ ctfAB::Em* [R], and MGCΔ*cac1502Δupp Δcac3535Δ ldhA::Em* [R] in batch culture at pH 4.5 in SM medium.

Deletion of the *ldhA* Gene in the MGCΔ*cac1502*Δ*upp*Δ*cac3535* to Create a Strain no Longer Producing Lactate

The *ldhA* genes (*CA_C0267*) encodes a lactate dehydrogenase involved in the last step of l-lactate formation [15]. To delete*ldhA*, the *ldhA* replacement cassette was cloned into the *Bam*HI site of the pCons::upp to generate the plasmid pREPldhA::upp. The plasmid pREPldhA::upp was used to transform the *C. acetobutylicum* MGCΔ*cac1502*Δ*upp*Δ*cac3535*strain by electroporation without previous in vivo methylation and cell containing the plasmid were selected on RCA plate with erythromycin at 40 µg/ml. To select for Δ*ldhA::Em* [R] integrants that lost the pREPldhA::upp plasmid, erythromycin- and 5-FU-resistant clones were selected on RCA plates containing erythromycin and 5-FU from 100 µl of a liquid culture of the MGC*cac1502*Δ*upp*Δ*cac3535* (pREPldhA::upp) strain. Approximately 500 colonies were obtained, and 50 of them were replica plated on both RCA with erythromycin and RCA with thiamphenicol. Most of the clones (80 %) were resistant to erythromycin and sensitive to thiamphenicol. Four clones were checked by PCR analysis (with primers LDH-0 and LDH-5 located outside of the *ldhA* replacement cassette and primers LDH-D and LDH-R located inside of *ldhA*). All four clones had the correct phenotype, and one of the clones was selected as the MGC Δ*cac1502*Δ*upp*Δ*cac3535*Δ*ldhA::mls* [R] strain. The fermentation profile of this strain was compared to the MGC Δ*cac1502*Δ*upp*Δ*cac3535* control strain during batch fermentation at pH 4.5 (Fig. 5). The production of l-lactate was totally abolished proving that *ldhA* encodes the main l-lactate dehydrogenase of *C. acetobutylicum*.

DISCUSSION

We developed a simple and efficient method to create mutations in the *Clostridium acetobutylicum* chromosome. This method is based on the use of (1) a replicative plasmid, (2) a deletion cassette containing both DNA sequences with homology to the flanking region of the target gene (to delete it) and an antibiotic resistance gene surrounded by FRT sequences (as an excisable marker), and (3) the *upp* gene, which encodes the uracil–phosphoribosyl-transferase, as a counter-selectable marker.

A plasmid that replicates via a rolling circle mechanism was more efficient in terms of double cross over frequency than a plasmid that replicates through a theta mechanism. This result is in agreement with previous findings in *Bacillus subtilis*showing that plasmid replication through a rolling circle mechanism favors recombination between homologous sequences [16, 17].

The deletion cassette can be rapidly constructed through a three-step procedure using pre-constructed building blocks. After a fusion PCR and TOPO cloning of the product, a predesigned antibiotic resistance gene surrounded by two FRT sites in direct repeats is inserted. The *upp* gene is located on the plasmid outside of the deletion cassette. This allows the positive selection of clones that have lost the plasmid and integrated the deletion cassette by a double recombination event. We demonstrate here that this event occurs at a frequency of 10^{-5}, which means that without the selection procedure, it would be much more difficult to isolate the correct deletion mutant by replica plating alone. Once the deletion cassette is integrated into the chromosome, the expression of the *flp* recombinase allows (1) the excision of the antibiotic marker for a clean in-frame deletion of the targeted gene (without polar effect) and (2) consecutive gene deletions. Such a strategy was previously applied to marker-less gene deletion in *E. coli* [18] and *Mycobacterium smegmatis*[19]. The plasmid expressing the FLP recombinase-encoding gene was further improved by coexpressing the *upp* gene to use it as a positive selection for the plasmid loss after excision of the *MLS* ^{R} marker. A similar tool was developed by Al-Hinai et al. [5] using a plasmid (1) that expresses the FLP recombinase-encoding gene and (2) that has an inducible segregational instability to promote the plasmid loss.

In this study *Cac824*II (encoded by *CA_C3535*), the second type II restriction enzyme of *C. acetobutylicum* predicted by REBASE [20], was biochemically characterized and it was demonstrated that it is an isoschizomer of *Acu*I [13] recognizing the 5'-CTGAAG-3' sequence. It was also shown that *Cac824*II methylase (encoded by *CA_C3534*) protects DNA against restriction by *Cac824*II and *Acu*I by probably methylating one of the adenine in the 5'-CTGAAG-3' sequence. Two *Cac824*II restriction sites are present (in the *ampR* gene and in the colE1 origin of replication) in most the shuttle vector use to transform *C. acetobutylicum* and it was then justified to construct a marker-less strain deleted from CA_C3535. The transformation efficiency of MGC*Δcac1502ΔuppΔcac3535* for unmethylated pCons2.1 was much higher (~eightfold higher) than that of MGC*Δcac1502* and it will be an interesting strain to develop new genetic tools based on suicide vectors [20].

The restriction-less marker-less strain and the method was successfully used to delete two genes (*ctfAB*) on the pSOL1 megaplasmid and one gene (*ldhA*) on the chromosome to get strains no longer producing acetone or l-lactate. This work demonstrate that (1) *ctfAB* encode an acetoacetyl-CoA:acetate CoA-transferase that coupled acetone formation to acetate consumption and (2) *ldhA* encodes the main lactate dehydrogenase in *C. acetobutylicum* although a second gene *ldhB*(*CA_C3552*) is also present [15]. A strain with clostron inactivated *ctfAB* genes was previously constructed [21]. From the physiological analysis of this mutant and with the help of a mathematical model [22], it was demonstrated that butyrate was mainly reconsumed by the phosphotransbutyrylase-buyrate-kinase pathway and not by the acetoacetyl-CoA:acetate CoA-transferase in agreement with the data presented in our study.

CONCLUSIONS

The restriction-less, marker-less strain and the genome modification method presented here become simple and convenient tools that are useful for research groups involved in functional genomic studies of *C. acetobutylicum* and for further metabolic engineering of this strain to produce bulk chemicals and biofuel. As a demonstration of the efficiency of the method, we constructed two strains unable to produce l-lactate or acetone. Furthermore, this method was successfully used by the Metabolic Explorer Company to develop and patent an industrial recombinant strain of *C. acetobutylicum* for *n*-butanol production [23] at high yield.

METHODS

Bacterial Strain, Plasmids And Oligonucleotides

The bacterial strain and plasmids used in this study are listed in Table 3. The specific oligonucleotides used for PCR amplification were synthesized by Eurogentec (Table 4).

Table 3: Bacterial strains and plasmids used in this study

Strain or plasmid	Relevant characteristics[a]	Source or reference[b]
Bacterial strains		
E. coli		
TOP10		Invitrogen
ER2275	RecA⁻ McrBC⁻	NEB
C. acetobutylicum		
ATCC824	Wild type	ATCC
MGC*Δcac1502*	Δ*CA_C1502*	This study
MGC*Δcac1502Δupp*	Δ*CA_C 1502*Δ *CA_C 2879*	This study

MGC*Δcacl502Δupp*Δcac3535	*ΔCA_C 1502ΔCA_C 2879ΔCA_C 3535*	This study
MGC*Δcacl502Δupp*Δcac3535ΔctfAB	*ΔCA_C 1502ΔCA_C 2879ΔCA_C 3535ΔCA_ P0162-3*	This study
MGC*Δcacl502Δupp*Δcac3535ΔldhA	*ΔCA_C 1502ΔCA_C 2879ΔCA_C 3535ΔCA_C 0267*	This study
Plasmids		
pAN1	Cm[r], φ3TI, p15A origin	[27]
pKD4	Ap[r] Km[r]	[18]
pETSPO	Cm[r] MLS[r]	[4]
pUC18	Ap[r]	Fermentas
pUC18-FRT-MLS2	Ap[r] MLS[r]	This study
pCons2-1	Cm[r]	This study
pCR-BluntII-TOPO	Zeo[r] Km[r]	Invitrogen
pCIP2-1	Cm[r]	This study
pREPcac15	Cm[r] MLS[r] *ΔCA_C1502*	This study
pCIPcac15	Cm[r] MLS[r] *ΔCA_C1502*	This study
pREPupp	Cm[r] MLS[r] *Δupp*	This study
pCP20	Ap[r] Cm[r] *FLP*	[29]
pSOS95	Ap[r] MLS[r], acetone operon, *repL* gene, ColE1 origin	[32]
pCLF1	Cm[r] *FLP*	This study
pCR4-TOPO-Blunt	Ap[r] Km[r]	Invitrogen
pCons::upp	Cm[r] MLS[r] *upp*	This study
pREPCAC3535::upp	Cm[r] MLS[r] *upp ΔCA_C3535*	This study
pREPctfAB::upp	Cm[r] MLS[r] *upp ΔctfAB*	This study
pREPldhA::upp	Cm[r] MLS[r] *upp ΔldhA*	This study

[a] *RecA* − homologous recombination abolished, *McrBC* − lacking methylcytosine-specific restriction system, *Cm* [r] chloramphenicol resistance, *Ap* [r] ampicillin resistance, *MLS* [r] macrolide lincosamide and streptogramin B resistance, *Zeo* [r] zeomycin resistance, *φ3TI* φ3TI methyltransferase, *repL* Gram-positive origin of replication from pIM13

[b] *NEB* New England BioLabs, *ATCC* American Type Culture Collection (Rockville, MD)

Table 4: Oligonucleotides used for PCR amplifications

Primer name	Oligonucleotide sequence
PKD4.1	ct*ggcgccc*tgagtgcttgcggcagcgtgagggg
PKD4.2	ag*cccgggg*atctcatgctggagttcttcgccc
FRT-MLSR-F	tac*aggcct*tgagcgattgtgtaggctggagc
FRT-MLSR-R	aac*aggcct*gggatgtaacgcactgagaagccc
FCONSAccI	ccggggtaccgtcgacctgcagcc
PCONS*Eco*RI	gaattccgcgagctcggtacccggc
CRI3-D	ccatcgatgggggtcatgcatcaatactatcccc
CRI4-R	gcttccctgttttaatacctttcgg
FLP1-D	aaaa*ggatcc*aaaaggagggattaaaatgccacaatttggtatattatgtaaaacac-cacct
FLP1-R	aaat*ggcgccc*gcgtacttatatgcgtctatttatgtaggatgaaaggta
REP-UPP-F	aaaacagctgggaggaatgaaataatgagtaaagttacac
REP-UPP-R	aaaacagctgttattttgtaccgaataatctatctccagc
CAC 1	aaa*ggatcc*atgcacactcataaatttactgtaggaagtctg
CAC 2	gggg*aggcct*aaaaagggggtcccaaataatatttgccatagtaaccacc
CAC 3	cccccttttt*aggcct*cccctcgaacttattagaatgattaagattccgg
CAC 4	aaa*ggatcc*tcattaaatttcctccattttaagcctgtc
CAC 0	gtgatataattttcctttaaatggaggaggatctg
CAC 5	gccgttaatagacattataattccattggc
CAC-D	gaattcttaaaaatatttggatcattaagcgg
CAC-R	gttgtattggaatctttgttattatttctccc
UPP 1	aaaa*ggatcct*cctgatctattaattcttgatgaaccc
UPP 2	gggg*aggcct*aaaaaggggggattgcataaataaaaagggctgaaaaata-aatttcag
UPP 3	cccccttttt*aggcct*cccccttatttcattcctccattgtattttttttctatttg
UPP 4	aaaa*ggatcc*gctattatgaataggttaaataagtcagctgg
UPP 0	aatacaagcaaagagaataggctatgtgcc
UPP 5	aatacaagcaaagagaataggctatgtgcc
UPP-D	ggcata*gaagtaacaagagaaatgcagc
UPP-R	ataatctatctccagcatctccaagacc
RM3535 1	aaaa*ggatcc*gcagctttctggaaggactacggcg
RM3535 2	gggg*aggcct*aaaaaggggggcatttacttatggtacggttcacccc
RM3535 3	cccccttttt*aggcct*ccccgtctttaaaaagtaatttatcaaaggcatcaaggc
RM3535 4	aaaa*ggatcc*ctaactctctaaacgttacaatagtaatgcgc

RM3535 0	cacattgtcatttataaaagtccctaggg
RM3535 5	gtagtaattccaacttcaactcttgccac
RM3535-D	cttagaatagctgatattgcttgcgg
RM3535-R	agcatctctcttaatgattctccgg
CTF1	aaaaggatcccagacactataatagctttaggtggtacccc
CTF2	ggggaggcctaaaaagggggattataaaaagtagttgaaatatgaaggttta-aggttg
CTF3	ccccctttttaggcctccccatatccaatgaacttagacccatggctg
CTF4	aaaaggatccgtgttataatgtaaatataaatataaataggactagaggcg
CTF0	taccaccttctttcacgcttggctgcgg
CTF5	tatttaaagaggcattatcaccagagcg
LDH1	aaaaggatccgctttaaaatttggaaagaggaagttgtg
LDH2	ggggaggcctaaaaaggggggttagaaatctttaaaaatttctctatagagcccatc
LDH3	ccccctttttaggcctccccggtaaaagacctaaactccaagggtggaggctaggtc
LDH4	aaaaggatcccccattgtggagaatattccaaagaagaaaataattgc
LDH0	cagaaggcaagaatgtattaagcggaaatgc
LDH5	cttcccattatagctcttattcacattaagc
Cac3535-d-*Spe*I	aaa*actagt*atgaatgatatattaaaatagctttgaaaaaattggttgac
Cac3535-R-*Bam*HI	aaaa*ggatcc*ctacaaattatatatatctgttaccaatgcctc

Restriction sites are in italic

Culture and Growth Conditions

C. acetobutylicum was maintained as spores in synthetic medium (SM) as previously described [24, 25] Spores were activated by heat treatment at 80 °C for 15 min. All *C. acetobutylicum* strains were grown in anaerobic conditions at 37 °C in SM, in *Clostridium* growth medium (CGM) [26] in 2YTG [27] or in reinforced clostridial medium (RCM) (Fluka). Solid media were obtained by adding 1.5 % agar to the liquid media. Media were supplemented, when required, with the appropriate antibiotic in the following concentrations: for *C. acetobutylicum*, erythromycin at 40 µg/ml and thiamphenicol at 50 µg/ml; for *E. coli*, erythromycin at 200 µg/ml and chloramphenicol at 30 µg/ml. Transformations of *C. acetobutylicum* were conducted by electroporation, as previously described [11]. 5-FU was purchased from Sigma, and stock solutions were prepared in DMSO (dimethyl sulfoxide).

DNA Manipulation Techniques

Total genomic DNA from *C. acetobutylicum* was isolated as previously described [27]. Plasmid DNA was extracted from *E. coli* with the QIAprep kit (Qiagen, France). Pfu DNA Polymerase (Roche) was used to generate PCR products for cloning, and Taq Polymerase (New England BioLabs) was used for screening colonies by PCR with standard PCR protocols employed for all reactions. DNA restriction and cloning were performed according to standard procedures [28]. Restriction enzymes and Quick T4 DNA ligase were obtained from New England BioLabs (Beverly, MA) and were used according to the manufacturer's instructions. DNA fragments were purified from agarose gels with the QIAquick gel purification kit (Qiagen, France).

Construction of pUC18-FRT-MLS2

Inverse PCR was performed using the pKD4 plasmid [18] as a template and oligonucleotides PKD4.1 and PKD4.2 as primers to amplify the plasmid region with the FRT sites but without the kanamycin resistance marker. This blunt end fragment was later ligated to the *MLS* ^{r} gene obtained after a *Hind*III digestion of the pETSPO plasmid [4] and Klenow treatment. The corresponding plasmid (pKD4-Ery1) was then used as a template to amplify by PCR the macrolide lincosamide streptogramin B resistance (*MLS* ^{r}) gene, functional in *Clostridia* and flanked by two FRT sites and two *Stu*I sites, using the oligonucleotides FRT-MLSR-F and FRT-MLSR-R as primers. This fragment was directly cloned into the *Sma*I digested pUC18 to generate the pUC18-FRT-MLS2 plasmid.

Construction of pCons2.1

Inverse PCR was performed using the pETSPO plasmid [4] as a template and oligonucleotides PCONSAccI (mutating a*Bam*HI site) and PCONS*Eco*RI as primers. The PCR product, containing a pIMP13 *B. subtilis* origin of replication functional in*Clostridia* (rolling circle mechanism of replication) and a *catP* gene conferring resistance to thiamphenicol was phosphorylated and ligated to yield the pCons0 plasmid. This plasmid was then digested with *Bam*HI to remove the *spoA*cassette, and the DNA fragment was purified and ligated to generate the pCons2-1 plasmid.

Construction of pCIP2-1

The pIMP13 origin of replication from pCons2-1 was replaced by the origin of replication of the pSOL1 megaplasmid. The origin of replication of pSOL1 was amplified by PCR using *C. acetobutylicum* total DNA as a template and oligonucleotides ORI3-D and ORI3-R as primers. This PCR product was cloned into the pCR-BluntII-TOPO vector, and the resulting plasmid was digested by *Eco*RI to obtain the 2.2 kb *Eco*RI fragment containing the origin of replication of pSOL1. The pCons2-1 plasmid was digested by *Eco*RI, and the 2.4 kb fragment was ligated to the 2.2 kb *Eco*RI fragment to generate the plasmid pCIP2-1.

Construction of pREPcac15

Two DNA fragments surrounding *cac*1502 were amplified by PCR using *C. acetobutylicum* total DNA as the template and two pairs of oligonucleotides as primers. Using the primers pairs CAC 1 and CAC 2 or CAC 3 and CAC 4, 1493 and 999-bp DNA fragments were obtained, respectively. Both primers CAC 1 and CAC 4 introduce a *Bam*HI site, whereas primers CAC 2 and CAC 3 have complementary 5′ extended sequences that introduce a *Stu*I site. DNA fragments CAC 1–CAC 2 and CAC 3–CAC 4 were joined in a PCR fusion with primers CAC 1 and CAC 4, and the resulting fragment was cloned into the pCR4-TOPO-Blunt vector to generate pTOPO::cac15. At the unique *Stu*I site of pTOPO::cac15, the 1372-bp *Stu*I fragment of pUC18-FRT-MLS2 carrying the antibiotic resistance *MLS* ^{r} gene with FRT sequences on both sides was introduced. The *cac*1502 replacement cassette obtained after *Bam*HI digestion of the resulting plasmid was cloned into the *Bam*HI site of the pCons2-1 to generate the plasmid pREPcac15.

Construction of pCIPcac15

The *cac*1502 replacement cassette above was cloned into the *Bam*HI site of the pCIP2-1 to generate the plasmid pCIPcac15.

Construction of pREPupp

Two DNA fragments upstream and downstream of *cac*2879 were amplified by PCR using total DNA from *C. acetobutylicum* as the template and two pairs of oligonucleotides as primers. With the primer pairs UPP 1–UPP 2 and UPP 3–UPP 4, 1103- and 1105-bp DNA fragments were obtained, respectively. Both primers UPP 1 and UPP 4 introduce a *Bam*HI site, whereas primers UPP 2 and UPP 3 have 5′ extended sequences that introduce a *Stu*I site. DNA fragments UPP 1–UPP 2 and UPP 3–UPP 4 were joined in a PCR fusion with primers UPP 1 and UPP 4, and the resulting fragment was cloned into pCR4-TOPO-Blunt vector to generate pTOPO::upp. At the unique *Stu*I site of pTOPO::upp, the 1372-bp *Stu*I fragment of pUC18-FRT-MLS2 carrying the antibiotic resistance MLS^r gene with FRT sequences on both sides was introduced. The *upp* replacement cassette obtained after *Bam*HI digestion of the resulting plasmid was cloned into the *Bam*HI site of the pCons2-1 to generate the plasmid pREPupp.

Construction of pCLF1

The *FLP1* gene was amplified by PCR using the pCP20 plasmid [29] as a template and oligonucleotides FLP1-D and FLP1-R as primers. These primers introduced *Bam*HI and *Sfo*I restriction sites on the ends of the PCR product. After a *Bam*HI–*Sfo*I double digestion, the PCR product was cloned into the *Bam*HI–*Sfo*I sites of the pSOS95 expression vector to generate the pEX-FLP1 plasmid. The 1585-bp *Sal*I fragment of pEX-FLP1 containing the *FLP1* expression cassette was cloned into the *Sal*I site of pCons2-1 to generate the pCLF1 plasmid.

Construction of pCons::upp

The *upp* gene with its own ribosome binding site (RBS) was amplified by PCR from *C. acetobutylicum* total DNA with the oligonucleotides REP-UPP-F and REP-UPP-R as primers. The 664-bp PCR product was digested by *Pvu*II and was cloned into pCons2.1, digested by *Bcg*I and treated with T4 DNA polymerase to generate the pCons::*upp* plasmid. In this way, the *upp*gene was located just downstream of the *catP* gene to construct an artificial operon with *upp* expressed under the control of the *catP* promoter.

Construction of pREPcac35::upp

Two DNA fragments upstream and downstream of *CA_C3535* were amplified by PCR using the total DNA from *C. acetobutylicum* as a template and two pairs of oligonucleotides as primers. With the primer pairs RM3535 1 and RM3535 2 or RM3535 3 and RM3535 4, 1044- and 938-bp DNA fragments were obtained, respectively. Both primers RM3535 1 and RM3535 4 introduce a *Bam*HI site, whereas primers RM3535 2 and RM3535 3 have 5' extended sequences that introduce a*Stu*I site. DNA fragments RM3535 1-RM3535 2 and RM3535 3-RM3535 4 were joined in a PCR fusion with primers RM3535 1 and RM3535 4, and the resulting fragment was cloned into the pCR4-TOPO-Blunt vector to generate pTOPO::cac3535. At the unique *Stu*I site of pTOPO::cac3535, the 1372-bp *Stu*I fragment of pUC18-FRT-MLS2 carrying the antibiotic resistance*MLS*ᵣ gene with FRT sequences on both sides was introduced. The *CA_C3535* replacement cassette obtained after *Bam*HI digestion of the resulting plasmid was cloned into the *Bam*HI site of the pCons::upp to generate the plasmid pREPcac3535::upp.

Construction of pREPctfAB::upp

Two DNA fragments upstream and downstream of *ctfAB* (*CA_P0162-CA_P0163*) were amplified by PCR using the total DNA from *C. acetobutylicum* as a template and two pairs of oligonucleotides as primers. With the primer pairs CTF 1 and CTF 2 or CTF 3 and CTF 4, 1144- and 1138-bp DNA fragments were obtained, respectively. Both primers CTF 1 and CTF 4 introduce a *Bam*HI site, whereas primers CTF 2 and CTF 3 have 5′ extended sequences that introduce a *Stu*I site. DNA fragments CTF 1-CTF 2 and CTF 3-CTF 4 were joined in a PCR fusion with primers CTF 1 and CTF 4, and the resulting fragment was cloned into the pCR4-TOPO-Blunt vector to generate pTOPO::ctfAB. At the unique *Stu*I site of pTOPO::ctfAB, the 1372-bp *Stu*I fragment of pUC18-FRT-MLS2 carrying the antibiotic resistance *MLS* ʳ gene with FRT sequences on both sides was introduced. The *ldhA* replacement cassette obtained after *Bam*HI digestion of the resulting plasmid was cloned into the *Bam*HI site of the pCons::upp to generate the plasmid pREPctfAB::upp.

Construction of pREPldhA::upp

Two DNA fragments upstream and downstream of *ldhA* (*CA_C0267*) were amplified by PCR using the total DNA from *C. acetobutylicum* as a template and two pairs of oligonucleotides as primers. With the primer pairs LDH 1 and LDH 2 or LDH 3 and LDH 4, 1135- and 1161-bp DNA fragments were obtained, respectively. Both primers LDH 1 and LDH 4 introduce a *Bam*HI site, whereas primers LDH 2 and LDH 3 have 5′ extended sequences that introduce a *Stu*I site. DNA fragments LDH 1-LDH 2 and LDH 3-LDH 4 were joined in a PCR fusion with primers LDH 1 and LDH 4, and the resulting fragment was cloned into the pCR4-TOPO-Blunt vector to generate pTOPO::ldhA. At the unique *Stu*I site of pTOPO::ldhA, the 1372-bp *Stu*I fragment of pUC18-FRT-MLS2 carrying the antibiotic resistance *MLS* ʳ gene with FRT sequences on both sides was introduced. The *ldhA* replacement cassette obtained after *Bam*HI digestion of the resulting plasmid was cloned into the *Bam*HI site of the pCons::upp to generate the plasmid pREPldhA::upp.

Construction of pCLF::upp

The 1585-bp *Sal*I fragment of pEX-FLP1 containing the *FLP1* expression cassette was cloned into the *Sal*I site of pCons::upp to generate the pCLF::upp plasmid.

Cac3535 Expression and Purification

For the general cloning methods of restriction endonuclease genes in *E. coli*, the first step to clone and express the recombinant *CA_C3535* gene into *E. coli* was to pre-protect the host genomic DNA against the restriction activity of the *Cac*3535 bi-functional enzyme. The *CA_C3534* methylase-encoding gene was thus amplified by PCR with Phusion DNA polymerase using *C. acetobutylicum* ATCC824 total genomic DNA as the template and *Cac*3534-d-*Age*I and *Cac*3534-R-*Pvu*I as primers. After digestion with *Age*I and *Pvu*I, the resulting 1748-bp fragment was then cloned into pAH105 [30] a pSC101 derivative, that has been previously digested with *Age*I and *Pac*I, resulting in the pSC-CAC3534 plasmid. In this construct, the *CA_C3534* gene expression was placed under the control of the pGI 1.6 promoter [31].

The *E. coli* BL21-AI strain (Invitrogen) was then transformed by the pSC-CAC3534 plasmid to give the BL21-AI-3534 strain. This strain, with host genomic DNA protected against the restriction activity of the *Cac*3535 bi-functional enzyme, was finally used as the host strain for the *CA_C3535* gene over-expression using the T7-based expression system (see below). The *Cac*3535 protein was expressed in *E. coli* BL21 AI-3534 and was purified using the Profinity eXact Protein Purification System, following the recommendations of the manufacturer (Biorad). The *CA_C3535* gene was amplified by PCR with Phusion DNA polymerase using *C. acetobutylicum* ATCC824 total gDNA as the template and *Cac*3535-d-*Spe*I and *Cac*3535-R-BamHI as primers. The resulting 3002 bp fragment was cloned into the Zero Blunt TOPO vector (Invitrogen) to generate the TOPO-CAC3535 plasmid. After verification by DNA sequencing, the 2988-bp *Spe*I-*Bam*HI fragment from the latter plasmid was then introduced into the pPAL7 vector previously digested with the same enzymes to give the final pPAl-3535-I_2.4 plasmid.

After transformation, E. coli BL21-AI-3534 cells harboring the pPAl-3535-I_2.4 plasmid were grown in TB medium in the presence of 50 µg/ml carbenicillin and 100 µg/ml Spectinomycin at 37 °C to an OD550 ~ 0.45 and were then induced with 500 µM IPTG for 4 h at 37 °C. After centrifugation, the cell lysate was obtained by sonicating the resuspended pellet in bind/wash buffer (0.1 M sodium phosphate buffer, pH 7.2).

The tag-free *Cac*3535 protein was prepared using the Profinity eXact protein purification system, according to the standard protocol. After the Profinity Exact mini-spin column was bound by the protein and washed, the proteolytic activity of the affinity matrix was activated by applying two column volumes of room temperature 0.1 M sodium phosphate buffer, pH 7.2, containing 0.1 M sodium fluoride. The column was incubated for 30 min to allow for the cleavage of the tag from the protein; then, the tag-free protein was released from the mini-spin column by centrifugation. The tag-free *Cac*3535 purified protein retains a Thr-Ser linker at its N-Terminus, ensuring optimal binding and cleavage during the purification steps ("Imprecise Fusion protein").

ABBREVIATIONS

5-FU: 5-fluorouracil

CGM: *Clostridium* growth medium

DMSO: dimethyl sulfoxide

FLP: flippase

FRT: flippase recognition target

MLS [r]: the macrolide lincosamide streptogramin B resistance gene

PCR: polymerase chain reaction

RBS: ribosome binding site

RCM: reinforced clostridial medium

SM: synthetic medium

Th [R]: thiamphenicol resistance gene

UPRTase: uracil phosphoribosyl-transferase

AUTHORS' CONTRIBUTIONS

CC, IMS, and PS conceived the study; CC performed the initial deletion of CA_C1502 and the biochemical characterization of *Cac*824II, NPTN optimized the method for efficiently selecting clones with gene deletion using a negative marker and performed the deletion of CA_C3535, MGP made the initial constructs for gene deletion using a replicative plasmid with a θ or a rolling circle replication mechanism, FSP made the construct for the *upp* deletion, and JL made the construct for the expression of the FLP recombinase on a plasmid with a negative selection marker; CR performed the deletion of *ctfAB* and*ldhA*; PS drafted the manuscript together with CC and supervised the work. All authors read and approved the final manuscript.

ACKNOWLEDGEMENTS

This work was financially supported by the European Community's Seventh Framework Program "CLOSTNET" (PEOPLE-ITN-2008-237942) (to TN) and by Metabolic Explorer Company.

REFERENCES

1. Nolling J, Breton G, Omelchenko MV, Makarova KS, Zeng Q, Gibson R, Lee HM, Dubois J, Qiu D, Hitti J, et al. Genome sequence and comparative analysis of the solvent-producing bacterium *Clostridium acetobutylicum*. J Bacteriol. 2001;183(16):4823–38.

2. Green EM, Boynton ZL, Harris LM, Rudolph FB, Papoutsakis ET, Bennett GN. Genetic manipulation of acid formation pathways by gene inactivation in *Clostridium acetobutylicum* ATCC 824. Microbiology. 1996;142(Pt 8):2079–86.

3. Green EM, Bennett GN. Inactivation of an aldehyde/alcohol dehydrogenase gene from *Clostridium acetobutylicum* ATCC 824. Appl Biochem Biotechnol. 1996;57–58:213–21.

4. Harris LM, Welker NE, Papoutsakis ET. Northern, morphological, and fermentation analysis of spo0A inactivation and overexpression in*Clostridium acetobutylicum*ATCC 824. J Bacteriol. 2002;184(13):3586–97.

5. Al-Hinai MA, Fast AG, Papoutsakis ET. Novel system for efficient isolation of *Clostridium* double-crossover allelic exchange mutants enabling markerless chromosomal gene deletions and DNA integration. Appl Environ Microbiol. 2012;78(22):8112–21.

6. Liu CC, Qi L, Yanofsky C, Arkin AP. Regulation of transcription by unnatural amino acids. Nat Biotechnol. 2011;29:164–8.

7. Heap JT, Ehsaan M, Cooksley CM, Ng YK, Cartman ST, Winzer K, Minton NP. Integration of DNA into bacterial chromosomes from plasmids without a counter-selection marker. Nucleic Acids Res. 2012;40(8):e59.

8. Heap JT, Pennington OJ, Cartman ST, Carter GP, Minton NP. The ClosTron: a universal gene knock-out system for the genus*Clostridium*. J Microbiol Methods. 2007;70(3):452–64.

9. Shao L, Hu S, Yang Y, Gu Y, Chen J, Yang Y, Jiang W, Yang S. Targeted gene disruption by use of a group II intron (targetron) vector in*Clostridium acetobutylicum*. Cell Res. 2007;17(11):963–5.

10. Soucaille P, Figge R, Croux C, Explorer M. Process for chromosomal integration and DNA sequence replacement in *Clostridia*. International Patent Application PCT/EP2006/066997. 2006.

11. Mermelstein LD, Papoutsakis ET. In vivo methylation in *Escherichia coli* by the *Bacillus subtilis* phage phi 3T I methyltransferase to protect plasmids from restriction upon transformation of *Clostridium acetobutylicum* ATCC 824. Appl Environ Microbiol. 1993;59(4):1077–81.

12. Fabret C, Ehrlich SD, Noirot P. A new mutation delivery system for genome-scale approaches in *Bacillus subtilis*. Mol Microbiol. 2002;46(1):25–36.

13. Samuelson J, Xu S, O'Loane D, New England Biolabs I. Method for cloning and expression of *Acu*I restriction endonuclease and *Acu*I methylase in *E. coli*. US patent No.7,011,966. 2006.

14. Janulaitis A, Vaisvila R, Timinskas A, Klimasauskas S, Butkus V. Cloning and sequence analysis of the genes coding for Eco57I type IV restriction-modification enzymes. Nucleic Acids Res. 1992;20(22):6051–6.

15. Yoo M, Bestel-Corre G, Croux C, Riviere A, Meynial-Salles I, Soucaille P. A quantitative system-scale characterization of the metabolism of *Clostridium acetobutylicum*. MBio 2015;6(6):e01808–15.

16. Noirot P, Petit MA, Ehrlich SD. Plasmid replication stimulates DNA recombination in *Bacillus subtilis*. J Mol Biol. 1987;196(1):39–48.

17. Petit MA, Mesas JM, Noirot P, Morel-Deville F, Ehrlich SD. Induction of DNA amplification in the *Bacillus subtilis* chromosome. EMBO J. 1992;11(4):1317–26.

18. Datsenko KA, Wanner BL. One-step inactivation of chromosomal genes in *Escherichia coli* K-12 using PCR products. Proc Natl Acad Sci USA. 2000;97(12):6640–5.

19. Stephan J, Stemmer V, Niederweis M. Consecutive gene deletions in *Mycobacterium smegmatis* using the yeast FLP recombinase. Gene. 2004;343(1):181–90.

20. Sillers R, Chow A, Tracy B, Papoutsakis ET. Metabolic engineering of the non-sporulating, non-solventogenic *Clostridium acetobutylicum* strain M5 to produce butanol without acetone demonstrate the robustness of the acid-formation pathways and the importance of the electron balance. Metab Eng. 2008;10(6):321.

21. Cooksley CM, Zhang Y, Wang H, Redl S, Winzer K, Minton NP. Targeted mutagenesis of the *Clostridium acetobutylicum* acetone-butanol-ethanol fermentation pathway. Metab Eng. 2012;14(6):630–41.

22. Millat T, Voigt C, Janssen H, Cooksley CM, Winzer K, Minton NP, Bahl H, Fischer RJ, Wolkenhauer O. Coenzyme A-transferase-independent butyrate re-assimilation in *Clostridium acetobutylicum*-evidence from a mathematical model. Appl Microbiol Biotechnol. 2014;98(21):9059–72.

23. Soucaille P. Metabolic engineering of *Clostridium acetobutylicum* for enhanced production of n-butanol. International patent application WO2008052973.

24. Vasconcelos I, Girbal L, Soucaille P. Regulation of carbon and electron flow in *Clostridium acetobutylicum* grown in chemostat culture at neutral pH on mixtures of glucose and glycerol. J Bacteriol. 1994;176(5):1443–50.

25. Peguin S, Goma G, Delorme P, Soucaille P. Metabolic flexibility of *Clostridium acetobutylicum* in response to met. Appl Microbiol Biotechnol. 1994;42(4):611–6.

26. Wiesenborn DP, Rudolph FB, Papoutsakis ET. Thiolase from *Clostridium acetobutylicum* ATCC 824 and its role in the synthesis of acids and solvents. Appl Environ Microbiol. 1988;54(11):2717–22.

27. Mermelstein LD, Welker NE, Bennett GN, Papoutsakis ET. Expression of cloned homologous fermentative genes in *Clostridium acetobutylicum* ATCC 824. Biotechnology (NY). 1992;10(2):190–5.

28. Sambrook J, Fritsch EF, Maniatis T. Molecular cloning: a laboratory manual. 2nd ed. NY: Cold Spring Harbor Laboratory Press; 1989.

29. Cherepanov PP, Wackernagel W. Gene disruption in *Escherichia coli*: TcR and KmR cassettes with the option of Flp-catalyzed excision of the antibiotic-resistance determinant. Gene. 1995;158(1):9–14.

30. Payne MS, Picataggio SK, Hsu AK, Nair RV, Valle F, Soucaille P, Trimbur DE, Inc GI, Company EIDPDNA. Promoter and plasmid system for genetic engineering. 2012.

31. Meynial-Salles I, Cervin MA, Soucaille P. New tool for metabolic pathway engineering in *Escherichia coli*: one-step method to modulate expression of chromosomal genes. Appl Environ Microbiol. 2005;71(4):2140–4.

32. Raynaud C, Sarcabal P, Meynial-Salles I, Croux C, Soucaille P. Molecular characterization of the 1,3-propanediol (1,3-PD) operon of*Clostridium butyricum*. Proc Natl Acad Sci USA. 2003;100(9):5010–5.

Chapter 9

REWRITING THE BLUEPRINT OF LIFE BY SYNTHETIC GENOMICS AND GENOME ENGINEERING

Narayana Annaluru, Sivaprakash Ramalingam and Srinivasan Chandrasegaran

Department of Environmental Health Sciences, Bloomberg School of Public Health, Johns Hopkins University, 615 North Wolfe Street, Baltimore, MD 21205, USA

ABSTRACT

Advances in DNA synthesis and assembly methods over the past decade have made it possible to construct genome-size fragments from oligonucleotides. Early work focused on synthesis of small viral genomes, followed by hierarchical synthesis of wild-type bacterial genomes and subsequently on transplantation of synthesized bacterial genomes into closely related recipient strains. More recently, a synthetic designer version of yeast *Saccharomyces cerevisiae* chromosome *III* has been generated, with numerous changes from the wild-type sequence without having an impact on cell fitness and phenotype, suggesting plasticity of the yeast genome. A project to generate the first synthetic yeast genome - the Sc2.0 Project - is currently underway.

INTRODUCTION

Biology is now undergoing a rapid transition from the age of deciphering DNA sequence information of the genomes of biological species to the age of synthetic genomes. Scientists hope to gain a thorough mastery of and deeper insights into biological systems by rewriting the genome, the blueprint of life. This transition demands a whole new level of biological understanding, which we currently lack. This knowledge, however, could be obtained through synthetic genomics and genome engineering, albeit on a trial and error basis, by redesigning and building naturally occurring bacterial and eukaryotic genomes whose sequences are known.

Synthetic genomics arguably began with the report from Khorana's laboratory in 1970 of the total synthesis of the first gene, encoding an artificial yeast alanine tRNA, from deoxyribonucleotides. Since then, rapid advances in DNA synthesis techniques, especially over the past decade, have made it possible to engineer biochemical pathways, assemble bacterial genomes and even to construct a synthetic organism [1–11]. Genome editing approaches for genome-wide scale alteration that are not based on total synthesis of the genome are also being pursued and have proved powerful; for example, in the production of a reduced-size genome version of *Escherichia coli* [4] and engineering of bacterial genomes to include many different changes simultaneously [8].

Progress has also been made in synthetic genomics for eukaryotes. Our group has embarked on the design and total synthesis of a novel eukaryotic genome structure - using the well-known model eukaryote *Saccharomyces cerevisiae* as the basis for a designer genome, known as 'Sc2.0'. The availability of a fully synthetic genome will allow direct testing of evolutionary questions that are not otherwise approachable. Sc2.0 could also play an important practical role, since yeasts are the pre-eminent organisms for industrial fermentations, with a wide variety of practical uses, including production of therapeutic proteins, vaccines and small molecules through classical and well-developed industrial fermentation technologies.

This article reviews the current status of synthetic genomics, starting with a historical perspective that highlights the key milestones in the field (Fig. 1) and then continuing with a particular emphasis on the total synthesis of the first functional designer eukaryotic (yeast) chromosome, *synIII*, and the Sc2.0 Project. Genome engineering using nuclease-based genome editing tools such as zinc finger nucleases, transcription activator-like effector nucleases and RNA-guided CRISPR-Cas9 is not within the scope of this minireview (Box 1). Recent advances in gene synthesis and assembly methods that have accelerated the genome synthesis efforts are discussed elsewhere [12].

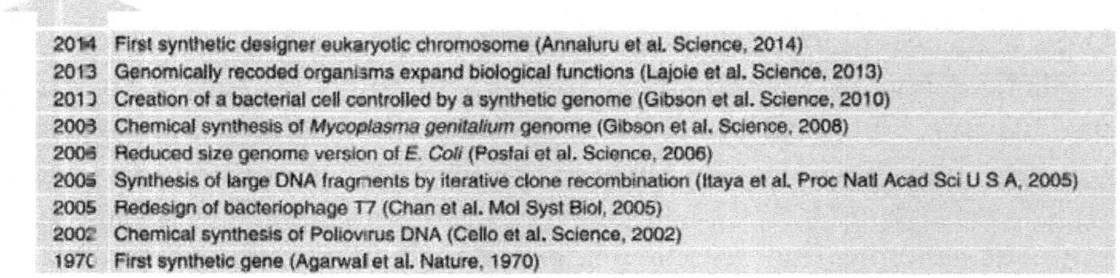

2014	First synthetic designer eukaryotic chromosome (Annaluru et al. Science, 2014)
2013	Genomically recoded organisms expand biological functions (Lajoie et al. Science, 2013)
2010	Creation of a bacterial cell controlled by a synthetic genome (Gibson et al. Science, 2010)
2008	Chemical synthesis of *Mycoplasma genitalium* genome (Gibson et al. Science, 2008)
2006	Reduced size genome version of *E. Coli* (Postai et al. Science, 2006)
2005	Synthesis of large DNA fragments by iterative clone recombination (Itaya et al. Proc Natl Acad Sci U S A, 2005)
2005	Redesign of bacteriophage T7 (Chan et al. Mol Syst Biol, 2005)
2002	Chemical synthesis of Poliovirus DNA (Cello et al. Science, 2002)
1970	First synthetic gene (Agarwal et al. Nature, 1970)

Figure. 1: Timeline of publication milestones for synthetic genomics.

Chemical Synthesis of Poliovirus cDNA (2002)

Viruses can be viewed as both chemical and 'living' entities. Since viral genomes are small, scientists wondered if it is possible to synthesize an infectious agent by in vitro chemical-biochemical means solely based on instructions from a known sequence. Poliovirus is an enterovirus of the Picornaviridae family and its sequenced genome comprises a single-stranded RNA genome of 7.5 kb in length. It replicates naturally in humans with high efficiency, occasionally causing the paralyzing and lethal poliomyelitis. The chemical synthesis of full-length Mahoney poliovirus cDNA *wt* PV1(M) by assembling oligonucleotides was first reported by Cello et al. [13]. The hierarchical strategy for synthesizing the genome of poliovirus involved three steps: (1) DNA fragments of 0.4-0.6 kbp length with overlapping complementary sequences at their termini were produced from purified oligonucleotides of approximately 70 nucleotides; (2) the 0.4-0.6 kbp fragments were then ligated into a plasmid vector to yield three larger DNA segments; (3) the assembly of a full-length cDNA carrying a phage T7 RNA polymerase promoter at the 5' end was achieved from these three large overlapping DNA segments by cloning into a plasmid vector, using unique restriction sites. Several clones were sequenced to identify either the correct DNA segments or the segments containing small numbers of errors that could be eliminated, either by combining the error-free portions of segments by using an internal cleavage site or by standard site-directed mutagenesis. Nucleotide substitutions were engineered into the synthesized viral genome sPV1(M) cDNA as genetic markers to distinguish it from the wild-type sequence [13]. *De novo* synthesis of poliovirus from transcript RNA of sPV1(M) cDNA in a cell-free extract of uninfected HeLa cells indicated that the input synthetic RNA was translated and replicated in the cell-free extract and that newly synthesized RNA was encapsulated into newly synthesized coat proteins, resulting in infectious poliovirus [13]. This elegant work clearly established that it was possible to synthesize the genome of an infectious agent by in vitro chemical-biochemical means based on a known sequence.

Refactoring Bacteriophage T7 (2005)

Evolution by natural selection gives rise to complicated biological systems that are difficult to understand and manipulate. Wild-type T7, an obligate lytic phage that infects *E. coli*, is one such natural biological system. The T7 genome comprises a 39,937 bp linear double-stranded DNA molecule. It is an excellent model organism for discovering the primary genetic components of a natural biological system. The 57 genes coding for 60 proteins have been identified, of which only 35 have a known function. Of the 25 non-essential proteins, only 12 are conserved across the T7-like phage family. Driven by a desire to better understand how the different parts that comprise bacteriophage T7 work together to encode a functioning whole, scientists wanted to refactor the genome to a more structured design that is easy to manipulate and study.

Chan et al. [14] reported the redesign of bacteriophage T7 by improving its internal structure for future use, while simultaneously maintaining external system function; that is, physically separating the primary genetic elements that are essential for the functioning of the bacteriophage from the overlapping genetic elements that are non-essential for the viability of the phage. The T7.1 design goals were as follows. First, define a set of components that function during T7 development and for each element choose an exact DNA sequence to encode the element function. Second, avoid overlap between DNA sequences that encode different element functions. Third, assign only one function to the DNA sequence that encodes each element. Fourth, incorporate unique restriction sites for precise and independent manipulation of each element. Fifth, construct the T7.1 genome. Sixth, refactor the T7.1 genome to encode a viable bacteriophage. Each functional genetic element was defined, for example, as a promoter, protein-coding domain, ribosome binding site and so on. The authors replaced 11,515 bp of the 5′ part of the 39,937 bp wild-type bacteriophage T7 genome with 12,179 bp of engineered DNA using both synthetic DNA fragments and PCR-amplified T7 fragments, which contained all genetic elements of the 5′ end plus restriction enzyme sites. The resulting partially synthetic genome encoded a viable bacteriophage that appeared to maintain key features of the original while being simpler to model and making it easier to manipulate each genetic element encoding a function. This important work established that large regions of genomes encoding natural biological systems can be systematically redesigned and built anew.

Synthesizing large DNA Constructs by Iterative Clone Recombination (2005)

While the smaller viral genomes, such as T7, are amenable to assembly by standard recombinant DNA techniques using synthetic or PCR-amplified precursor DNA fragments (see above), the assembly of larger bacterial genomes relies on recombination of the precursor DNA fragments in vivo in a host organism. For this approach to be successful, one has to be aware of the incompatibilities between the donor and the recipient host organism. Studies have shown that microbial genomes can be assembled in only evolutionarily divergent hosts (for example *Synechocystis* PCC6803 in *Bacillus subtilis*, or *Mycoplasma genitalium* in *S. cerevisiae*). In such instances the donor DNA remains transcriptionally silent without interfering with the viability of the host. The group of Itaya in Japan has used this approach to assemble a bacterial genome by serial integration of precursor DNA fragments directly into the *B. subtilis* genome. They cloned almost all of the 3.57 Mbp genome of the donor *Synechocystis* PCC6803 (a common and highly studied cyanobacterium) as a set of four separate fragments of approximately 800–900 kbp in a stepwise serial integration of PCR-generated precursor DNA fragments into the recipient *B. subtilis* genome [15]. This work showed that very large non-synthetic constructs could be produced from bacterial genomic DNA using in vivo methods. However, the resolution and activation of the synthetic donor genome is yet to be done.

Later, the Itaya group used the same approach to rebuild the full length mouse mitochondrial and rice chloroplast genomes from PCR-amplified precursors and recover the final synthetic DNA product as a circular episome [16]. Similarly, Holt et al. [17] achieved the reassembly of a fragmented donor genome of *Haemophilus influenzae* in a sequential manner into *E. coli*. This group used lambda Red recombination, which is an efficient system for *E. coli* chromosome engineering that uses electroporated linear DNA and a defective lambda phage to supply the functions needed for recombination. Using this technique, this group rebuilt two non-contiguous regions of *H. influenzae* genome totaling 190 kbp (approximately 10.4 % of the *H. influenzae* genome) as episomes in an *E. coli* host. However, both groups found that the bacterial recipient strains could not tolerate some sections of the donor genome, such as the rRNA operons and toxic genes.

Chemical Synthesis of *Mycoplasma genitalium* Genome (2008)

The J Craig Venter Institute has pursued complete synthesis and assembly of a whole bacterial (*M. genitalium*) genome from chemically synthesized oligonucleotides. They reported successful synthesis and assembly of a 582,970 bp *M. genitalium* genome, a culmination of about 10 years of work [5]. In this case, the final complete donor *M. genitalium* genome was assembled in the recipient host *S. cerevisiae* (yeast). The synthetic genome was essentially the wild-type *M. genitalium* G37 sequence except for the disruption of the gene M408 with an antibiotic marker to block pathogenicity and allow for selection. A few watermarks were inserted at intergenic sites in order to identify the genome as synthetic. The hierarchical synthesis of the *M. genitalium* genome was done in three steps: (1) overlapping 5–7 kbp DNA fragments were assembled from chemically synthesized oligonucleotides; (2) the 5–7 kbp fragments were joined by in vitro recombination to yield intermediate 24 kbp, 72 kbp and 144 kbp fragments that were cloned into bacterial artificial chromosomes in *E. coli*; (3) the complete synthetic genome was assembled by homologous recombination in the yeast *S. cerevisiae*. Although a clone with the correct sequence was identified, Gibson et al. [5] did not demonstrate that the synthesized genome encodes a living bacterium. However, in subsequent work this was shown by the same group for a synthesized *Mycoplasma mycoides* genome (below) [10]. This impressive work established that chromosome-size DNA molecules could be constructed from chemically synthesized pieces.

Synthesis and Assembly of the *Mycoplasma mycoides* Genome (2010)

Subsequently, Gibson et al. [10] reported the creation of a bacterial cell controlled by a chemically synthesized genome. A 1.08 Mbp *M. mycoides* genome was synthesized from known genome sequence; it was then transplanted into a closely related *Mycoplasma capricolum* recipient cell to form a new *M. mycoides* cell that was controlled solely by the synthetic genome. The chemically synthesized genome had several alterations compared with the wild-type CP001668, which included four watermark sequences, a designed 4 kbp gene deletion and nucleotide polymorphisms at 20 locations, 19 of which were from harmless mutations acquired during the assembly process. These 19 sequence alterations also served as polymorphic differences between the synthetic genome and the wild-type genome. The newly created cell had the expected phenotypic properties of *M. mycoides* and was capable of continuous self-replication [10].

The synthetic *M. mycoides* genome was assembled from 1,078 overlapping DNA cassettes in three steps: (1) DNA fragments of 1,080 bp, which were produced from overlapping synthetic oligonucleotides, were combined to form 109 larger DNA fragments of about 10 kbp; (2) these were then recombined in pools of 10 to create 11 DNA segments of about 100 kbp in length; (3) the 11 segments were recombined to form the complete *M. mycoides* genome. All assemblies were carried out by in vivo homologous recombination in yeast, except for two constructs that were enzymatically pieced together in vitro. The designed sequence was 1,077,947 bp in length.

The study also showed that a single base pair deletion in the essential gene *dnaA* could render the synthetic *M. mycoides* genome inactive, whereas large genome insertions and deletions in non-essential parts of the genome had no observable effect on viability. This foundational work provided a proof-of-principle experiment for producing cells based on computer-designed genome sequences, even though the synthetic genome had only very limited modifications from the naturally occurring *M. mycoides* genome.

Minimal Bacterial Genome

The vast differences that exist in the genome sizes of bacterial species begs the question, 'What is the minimal set of genes or the minimal genome [18] that is needed for cellular life?' A corollary to this question is, 'What is the minimal set of genes shared by all bacterial species through evolution?' Using gene deletion methods, several groups have successfully produced smaller and increasingly stable, streamlined bacterial genomes. These studies, using what is known as the top-down approach, have shown that large proportions of bacterial genomes can be deleted without any major growth defects. Research on *E. coli* laboratory strain MDS42 has shown that almost 15.3 % of the genome could be eliminated without affecting its growth characteristics [4, 19, 20]. The deleted genes include the transposable elements and horizontally derived genes that have important roles under special environmental conditions. Further work has shown that as much as 22 % of the MDS42 genome could be eliminated without any major growth defects. Other groups have also reported successful genome reduction efforts in *Schizosaccharomyces pombe*, *B. subtilis* and *E. coli* [21–23]. *M. genitalium* is a bacterium with the smallest genome of any independently replicating cell; it encodes 485 protein coding genes of which 100 are non-essential when individually disrupted. The small size of mycoplasma genomes makes them a prime candidate for creating a minimal genome using the bottom-up approach of synthetic genomics [5]. The J Craig Venter Institute is working towards a minimal mycoplasma genome by exploring whether genes that can be disrupted individually without affecting the fitness could also be deleted globally. De novo genome synthesis offers the ability to simultaneously implement many directed changes to the natural genome by building and testing a variety of reduced genomes by genome transplantation in a closely related host strain. The bottom-up approach should make it possible to arrive at the minimal mycoplasma genome that enables cellular life.

Expanding the Genetic Code of *E. coli* (2013)

Church, Isaacs and colleagues have used other genome-editing methods to alter the genetic code on a genome-wide scale in *E. coli*, thereby rewriting the genetic program. One approach, multiplex automated genome engineering (MAGE), allows for introduction of multiply targeted, small mutations through oligonucleotide-directed allelic replacement in an iterative manner (Fig. 2a; refer to [8] for more details). A second technique, conjugative assembly genome engineering (CAGE), allows for step-wise transfer of individually engineered genomic modules into a single genome (Fig. 2b; refer to [24] for more details). A combination of CAGE and MAGE was used to construct a recoded *E. coli* genome with an expanded genetic code [8, 24]. The translation-termination of the three stop codons (TAG, TAA and TGA) of the *E. coli* genetic code is mediated by two release factors, RF1 and RF2. RF1 recognizes the termination codons TAA and TAG, whereas RF2 recognizes TAA and TGA. The authors reasoned that replacing all TAG codons with synonymous TAA codons would abolish genetic dependence on RF1 and permit the newly reassigned TAA codons to be recognized by RF2. After removal of all genomic TAG codons, the *prfA* gene that codes for release factor 1 (RF1) was deleted. The authors hypothesized that this would enable them to test and leverage the redundancy of the genetic code and to provide a blank TAG codon that could be cleanly reassigned to a new function. The TAG codon was reintroduced along with an orthogonal set of aminoacyl-tRNA synthase and tRNA to encode a non-standard amino acid. The engineered *E. coli* incorporated non-standard amino acids into its proteins and showed enhanced resistance to bacteriophage T7 [25]. The Church group also recoded 13 codons in 42 highly expressed essential genes in *E. coli*, indicating that codon usage is quite flexible [26]. Recently, two laboratories have redesigned essential enzymes of *E. coli* with an altered genetic code by changing TAG codons to TAA. This confers metabolic dependence on non-standard amino acids for survival as a means for biocontainment of genetically modified organisms [27, 28].

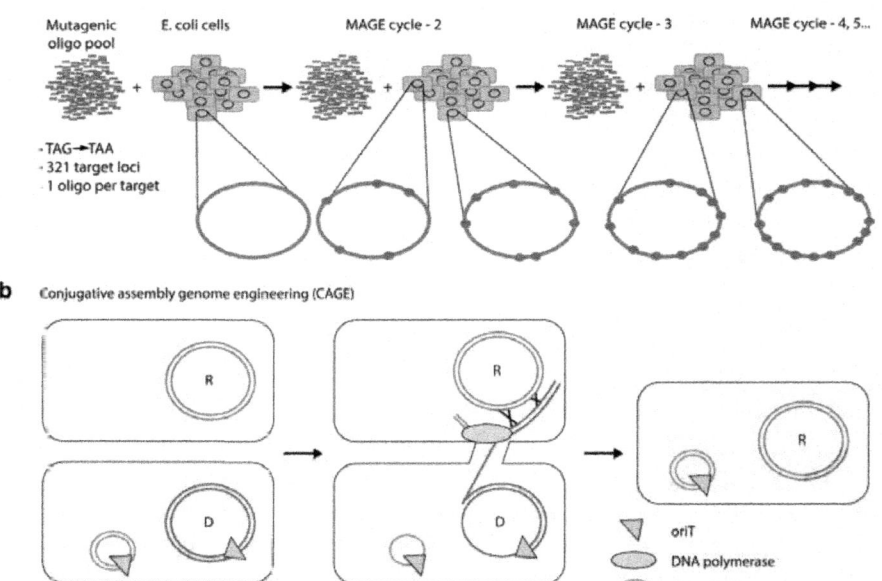

Figure. 2: Multiplex automated genome engineering (MAGE) and conjugative assembly genome engineering (CAGE). **a** Use of MAGE (refer to [8] for more details)

to replace all TAG codons with TAA in *E. coli*. **b** Use of CAGE (refer to [24] for more details) to incorporate a donor (D) into a recipient (R) genome. *oriT* in the donor genome serves as the transfer initiation point.

The First Synthetic Designer Eukaryotic Chromosome (2014)

The idea for designing and synthesizing a eukaryotic chromosome was initiated by our group in collaboration with Jef Boeke in 2005. The concept for hierarchically synthesizing a designer yeast chromosome was quite simple. First, design the synthetic chromosome incorporating all the desired changes based on the available wild-type chromosome sequence of *S. cerevisiae*. Second, compile the designed chromosome into pieces of about 10 kbp by including unique restriction sites at the 5′ and 3′ ends to enable further ligation of the 10 kbp pieces into segments of about 30–50 kbp. Synthesize these pieces of about 10 kbp using oligonucleotides from commercial vendors. Third, as yeast is highly recombinogenic, use an iterative strategy with alternating genetic markers to replace each 30–50 kbp segment of the wild-type sequence with the corresponding synthetic pieces, one at a time by homologous recombination in vivo in yeast.

The initial proof-of-principle experiment was performed in our laboratory by first designing and synthesizing a 30 kbp fragment of yeast chromosome *III* and then replacing the wild-type segment with the synthetic piece in yeast [29]. By 2007, the idea of synthesizing a eukaryotic chromosome had morphed into an ambitious project with the goal of rewriting wild-type *S. cerevisiae* Sc1.0 into a synthetic version, Sc2.0.

Design Principles for the Synthetic Yeast Genome (Sc2.0)

Suggestions for the types of changes to be incorporated into Sc2.0 were obtained by Boeke from the community of yeast researchers. Only conservative changes were included, as more drastic changes might result in 'dead' yeast. The synthetic yeast should have the same fitness as the wild type and grow normally; this is an obvious minimal requirement for Sc2.0. The three design principles for the synthetic yeast genome are as follows: (1) it should result in a (near) wild-type phenotype and fitness; (2) it should lack destabilizing elements to avoid the synthetic yeast genome from being unstable or undergoing rearrangements; (3) it should have genetic flexibility to facilitate future studies [30].

How does one design a Sc2.0 genome that will facilitate future studies? Yeast contains about 6000 genes and almost 5000 of these are non-essential when disrupted individually [31]. As such, all the non-essential genes were flanked with loxPsym sites. Once a synthetic chromosome or the Sc2.0 genome is built, in theory, one could expose the synthetic yeast strains to Cre recombinase for various time intervals and look for survivors. PCR-Tag analysis (see *synIII* construction) and sequencing of the genomes of survivors would reveal what combinations of non-essential genes have been deleted from the starting Sc2.0 genome, leaving the survivors viable.

synIII Design

After a successful proof-of-principle experiment involving the design of a synthetic 30 kbp chromosome III fragment that was used to replace the native sequence in yeast, the sequence of the whole native chromosome *III* was edited in silico using Biostudio [32] to incorporate a series of deletions, insertions and base substitution changes to produce the desired 'designer' sequence (Box 2 and Fig. 3a). The synthetic version of chromosome *III* (known as *synIII*) also encodes a built-in recombination system called SCRaMbLE (synthetic chromosome rearrangement and modification by loxP-mediated evolution) to enable removal of the non-essential parts of the chromosome, and therefore streamline it, by inducing genomic alterations of the *synIII* strain using Cre recombinase [32]. As the result of these alterations, *synIII* (272,871 bp) is about 13.8 % smaller than the native chromosome *III* (316,667 bp) [32].

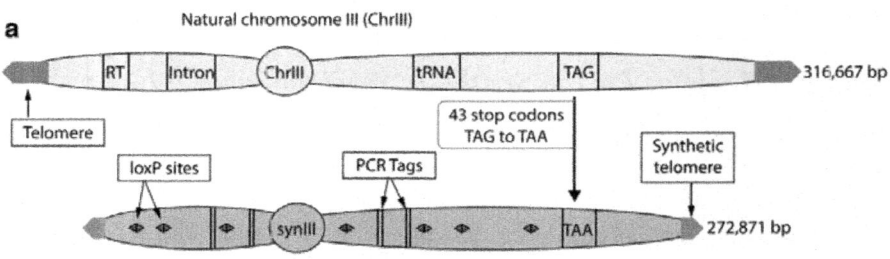

Synthetic chromosome III (synIII)

b

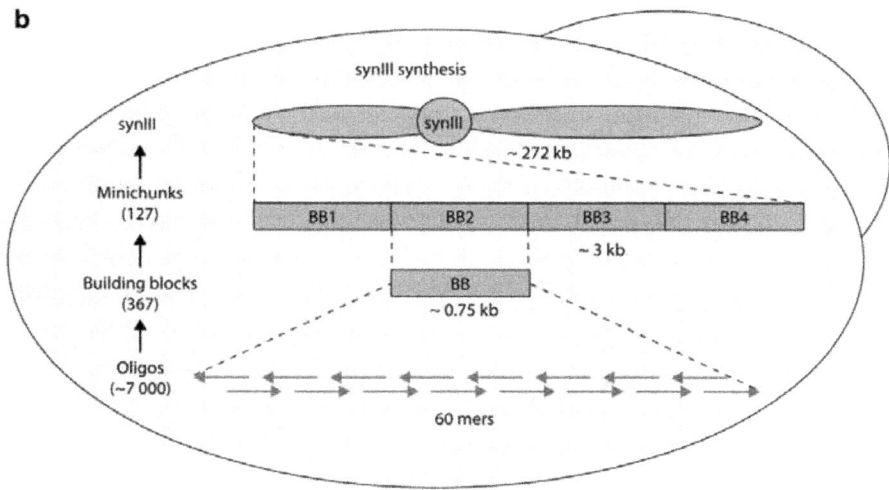

Figure. 3: *synIII* design and synthesis. **a** *synIII* design. Twenty-one retrotransposons (*RT*) and seven introns were removed. Forty-three TAG stop codons were changed to TAA stop codons. Ninety-eight loxPsym sites were introduced to enable SCRaMbLE analysis. The two natural telomeres were replaced with shorter universal telomere caps. A single copy of essential tRNA gene *SUP61*, which codes for tRNASer (CGA), was deleted and moved to a tRNA neochromosome. Numerous PCR-Tags were incorporated into *synIII* to distinguish it from the natural counterpart. As a result, *synIII* is about 13.8 % smaller than the native yeast chromosome *III* (Box 2). For the complete set of additions, deletions and other genome modifications to *synIII*, see Annaluru *et al.* [32]. **b** *synIII* synthesis.*synIII* was constructed in three steps (shown in the flow diagram on the left, from bottom to top). In step 1, 750 bp building blocks (*BB*) were synthesized from 60-mer oligonucleotides at Johns Hopkins University by undergraduate students in the Build-A-Genome course [33]. In step 2, three to five BB were assembled into 2–4 kb minichunks by homologous recombination in *Saccharomyces cerevisiae* [35]. Adjacent minichunks were designed to encode overlap of one BB to facilitate downstream assembly. In step 3, direct replacement of native yeast chromosome *III* with pools of synthetic minichunks was performed. Eleven iterative one-step assemblies and replacements of native genomic segments of yeast chromosome *III* were carried out using pools of overlapping synthetic DNA minichunks, encoding alternating genetic markers (*LEU2* or *URA3*), which enabled complete replacement of native *III* with *synIII* in yeast [32]. The number of oligonucleotides, BBs, and minichunks needed to construct *synIII* are shown in parentheses. *SynIII* is 272,871 bp long, compared with the 316,667 bp long native yeast chromosome *III*.

synIII Construction

The hierarchical workflow that was used to construct *synIII* (Fig. 3b) consisted of three major steps. In the first step, the 750 bp 'building blocks' (BBs) were produced starting from overlapping 60-mer to 79-mer oligonucleotides and assembled using standard PCR methods [33]. In a second step, the BBs were assembled into overlapping DNA 'minichunks' of approximately 2–4 kb using either the uracil-specific excision reaction [34] or cloning into a shuttle vector by homologous recombination in yeast *S. cerevisiae* [35–39]. In the USER approach, four to five BBs are used that each have a 5–13 bp sequence of the type $A(N)_3 T$ to $A(N)_{11} T$ that overlaps with their adjoining neighbors and a vector. These BBs are amplified using forward and reverse primers containing a single uracil instead of the T and are then treated with USER enzymes (a mixture of uracil DNA glycosylase and the DNA glycosylase-lyase endonuclease VIII) to generate complementary single-stranded ends. The BBs are then ligated and cloned into *E. coli* to recover recombinants containing the assembled 'minichunks'. The yeast homologous recombination cloning approach is much simpler, where four to five BBs each with 40 bp overlaps with their adjoining neighbors are assembled into a shuttle vector by direct transformation into the highly recombinogenic *S. cerevisiae*. This approach obviates the need for another round of PCR amplification of the BBs using primers containing uracil and the use of USER enzymes. Thus, as it turns out, all you need is yeast for minichunk assembly. In the third and final step, the adjacent minichunks for *synIII* were designed to overlap one another by one BB to facilitate further assembly in vivo by homologous recombination in yeast Using an average of 12 minichunks and alternating selectable markers in each experiment, the native sequence of *S. cerevisiae III* was systematically replaced by its *synIII* counterpart in 11 successive rounds of transformation. PCR-Tag analysis (Fig. 4) and sequencing confirmed the identity of *synIII* [32]. The fact that the numerous design changes to the DNA sequence of the chromosome *III* had little or no impact on cell fitness and phenotype suggests the very pliable nature of the yeast genome [32].

a   YCL061C.3_WT_F: TCTTTGCGTCTGCGTGAATGAAGAAGAG
         YCR061C.3_SYN_F: TCGTTGGGTTTGGGTAAAGCTGCTGCTA

         YCL061C.3_WT_R: AGACAATCCGCCAGAGTTGACTGGGAAC
         YCR061C.3_SYN_R: CGATAACCCACCTGAACTAACCGGTAAT

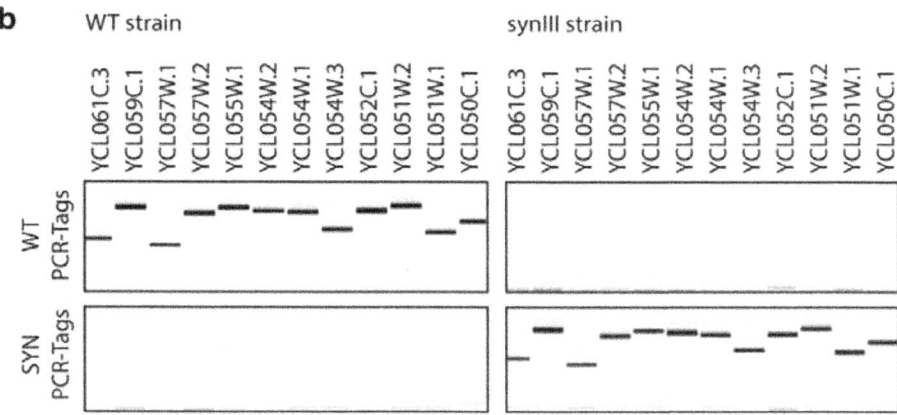

Figure. 4: PCR-Tag analysis of a *synIII* segment. **a** The YCL061C.3 locus-specific PCR-Tag forward (F) and reverse (R) primers for the wild type (WT) and *synIII* are shown. The changes between the two are shaded. PCR-Tags are short pairs of recoded segments used as genetic markers to verify introduction of a synthetic sequence and removal of native sequence. Pairs of 25–28 bp sequences about 500 bp apart were recoded with synonymous codons such that >33 % of the bases were changed; the first and last base PCR-Tag primers were coded to be different between the WT and *synIII* sequences. **b** Agarose gel profiles of PCR-Tag analysis of a WT DNA segment and the corresponding synthetic *synIII* segment (YCL061C.3 to YCL050C.1). A virtual gel image was generated using LabChip GX software version 4.0.1418.0.

International Consortium to Synthesize the Sc2.0 Genome

A group of international scientists has taken up the synthesis of the Sc2.0 genome. The Beijing Genome Institute in China was the first to agree to synthesize four of the yeast chromosomes. Since then laboratories from various other countries have also joined the Sc2.0 effort to synthesize the remaining yeast chromosomes. Each participating laboratory is required to sign an Agreement with Johns Hopkins University (now with New York University). This arrangement leaves the control of the Sc2.0 project to Boeke, who is a yeast expert. Such a central organization is needed for the coordination of a huge undertaking such as Sc2.0 and for the distribution of yeast strains, reagents and experimental protocols. Participating laboratories have to raise their own funds from their own country to synthesize the allotted chromosome.

What's Next for the Yeast Synthetic Genome?

The *synIII* chromosome is about 2.5 % of the yeast genome and the changes that were made were all conservative, although numerous. These sequence alterations have not reduced the fitness of the yeast, which is encouraging in terms of the potential for future modifications. There are about 98 loxPsym sites in *synIII*, which scales to about 4000 loxPsym sites for the entire Sc2.0 genome. It is not yet clear how all of these loxPsym sites along with all the other modifications will ultimately affect the stability of the Sc2.0 genome and the viability of the synthetic yeast cell. The results from *synIII* are encouraging and the synthesis of a few more chromosomes will give us a better idea. Boeke's laboratory is working on the assembly of the *synVI* chromosome using fragments of approximately 10 kbp from commercial vendors. Our laboratory is in the process of completing the assembly of the *synIX* chromosome. With the experience gained from the synthesis and assembly of *synIII*, we estimate that the construction of a chromosome about 1 Mbp could be done in 2–3 years.

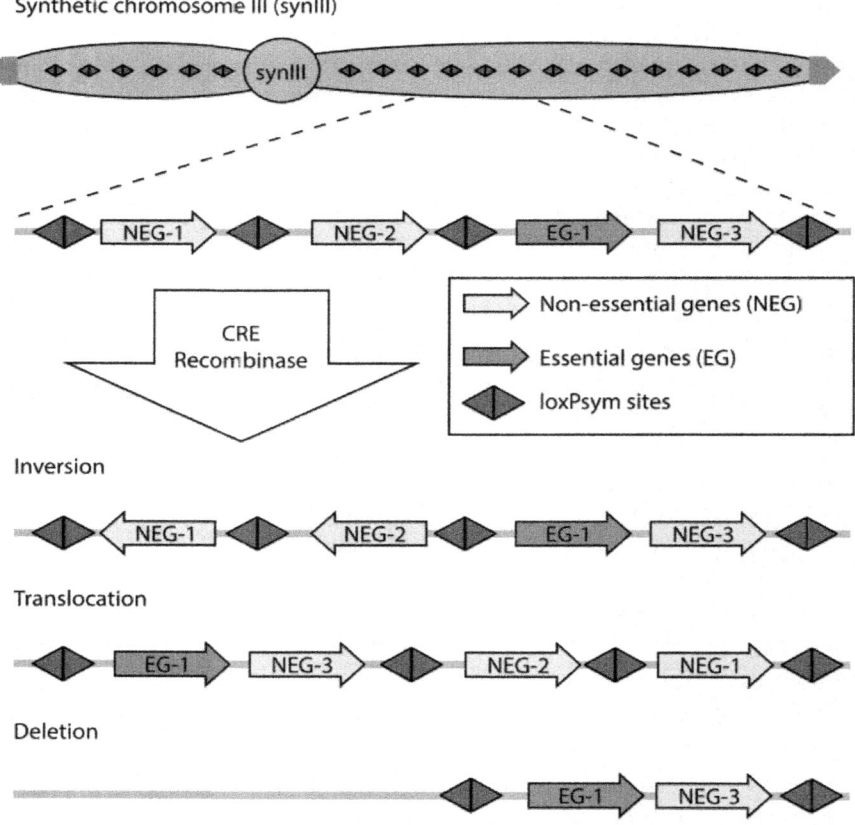

Figure. 5: Synthetic chromosome rearrangement and modification by loxP-mediated evolution (SCRaMbLE) of the *synIII* strain. Examples of inversion, translocation and

deletion products resulting from Cre recombinase treatment of *synIII* strain are shown.

Once the Sc2.0 genome is built, an important focus will be to determine the minimal eukaryotic (yeast) genome. If two or more genes perform a similar function, can one be deleted? Which combinations of the 5000 yeast non-essential genes that are dispensable individually can be simultaneously removed? If we possess this knowledge, we will be able to achieve further reduction in the size of the Sc2.0 chromosomes and the genome. The plan is to use SCRaMbLE analysis to arrive at the minimal yeast genome (Fig. 5). This approach involves exposing Sc2.0 to Cre recombinase for various time intervals and looking for survivors. We reason that PCRTag analysis and sequencing the genomes of the survivors would reveal what combinations of non-essential genes have been deleted from the starting synthetic genome, leaving the survivors viable.

This pathway to the minimal yeast genome would represent a 'top down' approach since we start from the entire newly designed Sc2.0 genome and progressively delete increasing parts of the genome. However, to complicate matters, the essential and non-essential genes of the synthetic yeast are interspersed with one another. Because of this intertwining, SCRaMbLEing of the Sc2.0 is likely to result in dead yeast most of the time. Only yeasts with small deletions are likely to survive, making it difficult to deduce the minimal genome. Furthermore, due to the inherent symmetry of loxPsym sites, when two such sites are brought together by a Cre recombinase, it could result in an insertion, a deletion, an inversion or a translocation (Fig. 3). Moreover, there is also the possibility of interchromosomal rearrangements through the loxPsym sites, in addition to the expected intrachromosomal deletions, inversions, insertions and rearrangements. Analysis of such widely variant genomes from a population of survivors would involve time-consuming costly experimentation and complicated data analysis to decipher the minimal yeast genome. This hurdle could be overcome to some extent by performing SCRaMbLE analysis at the level of intermediate yeast strains, each possessing an individual synthetic chromosome. Thus, one could delineate a set of 16 minimal chromosomes for yeast. All of the reduced yeast chromosomes could then be combined into a final yeast strain to form a minimal eukaryotic genome.

CONCLUSIONS AND FUTURE PERSPECTIVES

Recent literature reports make it clear that entire chromosomes and genomes can be designed, synthesized and incorporated into cells to produce synthetic organisms. However, to create a truly living, dividing synthetic cell from scratch by de novo synthesis, we need to know the minimal set of essential genes required for life and have a clear understanding of how each gene functions, and understand the regulatory mechanisms that are needed for harmonious gene function. It is very likely that the research on *Mycoplasma*, which have the smallest genomes among free-living cellular organisms, will be the first to lead to the delineation of the minimal set of genes required for life; this will be achieved either through stepwise iterative deletion of nonessential genes [40] or by de novo synthesis of several arbitrarily reduced genomes.

An international consortium of scientists is now working to synthesize the remaining 15 chromosomes of the yeast Sc2.0 genome, a model eukaryote. At this juncture, the best that one can say about Sc2.0 is that we are trying to rewrite Sc1.0 to Sc2.0, albeit with numerous conservative changes. The main stated purpose for designing and engineering of Sc2.0 is to improve our understanding of the evolution of eukaryotic genome structure and function.

Can the design rules be successfully applied across the entire yeast genome? Is the Sc2.0 genome worth doing? What are the potential industrial applications of Sc2.0? These are difficult questions to answer at this juncture. However, there is a critical need to develop alternative yeast strains as 'chassis' organisms for the production of pharmacologically important compounds such as artemisinin [41]. Will the 'streamlined' yeast strains resulting from in vitro evolution of the Sc2.0 genome be useful in this regard? Only time will tell. The total synthesis of a functional designer yeast chromosome represents an important milestone for eukaryotic synthetic biology. *synIII* work paves the way for other future synthetic genomics projects that seek to rewrite animal or plant chromosomes and genomes using specific design principles. The DNA synthesis technology and the genome engineering tools needed for such a major undertaking are currently available to scientists. However, it is too early to speculate about a minimal eukaryotic genome at this juncture.

In conclusion, recent breakthroughs in synthetic genomics have ushered in a new era for synthetic biology with the real potential to create a truly man-made living and dividing synthetic cell.

Box 1. Genome Engineering Using Programmable Nucleases

Genome engineering by genome-editing tools depends on cellular responses to targeted chromosomal double-strand breaks (DSBs). Except for mouse cells, mammalian cells are recalcitrant to gene targeting [42]. Only one in a million treated cells undergoes homologous recombination (HR). However, it was discovered that stimulation of both local mutagenesis and incorporation of homologous donor sequences can be achieved by generating targeted DSBs, which was demonstrated most clearly with rare-cutting endonucleases [43]. The generation of a targeted DSB remained the rate-limiting step in the development of HR technology for mammalian cells until the creation of zinc finger nucleases (ZFNs) by our laboratory, which ushered in the breakthrough in programmable nucleases [44–47].

ZFNs: the first truly targetable reagents were the ZFNs, which showed that predetermined DNA sequences could be addressed for cleavage by protein engineering. ZFNs are formed by fusing a zinc finger protein (ZFP) that comprises a tandem array of ZF motifs [48] to the FokI non-specific cleavage domain [44,45]. Each ZF motif recognizes a DNA site of 3–4 bp [49]. Studies of the ZFN cleavage mechanism established that the preferred substrates are inverted repeats [50]. Soon afterwards, ZFN-induced DSBs were shown to stimulate HR in cells [51–53]. Because ZF motifs interact with and influence the recognition of their neighbors, the selection methods used to generate highly specific ZFPs for desired target sites are quite laborious and time-consuming. The commercial pricing of ZFNs was prohibitively expensive, putting it beyond the reach of small laboratories.

Transcription activator-like effector nucleases (TALENs): TALENs are based on the fusion of a different class of DNA-binding modules, called bacterial transcription activator-like effectors (TALEs), to the FokI cleavage domain [54]. Each TALE motif recognizes a single base and appears not to influence the sequence recognition of its neighbors [55,56]. Therefore, TALENs are relatively easier to engineer than ZFNs and they expanded the targeting capability of programmable nucleases. The fact that ZFNs and TALENs have been used to modify genomic sequences of more than 40 different organisms and cell types attests to the success of this approach to genome engineering. However, although they are cheaper than ZFNs, the commercial pricing of TALENs was still too expensive for smaller laboratories.

RNA-guided CRISPR-Cas9: the second technology platform for inducing a targeted DSB in cellular genomes is the RNA-guided nucleases (RGNs), which are based on the type II prokaryotic CRISPR-Cas9 system [57–61]. Unlike ZFNs and TALENs, which use ZF and TALE motifs, respectively, for DNA sequence recognition, the CRISPR-Cas9 system depends on RNA-DNA recognition, and its natural function is to combat invaders of bacteria and archaea, a testament to nature's ability to solve problems several ways (compare restriction enzymes). The advantages of the CRISPR-Cas9 system are its ease of RNA design for new targets; the dependence on a single, constant Cas9 protein; and the ability to address many targets simultaneously with multiple guide RNAs. The CRISPR-Cas9 methodology is also very cheap and inexpensive, making it affordable for small laboratories. These have led to its wide adoption in research laboratories around the world.

These two technology platforms have equipped scientists with an unprecedented ability to modify cells and organisms almost at will, with wide-ranging implications across biology and medicine. However, both approaches have been shown to cut at off-target sites, with mutagenic consequences. Therefore, issues like efficacy, specificity and delivery are likely to drive selection of reagents for particular purposes. A word of caution about rushing to adopt CRISPR-Cas9 for human therapeutics and possibly for gene editing of human embryos: ease of design and use does not necessarily translate to safety. Therefore, human therapeutic applications of these technologies ultimately are likely to come down to risk versus benefit analysis and informed consent.

Box 2. Modifications in *synIII* Chromosome

- Elements removed: 10 transfer RNA genes, 21 Ty elements and/or derived long terminal repeats (LTRs), 7 introns, the silent mating loci *HML* and *HMR*, and subtelomeric sequences lying to the left of *YCL073C* and the right of *YCR098C* were removed [32]

- Elements relocated to extrachromosomal array: a single copy tRNA gene, *SUP61*, which codes for tRNASer (CGA) is essential to the yeast cell. Therefore, it was encoded in *trans* on a centromeric plasmid, which allowed deletion of the gene from *synIII* chromosome [32]

- Elements replaced: (1) TAG stop codons were replaced by TAA. Removal of the TAG stop codons from the synthetic genome will allow future genetic code manipulation [32]. (2) The telomeres were specified by a minimal 'universal telomere cap' comprising 305 bp of $T(G)_{1-3}$ sequence. (3) Single synonymous codons were used to incorporate unique restriction sites (or delete sites) to facilitate *synIII* assembly. (4) Short stretches of synonymous codons (fewer than ten codons) were recoded to generate 'PCR-Tags' that serve as the basis for PCR primer design [30]. PCR-Tags are used to distinguish wild-type from synthetic sequence by selective PCR amplification

- Elements introduced: symmetrical loxP (loxPsym) sites were inserted in the 3' UTR of all non-essential genes as well as at synthetic landmarks such as sites of LTR and tRNA deletion or flanking the centromere [32]. loxPsym sites lack the directionality of canonical loxP Cre recombinase sites and can align in two orientations. Therefore, both inversions and deletions are possible during SCRaMbLE using Cre recombinase [30]

- Elements not changed: gene order was preserved in *synIII* to prevent incorporation of a non-permissible configuration [32]. Induction of SCRaMbLE results in changes in gene order and chromosome structure

[30]. All recovered SCRaMbLEd yeasts will have viable genome structures

ABBREVIATIONS

BB: building block

bp: base pair

CAGE: conjugative assembly genome engineering

kb: kilobase

kbp: kilobase pair

MAGE: multiplex automated genome engineering

mpb: megabase pair

SCRaMbLE: synthetic chromosome rearrangement and modification by loxP-mediated evolution

TALEN: transcription activator-like effector nuclease

ZFN: zinc finger nuclease

ACKNOWLEDGEMENT

This work was supported by a grant from NSF (MCB 0718846). We thank Dr Hamilton O Smith for helpful comments and suggestions to improve the manuscript.

REFERENCES

1. Agarwal KL, Buchi H, Caruthers MH, Gupta N, Khorana HG, Kleppe K, et al. Total synthesis of the gene for an alanine transfer ribonucleic acid from yeast. Nature. 1970;227:27–34.

2. Menzella HG, Reid R, Carney JR, Chandran SS, Reisinger SJ, Patel KG, et al. Combinatorial polyketide biosynthesis by de novo design and rearrangement of modular polyketide synthase genes. Nat Biotechnol. 2005;23:1171–6.

3. Ro DK, Paradise EM, Ouellet M, Fisher KJ, Newman KL, Ndungu JM, et al. Production of the antimalarial drug precursor artemisinic acid in engineered yeast. Nature. 2006;440:940–3.

4. Posfai G, Plunkett 3rd G, Feher T, Frisch D, Keil GM, Umenhoffer K, et al. Emergent properties of reduced-genome *Escherichia coli*. Science.

2006;312:1044–6.

5. Gibson DG, Benders GA, Andrews-Pfannkoch C, Denisova EA, Baden-Tillson H, Zaveri J, et al. Complete chemical synthesis, assembly, and cloning of a *Mycoplasma genitalium* genome. Science. 2008;319:1215–20.

6. Itaya M, Fujita K, Kuroki A, Tsuge K. Bottom-up genome assembly using the *Bacillus subtilis* genome vector. Nat Methods. 2008;5:41–3.

7. Lartigue C, Glass JI, Alperovich N, Pieper R, Parmar PP, Hutchison 3rd CA, et al. Genome transplantation in bacteria: changing one species to another. Science. 2007;317:632–8.

8. Wang HH, Isaacs FJ, Carr PA, Sun ZZ, Xu G, Forest CR, et al. Programming cells by multiplex genome engineering and accelerated evolution. Nature. 2009;460:894–8.

9. Lartigue C, Vashee S, Algire MA, Chuang RY, Benders GA, et al. Creating bacterial strains from genomes that have been cloned and engineered in yeast. Science. 2009;325:1693–6.

10. Gibson DG, Glass JI, Lartigue C, Noskov VN, Chuang RY, Algire MA, et al. Creation of a bacterial cell controlled by a chemically synthesized genome. Science. 2010;329:52–6.

11. Benders GA, Noskov VN, Denisova EA, Lartigue C, Gibson DG, Assad-Garcia N, et al. Cloning whole bacterial genomes in yeast. Nucleic Acids Res. 2010;38:2558–69.

12. Gibson DG. Programming biological operating systems: genome design, assembly and activation. Nat Methods. 2014;11:521–6.

13. Cello J, Paul AV, Wimmer E. Chemical synthesis of poliovirus cDNA: generation of infectious virus in the absence of natural template. Science. 2002;297:1016–8.

14. Chan LY, Kosuri S, Endy D. Refactoring bacteriophage T7. Mol Syst Biol. 2005;1:2005.0018.

15. Itaya M, Tsuge K, Koizumi M, Fujita K. Combining two genomes in one cell: stable cloning of the Synechocystis PCC6803 genome in the *Bacillus subtilis* 168 genome. Proc Natl Acad Sci U S A. 2005;102:15971–6.

16. Itaya M, Fujita K, Kuroki A, Tsuge K. Bottom-up genome assembly using the *Bacillus subtilis* genome vector. Nat Methods. 2008;5:41–3.

17. Holt RA, Warren R, Flibotte S, Missirlis PI, Smailus DE. Rebuilding microbial genomes. Bioessays. 2007;29:580–90.

18. Glass JI, Assad-Garcia N, Alperovich N, Yooseph S, Lewis MR, Maruf

M, et al. Essential genes of a minimal bacterium. Proc Natl Acad Sci U S A. 2006;103:425–30.

19. Umenhoffer K, Feher T, Baliko G, Ayaydin F, Posfai J, Blattner FR, et al. Reduced evolvability of *Escherichia coli* MDS42, an IS-less cellular chassis for molecular and synthetic biology applications. Microb Cell Fact. 2010;9:38.

20. Csorgo B, Feher T, Timar E, Blattner FR, Posfai G. Low-mutation-rate, reduced-genome *Escherichia coli*: an improved host for faithful maintenance of engineered genetic constructs. Microb Cell Fact. 2012;11:11.

21. Ara K, Ozaki K, Nakamura K, Yamane K, Sekiguchi J, Ogasawara N. Bacillus minimum genome factory: effective utilization of microbial genome information. Biotechnol Appl Biochem. 2007;46:169–78.

22. Mizoguchi H, Mori H, Fujio T. *Escherichia coli* minimum genome factory. Biotechnol Appl Biochem. 2007;46:157–67.

23. Giga-Hama Y, Tohda H, Takegawa K, Kumagai H. *Schizosaccharomyces pombe* minimum genome factory. Biotechnol Appl Biochem. 2007;46:147–55.

24. Isaacs FJ, Carr PA, Wang HH, Lajoie MJ, Sterling B, Kraal L, et al. Precise manipulation of chromosomes in vivo enables genome-wide codon replacement. Science. 2011;333:348–53.

25. Lajoie MJ, Rovner AJ, Goodman DB, Aerni HR, Haimovich AD, Kuznetsov G, et al. Genomically recoded organisms expand biological functions. Science. 2013;342:357–60.

26. Lajoie MJ, Kosuri S, Mosberg JA, Gregg CJ, Zhang D, Church GM. Probing the limits of genetic recoding in essential genes. Science. 2013;342:361–3.

27. Mandell DJ, Lajoie MJ, Mee MT, Takeuchi R, Kuznetsov G, Norville JE, et al. Biocontainment of genetically modified organisms by synthetic protein design. Nature. 2015;518:55–60.

28. Rovner AJ, Haimovich AD, Katz SR, Li Z, Grome MW, Gassaway BM, et al. Recoded organisms engineered to depend on synthetic amino acids. Nature. 2015;518:89–93.

29. Wu J. Manipulating the eukaryotic genomes. PhD thesis. Johns Hopkins School of Public Health, Department of Environmental Health Sciences; 2007.

30. Dymond JS, Richardson SM, Coombes CE, Babatz T, Muller H, Annaluru

N, et al. Synthetic chromosome arms function in yeast and generate phenotypic diversity by design. Nature. 2011;477:471–6.

31. Goffeau A, Barrell BG, Bussey H, Davis RW, Dujon B, Feldmann H, et al. Life with 6000 genes. Science. 1996;274:563–7.

32. Annaluru N, Muller H, Mitchell LA, Ramalingam S, Stracquadanio G, Richardson SM, et al. Total synthesis of a functional designer eukaryotic chromosome. Science. 2014;344:55–8.

33. Dymond JS, Scheifele LZ, Richardson S, Lee P, Chandrasegaran S, Bader JS, et al. Teaching synthetic biology, bioinformatics and engineering to undergraduates: the interdisciplinary Build-a-Genome course. Genetics. 2009;181:13–21.

34. Annaluru N, Muller H, Ramalingam S, Kandavelou K, London V, Richardson SM, et al. Assembling DNA fragments by USER fusion. Methods Mol Biol. 2012;852:77–95.

35. Muller H, Annaluru N, Schwerzmann JW, Richardson SM, Dymond JS, Cooper EM, et al. Assembling large DNA segments in yeast. Methods Mol Biol. 2012;852:133–50.

36. Karas BJ, Tagwerker C, Yonemoto IT, Hutchison 3rd CA, Smith HO. Cloning the *Acholeplasma laidlawii* PG-8A genome in *Saccharomyces cerevisiae* as a yeast centromeric plasmid. ACS Synth Biol. 2012;1:22–8.

37. Larionov V, Kouprina N, Graves J, Chen XN, Korenberg JR, Resnick MA. Specific cloning of human DNA as yeast artificial chromosomes by transformation-associated recombination. Proc Natl Acad Sci U S A. 1996;93:491–6.

38. Kouprina N, Larionov V. TAR cloning: insights into gene function, long-range haplotypes and genome structure and evolution. Nat Rev Genet. 2006;7:805–12.

39. Lee NC, Larionov V, Kouprina N. Highly efficient CRISPR/Cas9-mediated TAR cloning of genes and chromosomal loci from complex genomes in yeast. Nucleic Acids Res. 2015;43, e55.

40. Suzuki Y, Assad-Garcia N, Kostylev M, Noskov VN, Wise KS, Karas BJ, et al. Bacterial genome reduction using the progressive clustering of deletions via yeast sexual cycling. Genome Res. 2015;25:435–44.

41. Paddon CJ, Keasling JD. Semi-synthetic artemisinin: a model for the use of synthetic biology in pharmaceutical development. Nat Rev Microbiol. 2014;12:355–67.

42. Mansour SL, Thomas KR, Capecchi MR. Disruption of the proto-oncogene int-2 in mouse embryo-derived stem cells: a general strategy

for targeting mutations to non-selectable genes. Nature. 1988;336:348–52.

43. Rouet P, Smih F, Jasin M. Introduction of double-strand breaks into the genome of mouse cells by expression of a rare-cutting endonuclease. Mol Cell Biol. 1994;14:8096–106.

44. Ramalingam S, Annaluru N, Chandrasegaran S. A CRISPR way to engineer the human genome. Genome Biol. 2013;14:107. doi:10.1186/gb-2013-14-2-107.

45. Kim YG, Cha J, Chandrasegaran S. Hybrid restriction enzymes: zinc finger fusions to Fok I cleavage domain. Proc Natl Acad Sci U S A. 1996;93:1156–60.

46. Kim YG, Chandrasegaran S. Chimeric restriction endonuclease. Proc Natl Acad Sci U S A. 1994;91:883–7.

47. Li L, Wu LP, Chandrasegaran S. Functional domains in Fok I restriction endonuclease. Proc Natl Acad Sci U S A. 1992;89:4275–9.

48. Miller J, McLachlan AD, Klug A. Repetitive zinc-binding domains in the protein transcription factor IIIA from Xenopus oocytes. EMBO J. 1985;4:1609–14.

49. Pavletich NP, Pabo CO. Zinc finger-DNA recognition: crystal structure of a Zif268-DNA complex at 2.1 A. Science. 1991;252:809–17.

50. Smith J, Bibikova M, Whitby FG, Reddy AR, Chandrasegaran S, Carroll D. Requirements for double-strand cleavage by chimeric restriction enzymes with zinc finger DNA-recognition domains. Nucleic Acids Res. 2000;28:3361–9.

51. Bibikova M, Carroll D, Segal DJ, Trautman JK, Smith J, Kim YG, et al. Stimulation of homologous recombination through targeted cleavage by chimeric nucleases. Mol Cell Biol. 2001;21:289–97.

52. Bibikova M, Beumer K, Trautman JK, Carroll D. Enhancing gene targeting with designed zinc finger nucleases. Science. 2003;300:764.

53. Bibikova M, Golic M, Golic KG, Carroll D. Targeted chromosomal cleavage and mutagenesis in Drosophila using zinc-finger nucleases. Genetics. 2002;161:1169–75.

54. Christian M, Cermak T, Doyle EL, Schmidt C, Zhang F, Hummel A, et al. Targeting DNA double-strand breaks with TAL effector nucleases. Genetics. 2010;186:757–61.

55. Moscou MJ, Bogdanove AJ. A simple cipher governs DNA recognition by TAL effectors. Science. 2009;326:1501.

56. Boch J, Scholze H, Schornack S, Landgraf A, Hahn S, Kay S, et al. Breaking the code of DNA binding specificity of TAL-type III effectors. Science. 2009;326:1509–12.

57. Doudna JA, Charpentier E. Genome editing. The new frontier of genome engineering with CRISPR-Cas9. Science. 2014;346:1258096.

58. Gasiunas G, Barrangou R, Horvath P, Siksnys V. Cas9-crRNA ribonucleoprotein complex mediates specific DNA cleavage for adaptive immunity in bacteria. Proc Natl Acad Sci U S A. 2012;109:E2579–86.

59. Jinek M, Chylinski K, Fonfara I, Hauer M, Doudna JA, Charpentier E. A programmable dual-RNA-guided DNA endonuclease in adaptive bacterial immunity. Science. 2012;337:816–21.

60. Mali P, Yang L, Esvelt KM, Aach J, Guell M, DiCarlo JE, et al. RNA-guided human genome engineering via Cas9. Science. 2013;339:823–6.

61. Cong L, Ran FA, Cox D, Lin S, Barretto R, Habib N, et al. Multiplex genome engineering using CRISPR/Cas systems. Science. 2013;339:819–23.

CITATION

CHAPTER 1

Vinayak S, Brooks CF, Naumov A, Suvorova ES, White MW, Striepen B. 2014. Genetic manipulation of the Toxoplasma gondii genome by fosmid recombineering. mBio 5(6):e02021-14. doi:10.1128/mBio.02021-14.

CHAPTER 2

Deyao Du, Lu Wang, Yuqing Tian, Hao Liu, Huarong Tan & Guoqing Niu, "Genome engineering and direct cloning of antibiotic gene clusters via phage φBT1 integrase-mediated site-specific recombination in Streptomyces," Scientific Reports 5, Article number: 8740 (2015), doi:10.1038/srep08740.

CHAPTER 3

Hu Z, Yang R-C (2014) Marker-Based Estimation of Genetic Parameters in Genomics. PLoS ONE 9(7): e102715. doi:10.1371/journal.pone.0102715

CHAPTER 4

Jones, M. L. et al. A versatile strategy for rapid conditional genome engineering using loxP sites in a small synthetic intron in Plasmodium falciparum. Sci. Rep. 6, 21800; doi: 10.1038/srep21800 (2016).

CHAPTER 5

Chandran et al.: TREC-IN: gene knock-in genetic tool for genomes cloned in yeast. BMC Genomics 2014 15:1180. doi:10.1186/1471-2164-15-1180.

CHAPTER 6

Karl J Clark, Daniel F Carlson, Linda K Foster, Byung-Whi Kong, Douglas N Foster and Scott C Fahrenkrug, "Enzymatic engineering of the porcine genome with transposons and recombinases," BMC Biotechnology20077:42, DOI: 10.1186/1472-6750-7-42.

CHAPTER 7

Dominic Pinel, David Colatriano, Heng Jiang, Hung Lee and Vincent JJ Martin, "Deconstructing the genetic basis of spent sulphite liquor tolerance using deep sequencing of genome-shuffled yeast," Biotechnology for Biofuels20158:53, DOI: 10.1186/s13068-015-0241-z.

CHAPTER 8

Christian Croux†, Ngoc-Phuong-Thao Nguyen†, Jieun Lee, Céline Raynaud, Florence Saint-Prix, Maria Gonzalez-Pajuelo, Isabelle Meynial-Salles and Philippe Soucaille, "Construction of a restriction-less, marker-less mutant useful for functional genomic and metabolic engineering of the biofuel producer Clostridium acetobutylicum," Biotechnology for Biofuels20169:23, DOI: 10.1186/s13068-016-0432-2.

CHAPTER 9

Narayana Annaluru, Sivaprakash Ramalingam and Srinivasan Chandrasegaran, "Rewriting the blueprint of life by synthetic genomics and genome engineering," Genome Biology201516:125, DOI: 10.1186/s13059-015-0689-y

INDEX

A

amplicons 119, 123, 127
antibody 108, 110
Apicomplexa 1, 2
apicoplast protein 17, 34
autonomously replicating sequence
(ARS) 113

B

Bayesian algorithms 57, 58

C

chloramphenicol acetyltransferase (CAT)
13
chromatin 4, 17, 19, 20
chromosome 220, 226, 229, 232, 233,
245
clamped homogeneous electric field
(CHEF) 22
Clostridium acetobutylicum 219, 220,
232, 244, 245, 246, 272
Clostridium growth medium (CGM)
237
clustered regularly interspaced short
palindromic repeat (CRISPR) 3
concatemer 147, 149, 150, 151, 156

conjugative assembly genome engineering (CAGE) 253, 254
copy number variation (CNV) 180
cosmids 3, 6, 8, 19, 28
cryptic polyketide (CPK) 46

D

deoxyribonucleotides 248
Dictyostelium 92
Diversity Array Technology (DArT) 63
downstream homology region (DHR)
118

E

enterovirus 249
Escherichia coli 248, 265, 267

F

fosmid 1, 2, 3, 4, 5, 6, 8, 10, 11, 13, 14,
17, 19, 20, 21, 22, 23, 24, 25, 29,
33, 271

G

gene of interest (GOI) 10
genes 91, 92, 93, 101, 104, 105, 110,
112